# Biofunctional Membranes

# Biofunctional Membranes

Edited by

D. Allan Butterfield

*University of Kentucky*
*Lexington, Kentucky*

Plenum Press • New York and London

Library of Congress Cataloging-in-Publication Data

On file

Proceedings of the International Conference on Biofunctional Membranes, held April 9 – 11, 1995, in Lexington, Kentucky

ISBN 0-306-45281-2


A Division of Plenum Publishing Corporation
233 Spring Street, New York, N. Y. 10013



10 9 8 7 6 5 4 3 2 1



Printed in the United States of America

To Marcia and Nyasha

**PREFACE**

Biofunctional membranes are entities in which a biomolecule, a collection of biomolecules, or cells are incorporated into a polymeric matrix cast in the form of porous membranes. The International Conference on Biofunctional Membranes, organized and hosted by the University of Kentucky Center of Membrane Sciences, was held in Lexington, Kentucky, April 9-11, 1995. Approximately 80 scientists and engineers from six countries attended the conference to present or hear the most current insights into the structure, function, and applications of biofunctional membranes, including catalysis (membrane-based enzyme bioreactors), bioseparations (affinity membranes), recognition and bioanalysis (biosensors and membrane-based imprinting), and medical applications (artificial organs, etc.). Most of the presenters at this meeting submitted papers that were reviewed by two independent reviewers. The papers that I accepted, based on the reviewers' recommendations, were compiled by me into the major application areas of biofunctional membrane cited above and constitute this book.

Many researchers in membrane sciences acknowledge that new applications of synthetic membranes will require greater understanding of the unique characteristics of biological molecules to impart desired properties to moieties that I have termed "biofunctional membranes." Properties such as specificity, selectivity, host-guest chemistry, to name a few, can be imparted to polymeric membranes by suitable incorporation of specific biomolecules or cells. The interface of biological and synthetic membranes is a particular research strength of the University of Kentucky Center of Membrane Sciences. Scientists and engineers from eight departments in six colleges within the university have strong research, educational, and programmatic interactions. As one of only a small number of academic membrane centers in the United States and relatively unique in our research efforts at the interface of biological and synthetic membranes, the University of Kentucky Center of Membrane Sciences is in a strong position to contribute to this rapidly expanding field of biofunctional membranes. Among our programmatic efforts are those involving technology transfer, bringing information quickly from the research laboratory to the membrane community. Accordingly, hosting the International Conference on Biofunctional Membranes and sharing the exciting information on biofunctional membranes presented is a natural consequence of our efforts in the center. I believe that this book will provide an invaluable resource for researchers who want to learn more of the theory and applications of biofunctional membranes and the prospects for the future of this rapidly emerging research area.

I would like to thank the National Science Foundation, the Kentucky EPSCoR Program, and the Vice President for Research and Graduate Studies of the University of Kentucky for their financial support for this conference. In particular, I want to express my gratitude to the dedicated Faculty Associates and graduate students of the Center of Membrane Sciences for their tireless efforts to make the University of Kentucky Center of Membrane Sciences a focus of cutting-edge science and engineering. Lastly, the excellent efforts of Ms. Lousetta Carlson, Staff Assistant for the Center of Membrane Sciences, and Ms. Monica Stoch and her staff of the UK Office of Conferences and Institutes, who ensured that the conference proceeded smoothly, are greatly appreciated.

D. Allan Butterfield, Ph.D
Professor of Chemistry, and
Director, Center of Membrane Sciences
University of Kentucky
Lexington, KY

## CONTENTS

### BIOMATERIALS



### BIOSENSORS



### SEPARATIONS AND RECOVERY

## BIOMIMETICS AND ENZYME BIOREACTORS



## PHARMACEUTICAL APPLICATIONS

## MEMBRANE FOULING

# SECTION I.

# BIOMATERIALS

# BIOMEMBRANES, ION CHANNELS AND NEW BIOMATERIALS

Parvez I. Haris, Richard J. Heaton, and Dennis Chapman

Interdisciplinary Research Centre in Biomedical Materials
Academic Department of Surgery
Royal Free Hospital School of Medicine
University of London
Rowland Hill Street
London, NW3 2PF
United Kingdom

## INTRODUCTION

The present consensus view of biomembranes is that of a dynamic structure consisting of a fluid lipid bilayer matrix into which membrane proteins are embedded through hydrophobic and electrostatic interactions. For some years we have been studying the structure of biomembranes and at the same time attempting to use our knowledge of biomembrane structure to develop new types of biomaterial. In this review we present some of our more recent studies of biomembranes and indicate the most recent developments of our studies on biomimetic biomaterials.

Biophysical techniques have been applied to obtain a wealth of information on the structure and dynamics of the lipid component of the biomembrane. A similar level of information on the proteins that span the lipid membrane are much more scarce. Information on the structure and dynamics of membrane spanning proteins is essential for gaining a better understanding of the diverse biological functions of these proteins such as signal transduction, ion transport etc. However, the large size and, more importantly, the membrane-bound nature of these proteins have proved to be major obstacles in their study using techniques normally available for protein structure determination. Thus, X-ray crystallography and NMR spectroscopy cannot be easily applied for the structural determination of membrane bound proteins. Furthermore, membrane proteins such as the voltage gated ion channels are not available in sufficient quantities for biophysical studies. We have been interested in searching for ways to overcome such problems and thereby obtain structural data on membrane bound proteins. We find the technique of Fourier transform infrared (FTIR) spectroscopy very useful for the study of membrane proteins and lipid-protein interactions. Another technique which we find useful in the study of complex macromolecular interaction is surface plasmon resonance (SPR)

*Biofunctional Membranes*, Edited by
D.A. Butterfield, Plenum Press, New York, 1996

## BASIC PRINCIPLES OF INFRARED SPECTROSCOPY

Chemical bonds undergo various forms of vibrations such as stretching, twisting and rotating. The energy of most molecular vibrations corresponds to that of the IR region of electromagnetic spectrum. There are two main types of spectroscopic methods based on the vibration of the atoms of a molecule, namely IR and Raman spectroscopy. Raman spectroscopy is sensitive to vibrations that modulate bond polarizability. Vibrations that lead to changes in the dipole moment of a molecular can be detected and measured using IR spectroscopy. The possible number of modes of vibrations is 3N-6 for a molecule consisting of N atoms (3N-5 if the molecule is linear). Thus for a biomolecule such as a protein there are many vibrations which can result in a complex spectrum. Fortunately, however, many of the vibrations can be localized to specific bonds or groupings, such as the C=O and O-H groups. This has led to the concept of characteristic group frequencies. Typical group frequencies of interest to biochemists include C=O, -COOH, COO-, O-H and S-H. There are many vibrational modes that do not represent a single type of bond oscillation but are strongly coupled to neighbouring bonds. For example, the IR spectrum of a protein is characterized by a set of absorption regions known as the amide modes.

## ANALYSIS AND ASSIGNMENT OF PROTEIN INFRARED SPECTRA

In IR spectra, the amide bands that arise from the vibration of the peptide groups provide information on the secondary structure of polypeptides and proteins. Analysis of the peptide group vibration in model compounds and in polypeptide systems allows assignment of these characteristic bands. Changes in the hydrogen bonding involved in the peptide linkages results in changes in the vibrational frequency of the different amide modes. The modes most widely used in protein structural studies are amide I, amide II and amide III. The amide I band arises principally from the C=O stretching vibration of the peptide group. The amide II band is primarily N-H bending with a contribution from C-N stretching vibrations. The amide III absorption is normally very weak in the IR, arising primarily from N-H bending and C-N stretching vibrations.

Quantitative estimation of protein secondary structure has been made by curve-fitting analysis of the amide I band. Other methods that do not rely on curve-fitting are also being developed. For example, we have recently reported a method which is useful for the reliable estimation of protein secondary structure from IR spectra (Lee et al. 1990). This method involves factor analysis of the IR spectra of 18 proteins whose crystal structures are known from X-ray studies. A good correlation is observed between the IR estimates and those calculated from X-ray data. This method gave standard errors of prediction of 3.9% for $\alpha$-helices, 8.3% for $\beta$-sheets and 6.6% for turns. This compares well with values obtained using other spectroscopic techniques such as CD spectroscopy. Nevertheless, the approach has some drawbacks, such as the influence of overlap of absorbance from amino acid side chains in the amide I region.

A particular feature of the technique of FTIR spectroscopy for the study of proteins is that high-quality spectra can be obtained with relative ease with very small amounts of protein (1mM). The most important advantage of FTIR spectroscopy for biological studies is that spectra of almost any biological material can be obtained in a wide variety of environments. Thus spectra of a protein can be obtained in single crystals, in aqueous solution, organic solvents, detergents micelles, lipid membranes etc. Figure 1 shows the second-derivative spectra of a soluble protein recorded in solution and in the crystalline state. Furthermore, there are no problems associated with background fluorescence, light scattering or the size of the molecule. FTIR spectroscopy, unlike fluorescence and resonance Raman spectroscopy, is not limited to providing information on the chromophores, as it can monitor absorption from all bonds of the biomolecule. Infrared spectroscopy provides a fast snap shot ($10^{-12}$s) of the biomolecule. Thus, interpretation of experimental data are not complicated by time scale averaging of anisotropic motions. The time scale of infrared spectroscopy complements those of NMR ($10^{-5}$s) and ESR ($10^{-8}$ s) spectroscopy. FTIR spectroscopy does not rely on the use of additional probe molecules, as is required with some

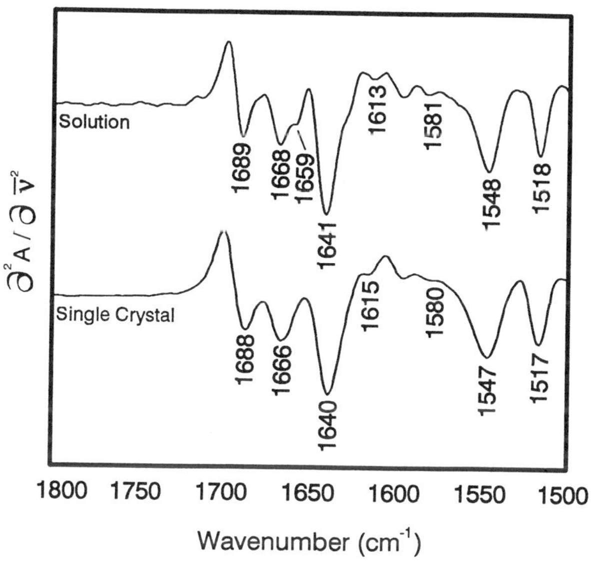


**Figure 1.** Second-derivative FTIR spectra of bovine pancratic ribonuclease A recorded in aqueous solution and in the crystalline state (Hadden et al. 1995).

spectroscopic methods. This technique has been used to study the structure of many different proteins. Figure 2 shows the temperature induced changes in the deconvolved FTIR spectra of a predominantly β-sheet (immunoglobulin G) and a predominantly α-helical (albumin) protein. In both proteins, thermal denaturation leads to a loss of native secondary structure and the appearance of aggregated β-sheet structures.

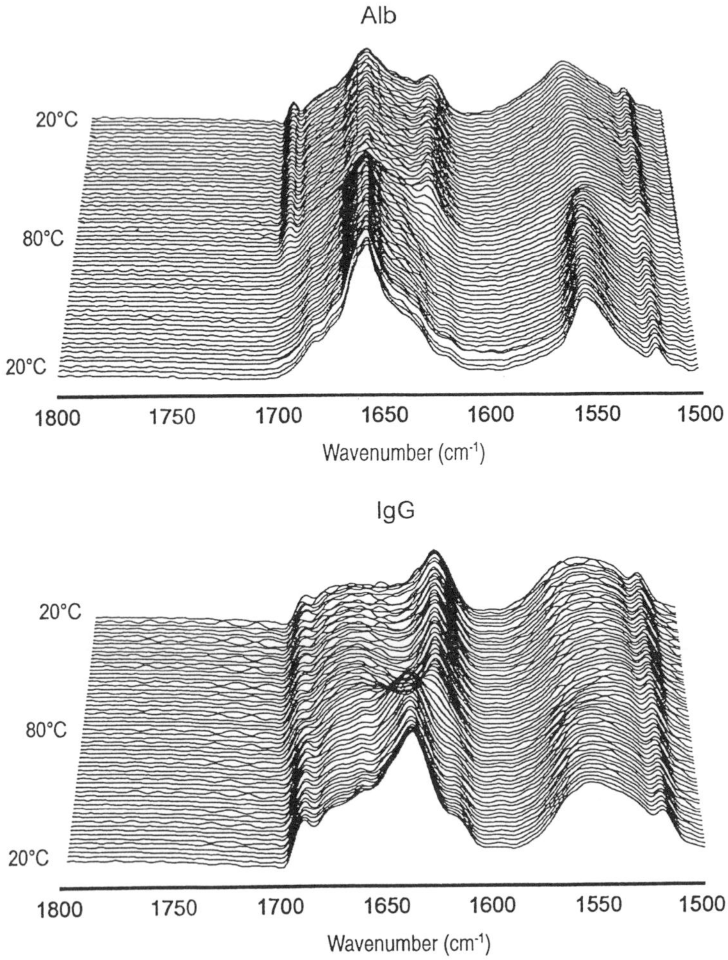


**Figure 2.** Pseudo-3D plots of the effect of temperature on deconvolved FTIR spectra of immunoglobulin G and albumin. Albumin is a predominantly α-helical protein with its amide I band near $1650cm^{-1}$, whereas immunoglobulin G is predominantly β-sheet and shows it's amide I band near $1630cm^{-1}$.

## THE STUDY OF MEMBRANE PROTEINS

We find the technique of FTIR spectroscopy very useful for the study of membrane proteins and lipid-protein interactions (for a review see Haris and Chapman, 1992). When FTIR spectroscopy is applied to the study of biomembranes, bands are observed arising from the lipids, proteins, and, if present, from the chromophores. The structures of a large number of membrane proteins have now been investigated. These studies provide the only experimental evidence presently available for the secondary structure of many of these membrane proteins. Using polarized FTIR spectroscopy, the orientation of the different secondary structures (e.g. the arrangement of the α-helices) within the lipid bilayer matrix of biomembrane systems has been determined. FTIR studies have also been used to investigate the structure of a range of peptides that interact with the membrane, for example, signal peptides and channel-forming peptides. Table 1 gives examples of some membrane proteins that have been studied using FTIR spectroscopy. Note that the vast majority of these membrane proteins show a similarity in their main amide I maximum frequency, which is consistent with the presence of α-helical structures. The exceptions to this (see Table 1) are the membrane proteins, bacteriorhodopsin (BRh) and porin. The latter displays its amide I maximum at 1631cm$^{-1}$, which is typical of a β-sheet structure, and the recently solved X-ray crystal structure supports this conclusion. The cryoelectron microscopic structure of BRh shows it to be predominantly α-helical. In this regard, the structures of BRh and porin represent two radically different folding arrangements of membrane proteins. The unusually high amide I maximum of 1662cm$^{-1}$ (1650-1658cm$^{-1}$ for normal α-helical structure; see Table 1) for BRh raises questions about the detailed structure of the helices in this protein. The present low resolution (3.5Å) of the electron microscopy data does not yet provide an explanation of the FTIR observation.

**TABLE I:** The amide I maximum frequency for a range of membrane proteins studied using Fourier transform infrared spectroscopy.

| Membrane Proteins and their source | Amide I Maxima in Wavenumbers (cm$^{-1}$) | Predominant Conformation |
|---|---|---|
| **ANIMAL** | | |
| Cytochrome C Oxidase | 1656 | α-helix |
| $Ca^{++}$-ATPase | 1656 | α-helix |
| $H^+/K^+$-ATPase | 1656 | α-helix |
| $Na^+/K^+$ATPase | 1656 | α-helix |
| Glucose Transporter | 1656 | α-helix |
| Rhodopsin | 1656 | α-helix |
| **PLANT** | | |
| Photosystem II reaction centre | 1656 | α-helix |
| **BACTERIA** | | |
| Bacteriorhodopsin from *Halobacterium halobium* | 1662 | α-helix |
| Reaction Centre from *Rhodopsedumonas sphaeroidis* | 1656 | α-helix |
| Porin from *Rhodobacter capsulatus* | 1629 | ß-sheet |

Recently, we have been interested in understanding the structural organisation of voltage-gated potassium channels. These are integral membrane proteins which conduct ions across biomembranes and are responsible for electrical excitability (for a review, see Pongs, 1992). They are thought to be constructed from four similar domains with each of the domains consisting of six membrane spanning segments, S1 to S6 (see Figure 3). Site-directed mutagenesis studies in conjunction with electrophysiological measurements have identified the functional role of several regions of this protein. For example, a highly conserved S4 transmembrane domain suggested to have a role in sensing the transmembrane electric potential has been identified (for reviews see Pongs, 1992, Franciolini, 1994). Mutations in S4 segment of voltage-dependent ion channels have been shown to affect both voltage-sensing and gating currents. The S4 segment contains a positively charged amino acid residue (lysine or arginine) at every third position. Mutagenesis studies have shown that these basic residues in the S4 region play a central role in the voltage-dependence of activation of $K^+$ channels.

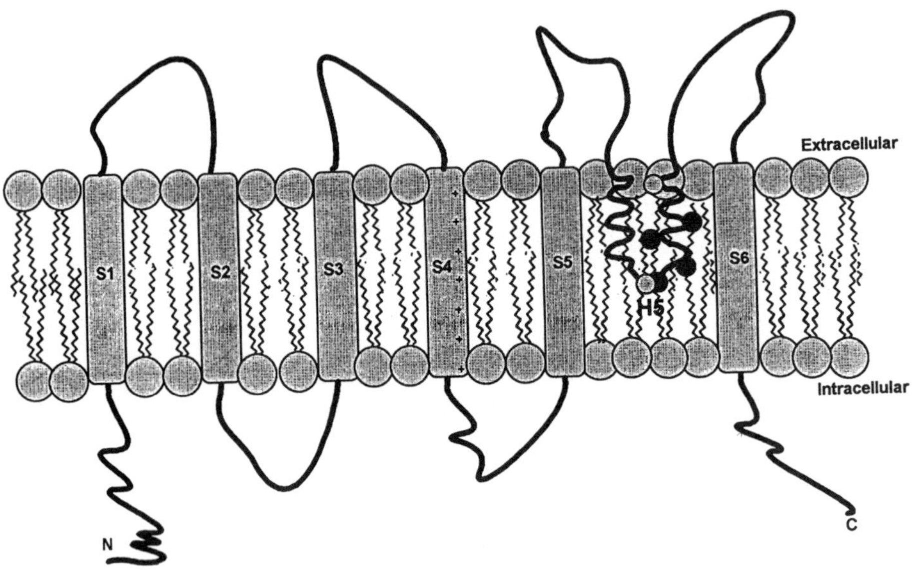


**Figure 3:** A schematic representation of the membrane spanning structure of a single-subunit of the voltage-gated potassium channel. Hydropathy profiles indicate the presence of six transmembrane stretches (S1-S6). The plus symbols in S4 indicate the presence of basic residues in this domain.

Further support indicating that S4 is involved in voltage sensing comes from experiments in which S4 domains are swapped. In these situations the voltage-dependence is conferred largely by the substituted S4 domain (Logothetis et al. 1993). Although mutagenesis studies have provided valuable information on the possible functional role of the S4 sequence, little is known about its structure. Voltage-gated ion-channels are not available in sufficient quantities from natural sources to characterise their structure using biophysical techniques. The problem of adequate sample for spectroscopic studies was overcome by dividing the primary protein structure of voltage-gated potassium channels into domains or motifs and chemically synthesising these separate domains and determining their structure by spectroscopy. We have synthesised peptides corresponding to the transmembrane segments (such as S2, S4 etc) as well as the proposed pore region (H5) of the *Shaker* potassium channel from *Drosophila*. The structure of these peptides in different solvents as well as in phospholipid micelles and bilayers was investigated using Fourier transform techniques. Interestingly, the conformation of the peptide segment that is postulated to form the pore of the ion channel was found to be α-helical (Haris et al. 1994a). This contrasts the proposed β-hairpin type arrangement suggested on the basis of mutagenesis studies. The sequence corresponding to the proposed voltage sensor region (S4) was also synthesised (Haris et al. 1994b).

In the absence of structural data various models have been proposed as to how S4 is involved in voltage-sensing (Guy and Seetharamulu, 1986, Durrel and Guy, 1992). In one such model it is postulated that the voltage-dependent conformational change involves a "helical screw" (Guy and Seetharamulu, 1986) movement of the charged S4 segment towards the extracellular surface. The S4 is in its "innermost" position in the protein structure when the channel is in the D1 deactivated state. Durell and Guy (1992) have described how the S4 can move from the $D_1$ to the open conformation by the "helical screw" mechanism. In the open conformation all the positively charged amino acids of S4 are on or near the extracellular surface, forming salt bridges with negatively charged residues in S1, S2, S3 and S5. The model of Durell and Guy indicate that a substantial (perhaps up to 90%) fraction of the S4 segment moves outside the lipid bilayer in the open state. This requires the peptide to be conformationally flexible, and the transfer from a membrane environment to a hydrophilic aqueous environment is likely to involve some rearrangement in the peptide structure. Results of our spectroscopic study (Haris et al. 1994b) provided information on such conformational changes for the S4 peptide which can be useful in rationalising the structural and functional role of the voltage-gated ion channels (see Figure 4). The peptide is predominantly α-helical in trifluoroethanol solvent, lysophosphatidylcholine micelles and in phospholipid membranes. However, upon transfer to an aqueous solution environment the peptide undergoes a major conformational rearrangement and adopts a random coil structure (see Figure 4).

An increase in α-helicity was detected with the peptide in DMPC bilayers at temperatures above the lipid phase transition. We attribute this increase in α-helical structure to greater penetration of the peptide into the lipid bilayer due to the increase in lipid fluidity. We envisage that as the lipid chains crystallise the peptide may be excluded from the lipid bilayer leading to a decrease in helical structure below the lipid phase transition temperature. The results of our FTIR spectroscopic study indicate that the S4 peptide has the conformational flexibility necessary for it to move from the lipid bilayer to the extracellular aqueous media. There is some evidence to indicate that S4 does get exposed to the extracellular medium. Mutating one positive charge (lysine) near the $NH_2$-terminal of S4 alters the influence of extracellular $Ca^{++}$ ions but not that of intracellular $Mg^{++}$ (Pusch, 1990).

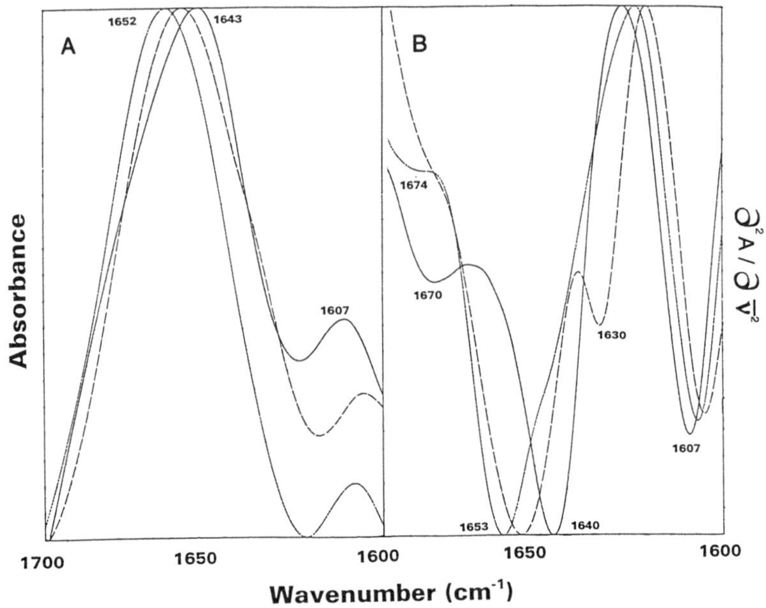


**Figure 4**: FTIR absorbance (A) and second-derivative (B) spectra of the S4 peptide in aqueous lysophosphatidyl choline micelles (- - - -); in DMPC vesicles (........); in aqueous solution ( ( ____ ). All the spectra were recorded for samples in phosphate buffer, (150mM sodium phosphate, pH 7.4) at 30°C.

This is interpreted to signify that this charge is exposed to the extracellular medium. Furthermore, antibodies to the S4 sequence of Repeat I in Na channels appear to bind on the extracellular surface when the membrane is depolarized (Schwartz et al. 1990). Our observation of a helix-coil transition for the S4 sequence is consistent with a recent theoretical study which suggest that such a major conformational change occurs during the closing and opening of the $Na^+$ channel (Leuchtag, 1993).

There are very few techniques available for investigating Interaction between biological macromolecules such as protein-protein interaction. We have recently shown using FTIR spectroscopy how a uniformly $^{13}C$-labelled protein can be used to aid the study of such protein-protein interactions (Haris et al. 1992). When two interacting unlabelled proteins are present, the overlap of absorbance of the amide I band is so great that it is impossible to distinguish the individual band from each protein. However, if one of the proteins, say B, is uniformly 13C-labelled, then its absorbance is clearly separated from the amide I absorbance of protein A. This is because $^{13}C$-labelling leads to a 45 cm-1 shift of the amide I band towards lower frequency. The secondary structure of each protein can be determined prior to and during their interaction with each other. The FTIR method has potential not only for protein-protein interaction studies, but also for protein-ligand interactions.

Another technique which is useful for investigating protein-protein interaction is the technique of surface plasmon resonance. We are using this technique in the study of receptor-ligand interaction with a view to develop biosensors. A brief introduction to the technique is given below.

## SURFACE PLASMON RESONANCE SPECTROSCOPY

A novel method for functional studies at a molecular level utilises the phenomenon of surface plasmon resonance (SPR) (Liedberg et al, 1993). SPR detects refractive index changes (see Figure 5) in a solution close to the interface of a metal and the solution.(Mayo et al, 1989) Analysis using this technique provides direct information on solute concentration, and, due to the real-time nature of the technique, association and dissociation kinetic data from the surface layer can be determined (O'Shannessy, 1994). The analysis is label free, allowing virtually any interactant to be studied.

This technique is particularly useful in investigating the interactions between biological species. For example, antibody-antigen interactions can be studied by immobilising one or other species to the surface and passing the complementary species over this layer (Liedberg et al, 1993). We are using the technique to study the interaction of biomolecules such as proteins with biomaterials (see Figure 6).

The surface plasmon resonance setup is shown above (Figure 5). Surface plasmons are transverse magnetic charge-density waves which propagate along the interface between a metal and

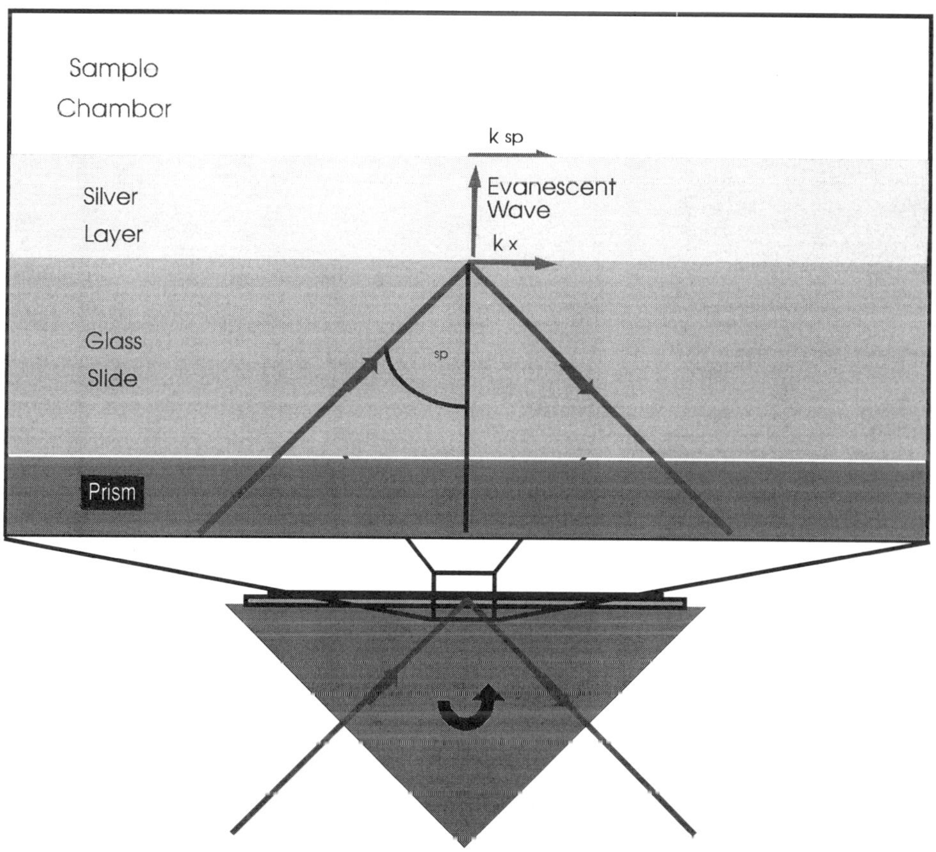


**Figure 5:** Optical setup for surface plasmon resonance spectrophotometry.

a dielectric (Raether, 1977). These surface plasmons can be set up by reflecting light from a glass-metal interface using attenuated total reflection. This results in an evanescent wave which penetrates the metal and, for thin metal films, can set up surface plasmons. At a particular angle of incidence, the energy from the light is coupled into the surface plasmons and resonance occurs, resulting in a minimum in reflected light.

The evanescent wave also penetrates someway into the test medium, and so changes in the optical properties close to the surface influence the resonance condition, which is seen by a change in the resonance angle. It is this change in angle due to changes in the optical conditions at the surface that allow SPR to monitor interactions.

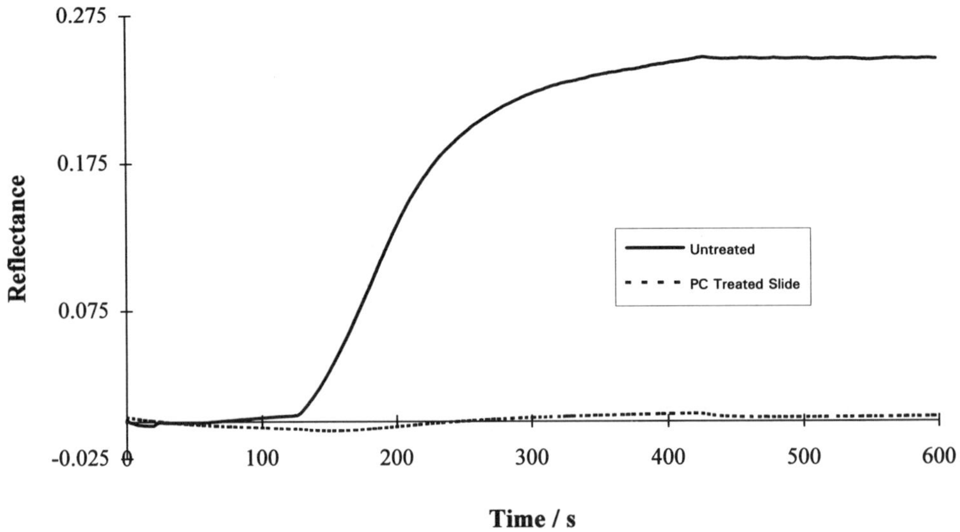


**Figure 6:** Surface plasmon resonance time-scan for albumin interaction with untreated and PC-treated surfaces.

## NEW BIOMATERIALS

Many materials used for fabricating medical devices produce adverse reactions when in contact with body tissues and fluids and especially blood. Such materials include metals and various polymers such as polyethylene, polyvinylchloride, polyurethane, cellulose and it's derivatives and materials such as silicone. Many of these materials have good mechanical properties but have unsatisfactory biocompatible characteristics. Indeed many of the materials commonly used in medical devices were found by empirical means without regard to consideration of their interaction with the biological fluids. In particular, the contact of materials with blood causes protein deposition as well as platelet adhesion and activation. To overcome these effects, anticoagulants such as heparin are sometimes used. This can lead to various complications including uncontrolled bleeding and can lead to a lowering of blood quality.

The search for satisfactory materials which can be placed in blood without causing blood to clot has let to many investigations but with little success. Negatively charged surfaces, positively charged surfaces and hydrophobic surfaces have all been tried. Andrade and Hlady (1986), commented "Virtually every physical and chemical characteristic of materials has been suggested as being important in blood coagulation and thrombosis".

In addition to problems with materials contacting blood, there are many other situations where adverse reactions occur between polymers and body fluids. One example of this is with Hydrogel contact lenses. These are known to suffer from the problem of protein deposition. The issue of biocompatibility (i.e. the property of interfacing with a biological system without modifying or adversely affecting its normal function) is being increasingly seen to be an important one which requires resolution. Associated with this requirement of biocompatibility is the related problem of obtaining surfaces which are protein resistant and which prevent cell attachment, e.g. the requirement to produce anti-fouling surfaces.

This feature, that proteins adsorb on to a variety of polymer and metal surfaces appears to be a key event in stimulating the blood coagulation process. Thus, when body fluids are in contact with the usual polymeric synthetic materials, protein adsorption takes place rapidly (in milliseconds). This means that a polymer when placed in blood or plasma has within a few minutes a coating of protein on its surface. The predominant adsorbed protein from blood is fibrinogen. Furthermore, fibrinogen is known to bind to platelets as well as to factors which can cause platelet activation. As well as this, fibrinogen is converted by thrombin to fibrin thereby producing an insoluble polymer. In this way clots and microemboli arise when a material is placed in blood.

We have introduced a new approach to this problem which has been shown to produce a satisfactory haemocompatible material (Chapman, 1979; Hayward and Chapman, 1984). This is based upon a mimicry of the outer lipid surface of the biomembranes of erythrocytes and platelet cells. The biomembranes of these cells are built upon a lipid matrix (usually a bilayer) into which the membrane proteins (integral proteins) are inserted (see Figure 3). It is known that the lipid bilayer matrix of red blood cells and platelet cells has lipid class asymmetry, i.e. the negatively charged phospholipids are predominantly located in the inner leaflet of the plasma membrane, while the zwitterionic phospholipids are found predominantly in the outer lipid leaflet. It was suggested some time ago that this lipid asymmetry may serve a biological purpose for the maintenance of a delicate balance between haemostasis and thrombosis (Zwaal and Bevers, 1983). It was consideration of this lipid class asymmetry of red blood cells and platelet cells that led us to the idea of mimicking the outer lipid surface of red blood cells and platelet cells so as to produce a new type of haemocompatible biomaterial (Hayward and Chapman, 1984). A coating of any one of these lipids, either phosphatidylcholine (see Figure 7) or sphingomyelin, provides a 90% mimicry of the outer lipid surfaces of erythrocytes.

We next began to realise that the total phospholipid structure of phosphatidylcholine lipids, including the fatty acid residues was not essential to producing these excellent haemocompatibility characteristics, rather it was the phosphorylcholine headgroup which is important. We therefore started the study derivatives of the phosphorylcholine headgroup to see how it could be attached to plastics and metal surfaces. These experiments also showed that improvements in haemocompatibility occur. For example, in collaboration with Biocompatibles Ltd, we synthesised physisorbable phosphorylcholine containing polymers for coating hydrophobic surfaces such as PVC, polyethylene, polypropylene, etc (see Chapman and Charles, 1992). These polymers, based on methacrylate chemistry, have high molecular weights and therefore form very stable coatings owing to their multipoint attachment. In addition, a variety of surface modifications have been used, such as plasma discharge and chemical procedures, so as to be able to attach or graft the PC polar groups to these materials. Materials such as celluloses and stainless steel have been successfully coated (Chapman and Charles, 1992).

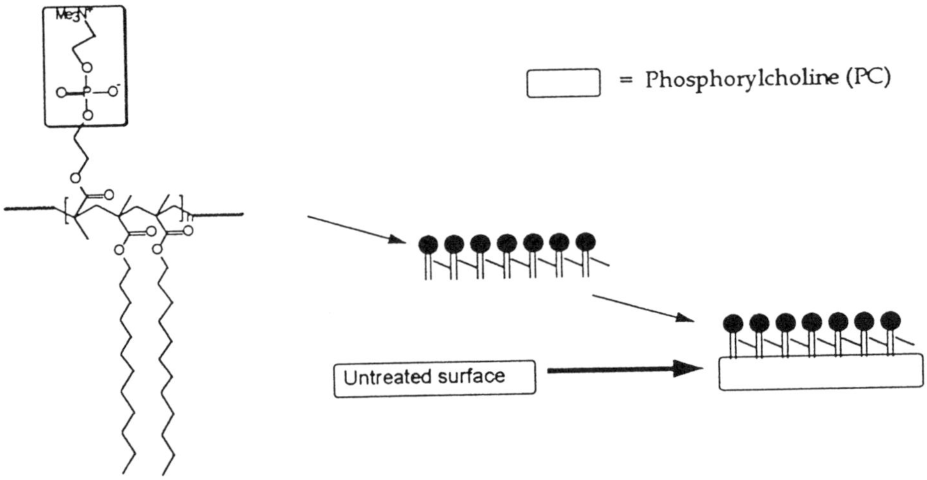


**Figure 7:** A schematic diagram showing the coating of surfaces with phosphatidylcholine lipids.

Functionally active phosphorylcholine derivatives have been synthesised, either as individual functionally active headgroups (Chapman and Durrani, 1984) or as methacrylate based polymers, and these molecules have been chemically attached to a variety of hydrophillic surfaces. Various surface modifications have been made to introduce partner functional groups where these groups are not already present, e.g. introducing acid groups into stainless steel for the subsequent reaction of aminophosphorylcholine derivatives.

New haemocompatible biomaterials have also been made either by introducing phosphorylcholine derivatives as plasticisers into polymers such as PVC or polyurethane (Valencia, 1985), or by copolymerizing phosphorylcholine monomers into the polymer backbone of polyurethanes and polyesters (Chapman and Valencia, 1984; Durrani, 1986).

These PC treated surfaces have been treated for fibrinogen adsorption, platelet adsorption, platelet activation and complement activation (see Chapman and Charles, 1992). Fukushima et al.

1983, have also pursued this concept of development of phosphorylcholine polymers produced from vinyl-containing monomer derivatives and have produced new materials and surface coatings which also show good compatibility with biological fluids. Using *in vitro* platelet adhesion and activation assays they also demonstrated that polymeric phosphorylcholine-based materials are non-thrombogenic.

Coagulation patterns in left heart bypass operations with bovine experiments (the left heart bypass is used for mechanical circulatory support). Experiments have been carried out by Segesser et al. (1992), in Zurich. Heparin coatings and phospholipid coatings were compared with systemic heparinization was absent. The coagulation studies included platelet counts, activated coagulation time, thrombin time, fibrinogen (factor 1), antithrombin III and fibrinopeptide A (FPA) measurements. With both the heparin and the lipid coating, no macroscopic red clots were observed. Segesser concluded that the antithrombotic properties of the phospholipid coating is similar and even superior to that of the bonded heparin. A particular feature noted by Segesser is that the fibrinopeptide A levels remain constant, i.e. the lipid coatings do not trigger the coagulation cascade. This is in contrast to the heparin bonded surfaces which show an increase with time of this peptide level. Segesser pointed out that there are situations where even bonded heparin is not recommended as occurs with patients with heparin induced thrombocytopenia.

Experiments have also been made by Hunter and Angelini (1993) with chest drainage tubes following open heart surgery. Occlusion of these tubes by thrombus is fairly common after such an operation. The experiments were carried out with 102 patients and the comparison was made with non-coated polyvinylchloride (PVC) chest drainage tubes. In these experiments, patients with PC coated tubes had, after removal, less post-operative supraventricular arrhythmias and a shorter hospital stay (8.4 vs 9.7 days).

From the many various *in vitro* and *in vivo* experiments that have been performed, it is clear that these new biomatieral surfaces containing well packed PC polar groups provide a new satisfactory haemocompatible biosurface. As we have seen, these results are in marked contrast to previous attempts to produce such haemocompatible surfaces using either positively or negatively charged surfaces or hydrophobic surfaces. The phosphorylcholine polar group on the other hand has a zwitterion structure, i.e. it contains within the same polar group both a positive and also a negative charge. It is in fact isoelectric over a wide pH range (3-10). It also binds large amounts, ~12 molecules, of water as indicated by a range of experiments including our own early studies using differential scanning calorimetry and also deuterium nmr spectroscopy.

Close packed arrays of these phosphorylcholine PC groups are protein resistant. This has been shown in experiments with fibrinogen and also with lysozyme. Some recent experiments with albumin were carried out in our laboratory using the surface plasmon resonance technique (see Figure 6). This shows the considerable reduction in protein adsorption which occurs after treating the metal surface so that it is covered with an array of PC-polar groups. We conclude that it is this protein resistant characteristic of the close packed arrays of the phosphorylcholine groups which is the key for its important and successful haemocompatible characteristics. A further important feature of these new haemocompatible materials is that because proteins do not bind to their surfaces, cell attachment is inhibited thereby reducing infection by bacteria.

The production of these new biomaterials provide polymers and coatings which have considerable potential for a range of medical applications. These include blood contacting devices, catheter, in-dwelling biosensors, stents and filtration membranes. The important characteristic of protein resistance has led Biocompatibles Ltd at Uxbridge to produce new types of hydrogel polymers which are synthesised so as to contain arrays of these PC groups. These are then used to make contact lenses. The PC contact lenses have new and improved characteristics. They show no protein deposition, do not readily "dry-out" and also have excellent oxygen permeability characteristics.


## ACKNOWLEDGEMENTS

We would like to thank the IRC in Biomedical Materials for financial support.


## REFERENCES


1. Andrade, J.D., Hlady, V. (1986) *Adv. Polym. Sci.* 79, 1.
2. Chapman, D. and Charles, S.A. (1992) *Chemistry in Britain*, 28, 253.
3. Chapman, D. and Durrani, A. (1984) *European patent*, 157469
4. Chapman, D. and Valencia, G.P. (1984) *European patent*, 199790
5. Durrani, A.A. (1986) *European patent* 275293
6. Durrel, S.R. and Guy, H.R. (1992) *Biophys. J.* 62, 238.
7. Franciolini, F. (1994) *Biochim. Biophys. Acta.* 1197, 227..
8. Fukushima, S., Kadoma, Y. and Nakabayashi, N. (1983) *Kobunshi Ronbunshu*, 40, 785.
9. Guy, H.R. and Conti, F. (1990) *Trends Neurosci.* 13, 201.
10. Guy, H.R. and Seetharamulu, P. (1986) *Proc. Natl. Acad. Sci. USA* 83, 508.
11. Hadden, J.M., Chapman, D. and Lee, D.C. (1995) *Biochim. Biophys. Acta.* (in press)
12. Haris, P.I. and Chapman, D. (1992) *Trends Biochem. Sci.* 17, 328.
13. Haris, P.I., Ramesh, B., Sansom, M.S.P., Kerr, I.D., Srai, K.S. and Chapman, D. (1994a) *Protein Enginnering,* 7, 255.
14. Haris, P.I., Ramesh, B., Braizer, S. and Chapman, D. (1994) *FEBS Lett.* 349, 371.
15. Haris, P.I., Robillard, G.T., van Dijk, A.A. and Chapman, D. (1992) *Biochemistry*, 31, 6279.
16. Hayward, J.A. and Chapman, D. (1984) *Biomaterials*, 5, 135.
17. Hunter, S. and Angelini, G.D. (1993) *Annals of Thoracic Surgery* 56, 1339.
18. Lee, D.C., Haris, P.I., Chapman, D. and Mitchell, R.C. (1990) *Biochemistry* 29, 9185.
19. Leuchtag, H.R. (1994) Biophys J. 66, 217.
20. Liedberg, B., Lundström, I. and Stenberg, E. (1993) Sensors and Actuators B, 11, 63.

21. Logothetis, D.E., Movahedi, S., Satler, C., Lindpainter, K. and Nadal-Ginard, B.(1992) Neuron. 8, 531.

22. Mayo, C.S. and Hallock, R.B. (1989), J. Immun. Methods, 120, 105.

23. Mulvey, D., King, G.F., Cooke, R.M., Doak, D.G., Harvey, T.S. and Campbell, I.D. (1989) FEBS Lett. 257, 113.

24. O'Shannessy, D.J. (1994) Current Opinion in Biotech. 5, 65.

25. Pongs, O. (1992) Physiol. Rev. 72, S69.

26. Pongs, O. (1993) J. Memb. Biol. 136, 1.

27. Pusch, M. (1990) Eur. Biopys. J. 18, 327.

28. Raether, H. (1977) in Physics of Thin Films, Academic Press 9, 230.

29. Schwartz, A., Palti, Y. and Meiri, H. (1990) J. Membr. Biol. 116, 117.

30. Segesser, L.V., Olan, M., Leskosek, B. and Turina, M. (1992) ASAIO Abstract, 38th Annual Meeting, p. 25.

31. Valencia, G.P. (1985), European Patent 247114

32. Waterhous, D.V., and Johnson, W.C. (1994) Biochemistry 33, 2121.

33. Zwaal, R.F.A. and Beavers, E.M. In Haemostasis Subcellular Biochemistry; Roodyn, D.B. Ed.; Plenum Press, New York, Vol. 9, 299.

# BIOFUNCTIONAL MEMBRANE SUBSTRATES BASED ON BLENDS OF POLYETHERIMIDE WITH POLY (VINYL PYRROLIDONE-CO-VINYL ACETATE)

Rajesh R. Shinde and Douglass S. Kalika

Department of Chemical and Materials Engineering
Center of Membrane Sciences
University of Kentucky, Lexington, KY 40506-0046

## INTRODUCTION

Amorphous engineering thermoplastics such as polysulfone, polyethersulfone, and polyetherimide have been used for the fabrication of a wide range of membrane materials via phase inversion methods. The high glass transition temperatures inherent to these materials lead to glassy membranes which display excellent mechanical integrity for a variety of applications. The application of engineering thermoplastics as microporous substrates for biofunctional membranes is typically precluded, however, owing to a lack of suitable functional attachment sites on the membrane surface. Further, the hydrophobic nature of these materials leads to difficulties with protein adsorption (*i.e. fouling*) and wettability.

A number of strategies have evolved for the chemical modification of engineering thermoplastics in an effort to enhance the hydrophilicity of the surface as well as to generate sites for the attachment of specific ligands; an overview of these methods is provided by Klein.[1] Direct chemical approaches include the development of copolymers which contain appropriate groups for derivitization, chemical substitution (*e.g. sulfonation*) of existing membrane materials, or covalent attachment to polymer end groups. Physical methods include the introduction of strongly-adsorbed hydrophilic coatings or the blending of hydrophilic alloying polymers. The various advantages and disadvantages of each strategy in terms of surface characteristics and the retention of functionality and bulk membrane properties is discussed in reference 1.

The blending of hydrophilic polymeric additives with hydrophobic engineering thermoplastics can be used to effectively modify membrane surface properties. Such additives have also been introduced to increase the viscosity of casting solutions and typically act as pore formers, thereby leading to increased permeability.[2,3] A model system in this regard is membranes based on blends of polyetherimide [PEI] with poly (N-vinyl pyrrolidone) [PVP] as described by Roesink et al.[4-7] PVP is a hydrophilic, water-soluble polymer which is miscible with PEI at all compositions. When PEI/PVP mixtures are solution cast into aqueous nonsolvent, a significant fraction of the PVP additive is retained in the resulting membrane matrix, with the PVP tending to segregate at the surface of the pores. Subsequent post-treatment leads to rupture of the thinner, primarily hydrophilic pore walls, and a correspondingly open, interconnected morphology is obtained with good toplayer characteristics. The key to retaining the desired hydrophilic surface properties in the microporous membrane is the relative compatibility of the hydrophobic and hydrophilic components: the favorable molecular-level interactions between the two constituents prevent the long-time leaching of the hydrophilic component.

For the work described here, an analogous blending approach is used for the establishment of biofunctional attachment sites on microporous PEI membranes. Specifically, poly (N-vinyl pyrrolidone-co-vinyl acetate) [PVP-PVAc] is introduced as an additive in order to generate hydroxyls on the membrane surface (see polymer structures, Figure 1). By using a copolymer based on N-vinyl pyrrolidone and vinyl acetate monomers, we take advantage of the favorable interaction of the vinyl pyrrolidone moiety with the PEI matrix. This interaction leads to a finite retention of the copolymer in the microporous substrate, and hence the availability of acetate groups for hydrolysis and ultimate functional attachment.

In this paper, we report the characteristics of the PEI-copolymer casting system, and the properties of the resulting microporous membranes. Binodal cloud point determinations are presented based on the inclusion of both PVP homopolymer and copolymer additives, and the corresponding membrane morphologies are elucidated as a function of casting composition. The relative retention of the PVP-PVAc copolymer in the membrane matrix is assessed, and the measurement of the surface hydroxyl population is described. For these membranes, adequate retention of the PVP-PVAc copolymer is observed, with levels of surface attachment which compare favorably with values reported based on polymer end group population.

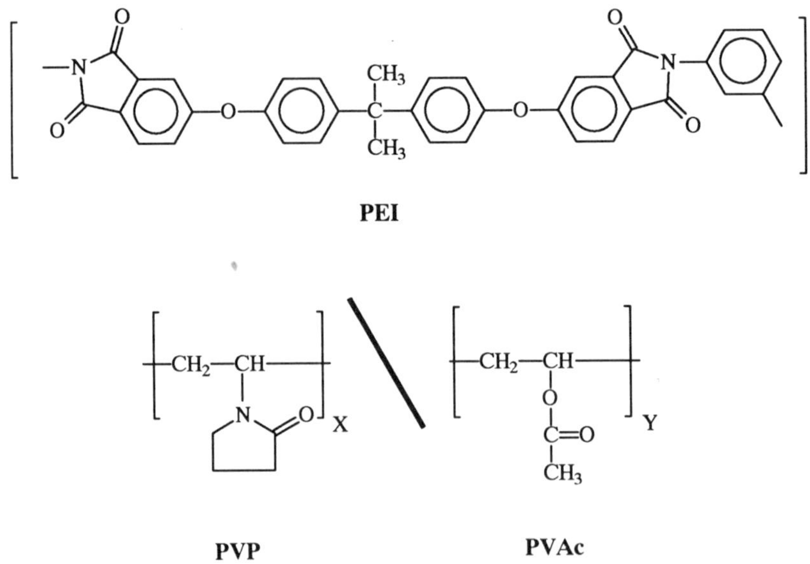


**Figure 1.** Repeat unit structures for polyetherimide [PEI], poly (N-vinyl pyrrolidone-co-vinyl acetate) [PVP-PVAc].

## MATERIALS AND EXPERIMENTAL METHODS

Polyetherimide (Ultem 1000) was obtained through the courtesy of GE Plastics, Mount Vernon, IN. Molecular weights as reported by the manufacturer were $M_n = 20,000$; $M_w = 59,000$. Poly (N-vinyl pyrrolidone) homopolymer (40,000 M.W.) and random poly (N-vinyl pyrrolidone-co-vinyl acetate) copolymers were purchased from Polysciences Inc., Warrington, PA. Two copolymer compositions were obtained: 60:40 (molar ratio) PVP-PVAc (65,000 M.W.) and 70:30 PVP-PVAc (100,000 M.W.). The 60:40 copolymer was provided in powder form, while the 70:30 copolymer was provided as a 50% solution in isopropanol. The 70:30 copolymer was recovered from solution by drying under vacuum prior to use. N-methyl-2-pyrrolidone (NMP) solvent was obtained from Sigma Chemical Co.

Cloud point determinations were accomplished by nonsolvent titration at 25°C in order to establish the binodal for the ternary system based on PEI/NMP/$H_2O$, and quaternary systems based on PEI/PVP/NMP/$H_2O$ and PEI/copolymer/NMP/$H_2O$. The titrations were carried out under inert atmosphere by the incremental addition of solvent/nonsolvent mixture; the cloud point corresponded to those solutions which remained turbid for at least thirty minutes based on visual inspection.

Phase inversion membranes were prepared by hand casting on a glass plate using a variable height casting blade (Paul Gardner, Pompano Beach, FL) with immediate quenching in the water bath; final membrane thickness was in the range of 50 to 125 μm. All membranes were prepared under inert ($N_2$) conditions in an atmosphere-controlled glove box. Dense (homogeneous) polymer blend films were prepared by casting in the glove box with solvent removal under vacuum. Dense film thickness was 20 to 30 μm.

Examination of membrane morphology was accomplished by scanning electron microscopy (SEM; Hitachi S800 scanning electron microscope). All samples were air dried and fractured in liquid nitrogen prior to examination.

The retention of PVP in the polymer blends was determined colorimetrically. Cast films were weighed on an analytical balance, and the outdiffusion of PVP upon quenching was monitored by periodic sampling of the nonsolvent water bath. PVP concentration in the aliquots was determined by addition of iodine reagent with measurement of absorbance at 432 nm against a distilled water blank.[8,9] At the end of the sampling period, the membrane was fully dried under vacuum. The final concentration of PVP (copolymer) in the membrane was determined either by comparison of the total outdiffusion of PVP with the initial cast film solution weight, or by comparison with the final membrane weight, assuming no loss of PEI during the fabrication process. Results by both material balance methods were in good agreement. Additional insight into the phase behavior and additive retention in the blend membranes was obtained by differential scanning calorimetry (Perkin-Elmer DSC-7). Membrane samples (5 mg) were heated in the DSC at 20°C/min across the range of the copolymer and PEI glass transitions.

As-cast membranes were hydrolyzed in aqueous sodium hydroxide (0.1 N and 1.0 N NaOH) in order to convert the copolymer vinyl acetate groups to hydroxyls. The population of accessible surface hydroxyl groups suitable for ligand attachment was assessed by linking lysine ($C_6H_{14}N_2O_2$) directly to the hydroxyls according to the procedure described by Feldhoff.[10] This method entails surface activation of the membranes using 1,1′ carbonyldiimidazole (CDI), with determination of the extent of lysine attachment (and corresponding -OH population) by ninhydrin assay. A series of well-defined controls were performed in order to account for the possibility of background response, and also non-covalent binding of the ligand to the membrane surface.

## RESULTS AND DISCUSSION

### Thermodynamics of the Casting System

Results of the cloud point determinations for the ternary PEI/NMP/$H_2O$ casting system and the quaternary PEI/PVP/NMP/$H_2O$ and PEI/copolymer/NMP/$H_2O$ systems are provided in Figures 2 and 3. In Figure 2, the binodal curve is plotted on a ternary phase diagram wherein the polymer apex point represents the total polymer content. The quaternary system data presented in Figure 2 correspond to a 1:1 weight ratio of PEI to additive. The results for the PEI/NMP/$H_2O$ and PEI/PVP/NMP/$H_2O$ systems are in good agreement with data reported by Roesink.[4] The introduction of PVP into the casting solution leads to a shift in the position of the binodal curve towards the polymer-solvent axis; i.e., less water is required to initiate the demixing process in the quaternary system based on the same overall polymer content. The introduction of PVP-PVAc copolymer into the casting solution in place of the PVP homopolymer leads to a further shift of the binodal towards the polymer-nonsolvent axis. Comparison of data based on inclusion of the 70:30 versus 60:40 PVP-PVAc copolymer composition indicates a progressive decrease in the amount of water required to start demixing with decreasing PVP content. These same data are plotted on a two-dimensional basis in Figure 3a, wherein the systematic decrease in nonsolvent required for demixing is clearly evident. The influence of PEI:copolymer ratio on the binodal characteristics for the 70:30 copolymer quaternary system is shown in Figure 3b. In this case, the presence of increasing amounts of the copolymer leads to a decrease in the quantity of water necessary to initiate demixing.

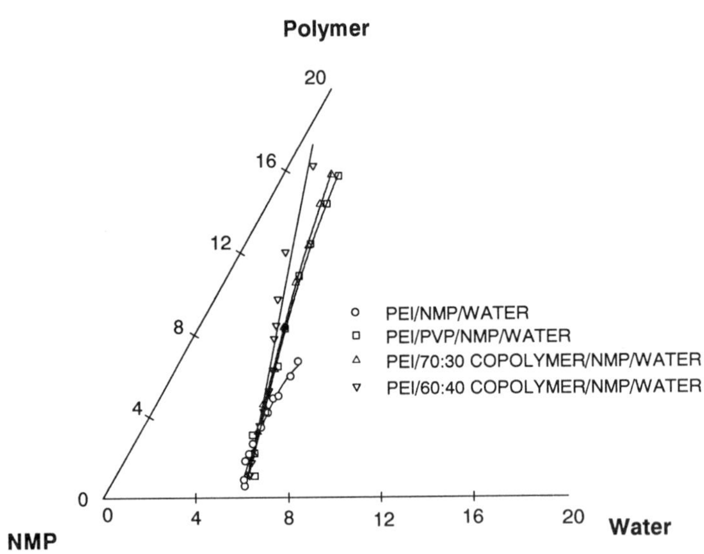


**Figure 2.** Cloud points for ternary PEI/NMP/$H_2O$ and quaternary PEI/PVP/NMP/$H_2O$ and PEI/copolymer/NMP/$H_2O$ casting systems. Weight percent basis, 25°C. For quaternary systems, PEI:additive ratio is 1:1.

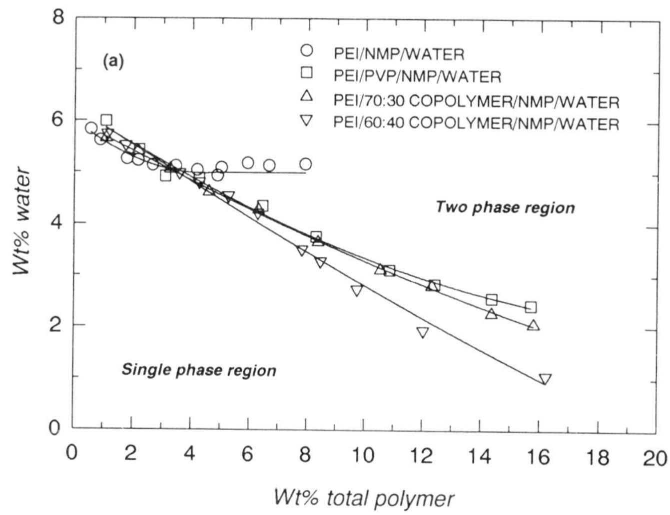


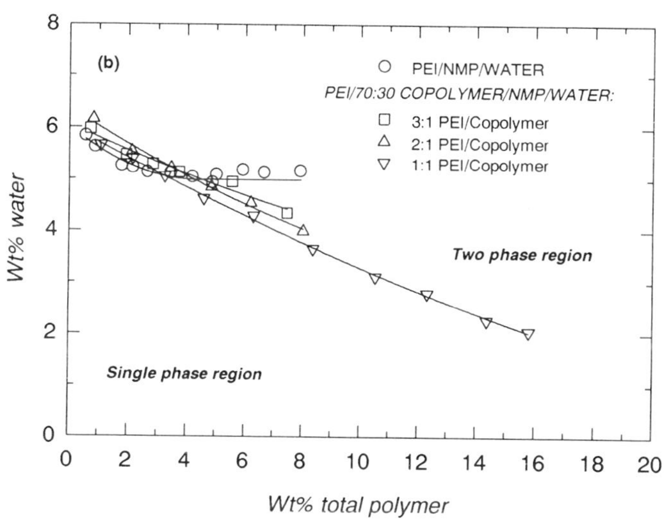


**Figure 3.** Cloud points for ternary PEI/NMP/$H_2O$ and quaternary PEI/PVP/NMP/$H_2O$ and PEI/copolymer/NMP/$H_2O$ casting systems, 25°C. Two-dimensional basis. **(a)** Wt% water *versus* wt% total polymer content for systems with different additives; PEI:additive ratio is 1:1; **(b)** Wt% water *versus* wt% total polymer content for PEI/70:30 PVP-PVAc/NMP/$H_2O$ casting system with varying PEI:copolymer ratio.

Examination of Figure 2 reveals that for the quaternary system based on the 60:40 PVP-PVAc copolymer, the binodal approaches the polymer-solvent axis at high overall polymer contents. This behavior indicates a limitation in the range of initial membrane casting solutions which can be prepared with the 60:40 copolymer additive. Specifically, single-phase casting solutions could only be obtained at overall polymer contents up to approximately 20 to 25 wt%. At these compositions, the resulting membranes were highly susceptible to the formation macrovoids (see discussion, below). Owing to this complication and the relatively limited range of accessible casting solutions for the 60:40 copolymer, subsequent studies in terms of membrane preparation and functionalization were focused on the quaternary system based on the 70:30 PVP-PVAc copolymer, for which a much a wider range of initial casting compositions was available.

**Membrane Morphology**

The membrane morphologies obtained via phase inversion for the quaternary system based on the 70:30 PVP-PVAc copolymer additive were examined as a function of initial casting composition by scanning electron microscopy. Membrane cross sections are shown in Figure 4 for the following casting compositions: 15/5/80 PEI/copolymer/NMP (Figure 4a), 15/10/75 PEI/copolymer/NMP (Figure 4b), and 15/15/70 PEI/copolymer/NMP (Figures 4c & 4d). At lower overall polymer solution content (Figures 4a & 4b) large, finger-like macrovoids are evident which appear to develop just below the membrane toplayer; the initiation point of the macrovoids is shifted downward with increasing polymer content. These macrovoids form under conditions of instantaneous liquid-liquid demixing, as discussed by Smolders,[11-13] which simultaneously lead to local regions of delayed demixing just below the incipient membrane toplayer. In the local regions of delayed liquid-liquid demixing, nucleated droplets of the polymer lean phase grow to large dimensions with minimal influx of nonsolvent. Macrovoid formation can be suppressed by avoiding conditions which lead to local delayed demixing, often by increasing overall polymer content in the casting solution or by the introduction of a polymeric additive which is soluble in the nonsolvent.

Virtually complete suppression of macrovoids was realized for membranes cast from the 15/15/70 PEI/copolymer/NMP solution due to the combined influence of higher overall polymer content and the presence of the additive. The photomicrograph shown in Figure 4c reveals only a few small macrovoids present towards the bottom of the membrane for a sample with thickness of approximately 125 μm; when thinner samples were fabricated (50 to 75 μm), no macrovoids were observed. Examination of the toplayer of the 15/15/70 membrane (Figure 4d) reveals a closely-packed nodular structure, with the size of the nodules increasing with distance from the top of the membrane. The nodular structure is characteristic of membranes containing a soluble polymeric additive, and has been attributed to spinodal decomposition owing in part to the rapid outdiffusion of the additive from the polymer rich phase into the polymer lean phase.[12]

**Retention of Polymeric Additive**

The retention of some fraction of the copolymer additive in the PEI matrix upon phase inversion membrane casting is essential for the establishment of functional attachment sites. For the 15/15/70 PEI/copolymer/NMP (70:30 PVP-PVAc copolymer) casting solution, the outdiffusion of copolymer into the nonsolvent was monitored by periodic sampling of the water quench bath after membrane coagulation. The concentration of PVP in the sample aliquots was determined colorimetrically, and the final composition of the membrane was established by mass balance. Typical results plotted as retained copolymer content *versus* time in the quench bath are provided in Figure 5. These data reveal that the outdiffusion of the copolymer is complete

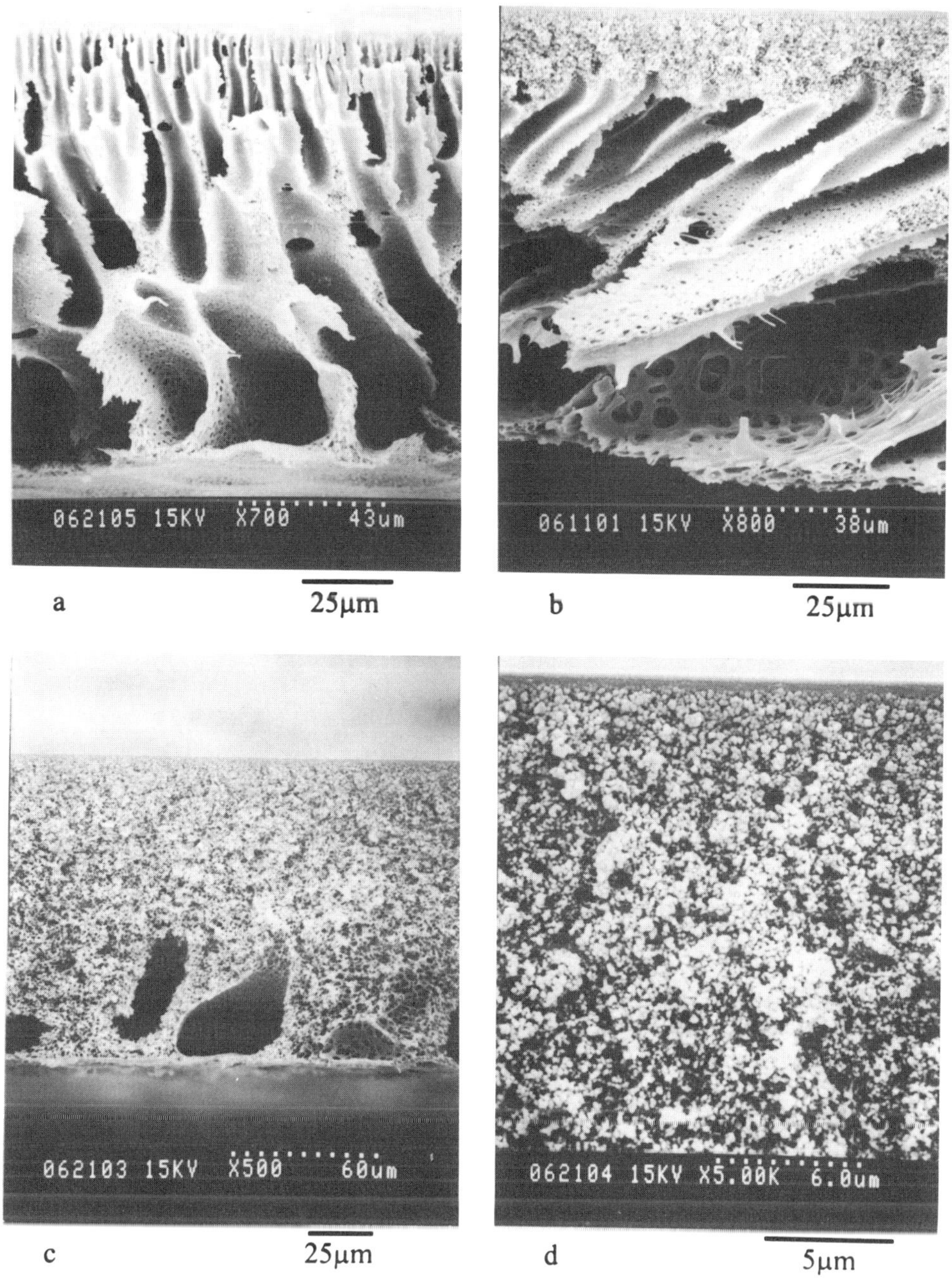


**Figure 4.** Scanning electron micrographs of phase inversion membranes based on PEI/70:30 PVP-PVAc/NMP/$H_2O$ casting system. **(a)** 15/5/80 PEI/copolymer/NMP casting composition; **(b)** 15/10/75 PEI/copolymer/NMP casting composition; **(c)** 15/15/70 PEI/copolymer/NMP casting composition; **(d)** 15/15/70 PEI/copolymer/NMP, toplayer only.

at very short times (less than 60 minutes), with a final copolymer content of approximately 5.5 wt% [membrane composition of 94.5% PEI, 4.1% PVP, and 1.4% PVAc]. The degree of retention observed here for the 70:30 copolymer is significantly less than that reported for blend membranes based on the PVP homopolymer, particularly for lower PVP molecular weights.[5] However, no additional leaching of copolymer was detected over long times in the quench bath, either at ambient conditions (25°C, 30 days), or at elevated temperature (80°C, 15 days).

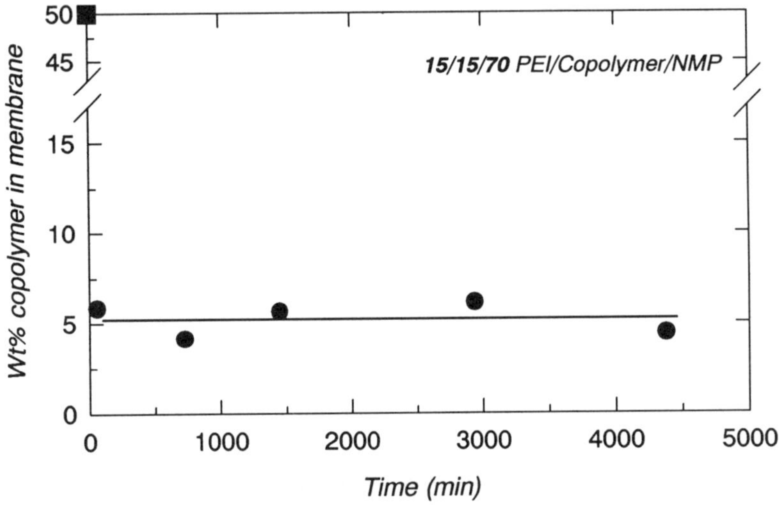


**Figure 5.** Weight percent 70:30 PVP-PVAc copolymer retained in membrane upon phase inversion casting (25°C) as determined by colorimetric methods. 15/15/70 PEI/copolymer/NMP casting composition.

Additional insight as to the retention and phase behavior of the copolymer additive in the membrane matrix was obtained by DSC. Figure 6 shows a heating sweep for a microporous membrane cast from 15/15/70 PEI/copolymer/NMP solution. Two distinct glass transitions are observed with increasing temperature which correspond to the presence of a copolymer-rich phase and a PEI-rich phase. The measured glass transition temperatures, taken as the midpoint of each transition event, are offset slightly as compared to the $T_g$'s determined for the individual constituents. This offset is indicative of a limited degree of miscibility between the two components. An attempt was made to quantify the amount of copolymer retained in the membrane based on the relative step change in heat capacity measured for the blend membrane as compared to a neat copolymer sample. When calculated on this basis, copolymer retention levels of approximately 8 to 10 wt% were indicated, in reasonable agreement with the values determined by colorimetric assay.

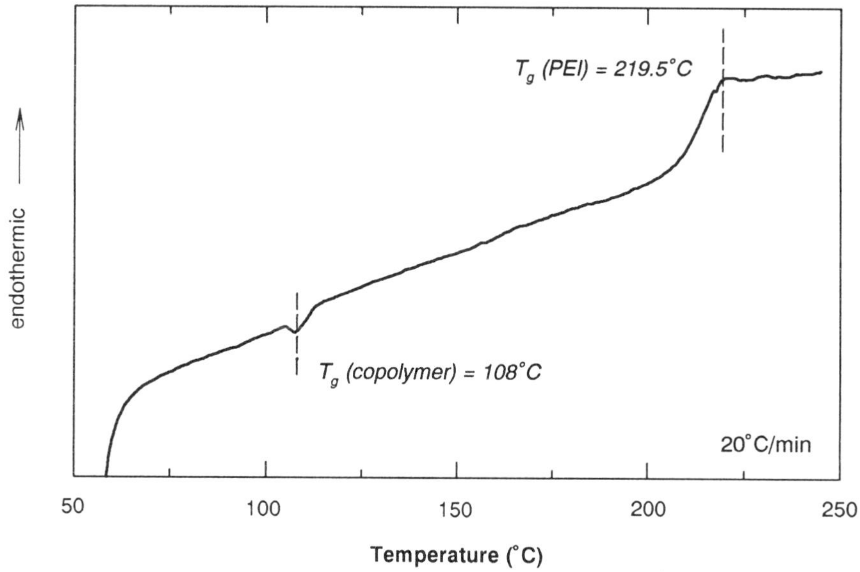


**Figure 6.** DSC sweep for air-dried phase inversion blend membrane based on 15/15/70 PEI/70:30 PVP-PVAc/NMP casting composition. Dashed lines indicate glass transition temperatures determined independently for 70:30 PVP-PVAc copolymer, PEI homopolymer.

## Determination of Surface Hydroxyl Population

The results presented above confirm that upon phase inversion casting, membranes prepared from 15/15/70 solutions of PEI, 70:30 PVP-PVAc copolymer, and NMP retain a finite fraction of the copolymer additive in an apparently phase-separated membrane matrix. In order to establish surface hydroxyl sites for subsequent ligand attachment, it was necessary to hydrolyze the vinyl acetate groups inherent to the copolymer. This hydrolysis was accomplished by exposing the membranes to aqueous sodium hydroxide; both 0.1 N NaOH and 1.0 N NaOH solutions were used. Membranes were either hydrolyzed directly from the quench bath, or were air dried prior to hydrolysis.

The population of surface hydroxyls was determined by attachment of lysine with subsequent ninhydrin assay for detection of amine groups. For each membrane evaluated, a separate control was performed in which the membrane was exposed to lysine without prior CDI surface activation. The response which was measured for these controls represented the extent of non-covalent binding of lysine to the membrane surface. The values reported below correspond to the net response for each membrane; i.e., the response measured for the CDI activated attachment less the response corresponding to non-covalent adsorption. In virtually all cases, the non-covalent response represented only a small fraction (< 15%) of the total response of the blend membranes.

The viability of the method for the detection of surface hydroxyl groups was confirmed by carrying out reference measurements on a cellulose flat sheet membrane, which contains native -OH sites. The values obtained for the CDI-activated membrane and corresponding control are reported in Table 1, with a net response of 248 μM amine per gram of membrane.

**Table 1.** Ninhydrin response (micromoles amine per gram) for lysine-functionalized cellulose flat sheet membrane and dense PEI and PEI/copolymer blend films. Non-activated control corresponds to materials exposed to lysine without prior CDI activation.

| | **Non-Activated Control** | **Activated Membrane** |
|---|---|---|
| | *(μM amine per gram membrane)* | |
| Cellulose Membrane | 4.7 | 253 |
| Dense PEI Film | 1.4 | 3.7 |
| 1:1 PEI/70:30 Copolymer Dense Film | 10.3 | 116 |

This value is in good agreement with results reported by Feldhoff.[10] Table 1 also contains results obtained for the activation of both a dense (non-porous) PEI film, and a dense blend film based on a 1:1 ratio of PEI and 70:30 copolymer. The blend film was prepared by solution casting from NMP in an inert atmosphere; solvent removal was accomplished by drying the film at 150°C under vacuum. Both films were exposed to hydrolysis in sodium hydroxide (1.0 N) for 4.5 hours prior to activation. The presence of hydroxyl groups on the surface of the 1:1 blend film is clearly demonstrated by the high degree of ninhydrin response as compared to the PEI film. This result thus indicates that the some fraction of the vinyl acetate moieties are accessible at the membrane surface, and have been successfully hydrolyzed to -OH.

Results for lysine attachment to the phase inversion membranes (15/15/70 PEI/copolymer/NMP casting solution) as a function of hydrolysis strength, hydrolysis time, and memrbrane post-treatment are provided in Figure 7. Data for the as-cast blend membranes hydrolyzed in 1.0 N NaOH (Figure 7a) indicate high degrees of response at relatively short hydrolysis times, with a modest, asymptotic increase observed over longer times. The air-dried membranes display a similar behvaior, although their net response is lower by a factor of two; the difference between the as-cast and the air dried membranes may reflect the rupture of pore walls (i.e., pore collapse) during the drying process, leading to a decrease in overall pore surface area. Also included in Figure 7a is the measured response of a PEI phase inversion membrane cast from 15/85 PEI/NMP solution and air dried. The PEI membrane displays a small, reproducible level of response which increases linearly with increasing hydrolysis time. This result is unexpected for the pure PEI membrane, and suggests that some type of hydrolytic attack is occurring at the membrane surface, resulting in the generation of hydroxyls. In this same regard, it is noted that exposure of the membranes to the 1.0 N NaOH environment for long times (> 12 to 24 hours) leads to embrittlement.

In order to reduce the severity of the hydrolysis environment, PEI/copolymer phase inversion membranes were hydrolyzed using 0.1 N NaOH. The corresponding response for a series of as-cast membranes as function of hydrolysis time is provided in Figure 7b. Although the timescale for membrane hydrolysis in the 0.1 N NaOH case is considerably longer as compared to those samples hydrolyzed in 1.0 N NaOH, the measured response is comparable. At long times (> 72 hours) the as-cast membranes have a hydroxyl surface population of approximately 125 μM per gram of membrane. Such levels are favorable as compared to the

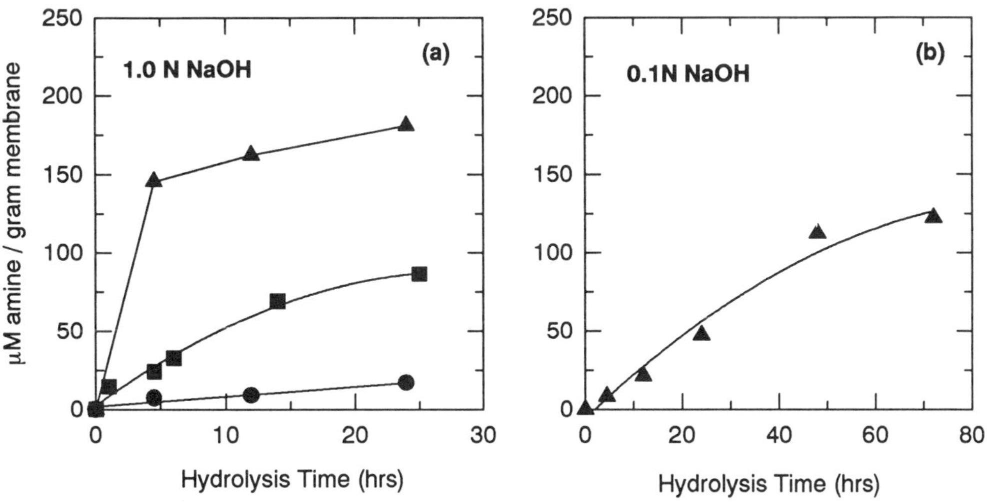


**Figure 7.** Results of ninhydrin assay for lysine-functionalized phase inversion membranes; 15/15/70 PEI/70:30 PVP-PVAc/NMP casting composition. Micromoles amine/gram of membrane *versus* hydrolysis time. **(a)** Hydrolysis in 1.0 N NaOH; **(b)** Hydrolysis in 0.1 N NaOH. (▲) as-cast membrane, (■) air-dried membrane, (●) PEI phase inversion membrane cast from 15/85 PEI/NMP.

end-group populations reported for polysulfone[1] (16 µM per gram) and microporous nylon[14] (21 µM per gram) hollow fiber membranes.

## CONCLUSIONS

The viability of a polymer blending approach for the development of microporous biofunctional substrate materials based on polyetherimide thermoplastic has been demonstrated. Poly (N-vinyl pyrrolidone-co-vinyl acetate) served as an effective additive for the establishment of hydroxyl groups on the surface of the phase inversion membranes. Homogeneous casting solutions comprised of PEI and the 70:30 PVP-PVAc copolymer in NMP were accessible across a broad range of compositions. For casting solutions with relatively low overall polymer content (below 30 wt% polymer), phase inversion led to the formation of large macrovoids extending from just below the membrane toplayer. These macrovoids were absent for casting solutions based on higher overall polymer content and higher levels of copolymer additive. In all cases, a nodular toplayer was observed owing to the onset of spinodal decomposition during the liquid-liquid demixing process.

Retention studies performed on membranes prepared from a 15/15/70 PEI/copolymer/NMP casting solution indicated an overall retention of approximately 5.5 wt% copolymer in the membrane matrix. DSC studies verified the retention of the copolymer additive, and revealed a phase-separated morphology in which the PEI and copolymer constituents displayed limited miscibility. Hydrolysis of the blend membranes was carried out in aqueous sodium hydroxide (0.1 N and 1.0 N NaOH) in order to convert the vinyl acetate groups of the copolymer additive to hydroxyls; the population of accessible -OH sites was assessed by covalently linking lysine to the membrane, followed by ninhydrin assay. Hydrolysis in 1.0 N NaOH led to an apparent saturation in ninhydrin response at relatively short times.

This was accompanied, however, by some level of hydrolytic attack on the PEI surface, as manifested by a PEI background response which increased monotonically with hydrolysis time. Hydrolysis in 0.1 N NaOH led to comparable levels of membrane response, although over longer timescales. For long hydrolysis times (> 72 hours), the as-cast membranes had a surface hydroxyl population of approximately 125 μM per gram. Continuing work on this system is focused towards optimization in terms of initial blend composition (PEI:copolymer ratio), membrane post-treatment, and the exploration of alternate hydrolysis methods.

## ACKNOWLEDGMENTS

This work was supported through a grant from the National Science Foundation (EHR-91-08764). We are pleased to acknowledge Dr. Pamela Feldhoff at the University of Louisville for her assistance in the development of the lysine attachment techniques used in this work. Also, Eric Pigman is acknowledged for his contributions to the colorimetric retention studies completed as part of the NSF Research Experience for Undergraduates Program at the UK Center of Membrane Sciences (DMR-92-00164).

## REFERENCES

1. E. Klein, E. Eichholz, and D.H. Yeager, "Affinity membranes prepared from hydrophilic coatings on microporous polysulfone hollow fibers", *J. Membrane Sci.*, **90**, 69 (1994).

2. I. Cabasso, E. Klein, and J.K. Smith, "Polysulfone hollow fibers. I. Spinning and Properties", *J. Appl. Polym. Sci.*, **20**, 2377 (1976).

3. L.Y. Lafreniere, F.D. Talbot, T. Matsuura, and S. Sourirajan, "Effect of polyvinylpyrrolidone additive on the performance of polyethersulfone ultrafiltration membranes", *Ind. Eng. Chem. Res.*, **26**, 2385 (1987).

4. H.D.W. Roesink, M.J. Otto, J.A. Ronner, M.H.V. Mulder, and C.A. Smolders, "Demixing phenomena and membrane formation in the four component system polyetherimide / polyvinylpyrrolidone / N-methylpyrrolidone / water", Ph.D. Thesis Chapter 2, University of Twente, Enschede, Netherlands (1989).

5. H.D.W. Roesink, Z. Borneman, M.H.V. Mulder, and C.A. Smolders, "Polyetherimide-polyvinylpyrrolidone blends as membrane material", Ph.D. Thesis Chapter 4, University of Twente, Enschede, Netherlands (1989).

6. H.D.W. Roesink, S. Oude Vrielink, M.H.V. Mulder, and C.A. Smolders, "Post-treatment of hydrophilic membranes prepared from polyetherimide-polyvinylpyrrolidone blends", Ph.D. Thesis Chapter 5, University of Twente, Enschede, Netherlands (1989).

7. H.D.W. Roesink, M.A.M. Beerlage, W. Potman, T. van den Boomgaard, M.H.V. Mulder, and C.A. Smolders, "Characterization of new membrane materials by means of fouling experiments: Adsorption of BSA on polyetherimide-polyvinylpyrrolidone membranes", *Colloids and Surfaces*, **55**, 231 (1991).

8. E. Schchori and J. Jagur-Grodzinski, "Polymeric alloys of poly (vinyl pyrrolidone) with a macrocylic polyether-polyamide", *J. Appl. Polym. Sci.*, **20**, 1665 (1976).

9. R.L. Larkin and R.E. Kupel, "Quantitative analysis of polyvinylpyrrolidone in atmosphere samples and biological tissues", *J. Amer. Ind. Hyg. Assoc.*, **26**, 558 (1965).

10. P.W. Feldhoff, "Comparison of coupling procedures for development of affinity membranes: Optimization of the CDI method", *in* Techniques in Protein Chemistry III, Academic Press, 1992.

11. C.A. Smolders, A.J. Reuvers, R.M. Boom, and I.M. Wienk, "Microstructures in phase-inversion membranes. Part 1. Formation of macrovoids", *J. Membrane Sci.*, **73**, 259 (1992).

12. R.M. Boom, I.M. Wienk, T. van den Boomgaard, and C.A. Smolders, "Microstructures in phase-inversion membranes. Part 2. The role of a polymeric additive" *J. Membrane Sci.*, **73**, 277 (1992).

13. A.J. Reuvers and C.A. Smolders, "Formation of membranes by means of immersion precipitation. Part II. The mechanism of formation of membranes prepared from the system cellulose acetate-acetone-water", *J. Membrane Sci.*, **34**, 67 (1987).

14. K. Kugel, A. Moseley, G.B. Harding, and E. Klein, "Microporous poly (caprolactam) hollow fibers for therapeutic affinity adsorption", *J. Membrane Sci.*, **74**, 115 (1992).

# SECTION II.

# BIOSENSORS

# BIOMIMETIC APPROACH TO THE DESIGN OF SELECTIVE OXOANION RECEPTORS FOR USE IN MEMBRANE-BASED POTENTIOMETRIC SENSORS

Richard S. Hutchins and Leonidas G. Bachas

Department of Chemistry and Center of Membrane Sciences
University of Kentucky
Lexington, KY 40506-0055

## INTRODUCTION

The need to selectively recognize and complex ions and molecules is common in many areas of science and industry. Nature's ability to selectively detect or sense specific compounds in a variety of ways and under many different conditions has long been the source of inspiration in the application and development of recognition-based chemistry. Biofunctional membranes present one such example where polymeric membranes can be designed with a desired function by using or mimicking natural chemical recognition systems. The components found in living organisms that have selective recognition features (i.e., proteins, enzymes, membrane systems, sensory neurons, etc.) have been the subject of much intense study. The rapid, selective, and sensitive response of natural sensory systems that employ recognition chemistry can serve as a model in the development of biosensors. This includes mimicking natural receptors,[1] antibiotics,[2,3] and even olfactory membranes[4] to accomplish desired tasks. One group has successfully immobilized double-helical DNA directly onto an electrode surface and obtained a biosensor that responded selectively to DNA-binding substances (via intercalation) as well as to magnesium ion.[5] The applications which await the development of novel biosensors cover a wide range of disciplines, including environmental, medical, industrial, and other venues of research.

## BIOSENSORS AND ION-SELECTIVE ELECTRODES

As shown in Figure 1, a biosensor usually contains two essential features: a biological recognition component and a transducer. The recognition component is typically the crucial element in the development of a biosensor.[6] It must be chosen to meet the selectivity, sensitivity, and response time criteria of the application for which the biosensor is intended. Although one approach towards achieving the desired recognition properties is often attempted by directly incorporating native receptors (or enzymes, antibodies, etc.), there is often a synthetic strategy involved in obtaining a receptor with all of the needed characteristics. Indeed, the rational design of synthetic receptors enables one to copy

*Biofunctional Membranes*, Edited by
D.A. Butterfield, Plenum Press, New York, 1996

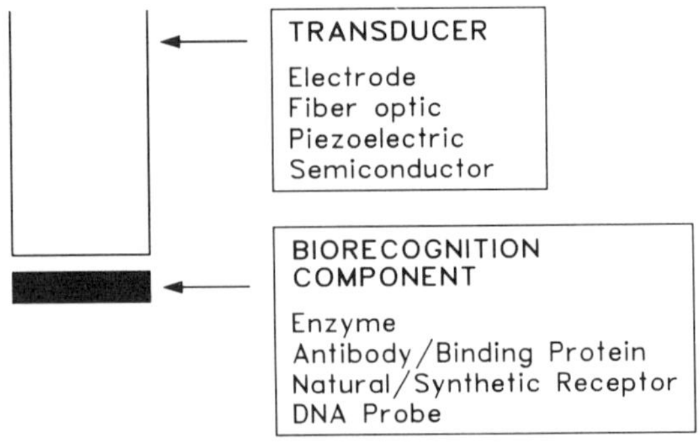


**Figure 1.** Schematic of a biosensor showing possible variations for each component.

selective chemical recognition abilities found in nature, while at the same time making it easier to control such parameters as hydrophobicity/hydrophilicity of the receptor, immobilization and/or attachment sites, etc. Once the recognition element is obtained, it must be amenable to some type of transducing element, which is the signal producing feature of the biosensor. The transducer enables the recognition process to be monitored by converting a signal observed following recognition (i.e., optical, electrical, etc.) to the desired mode of detection (voltage, current, etc.).

One type of sensors that have utilized biological recognition elements with a high degree of success are potentiometric liquid-membrane based ion-selective electrodes (ISEs). The process of preparation of such membranes for use in ISEs is outlined in Figure 2. The ionophore (Greek for "ion carrier") is the ingredient in the membrane that must be chosen to recognize and complex selectively the analyte of interest. The term polymer membrane-based ISEs is also used to describe these electrodes because the membranes are commonly formed by impregnating polymeric materials, such as plasticized poly(vinyl chloride), with the ionophore.

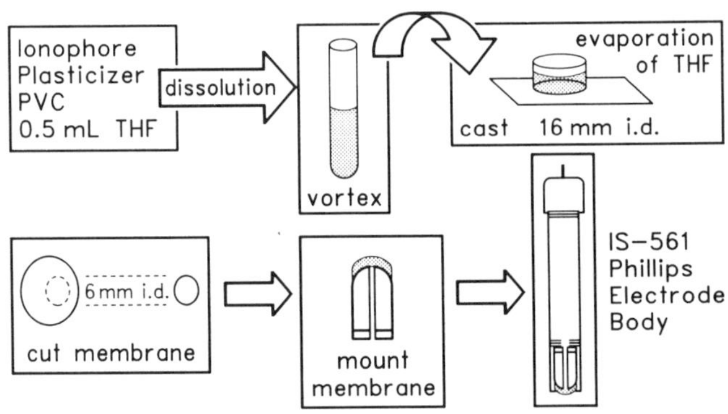


**Figure 2.** General preparation diagram for making and assembling polymer membrane-based ion-selective electrodes.

The development and commercialization of a highly selective potassium ISE, which incorporated the naturally occurring antibiotic valinomycin as the ionophore,[7] was soon followed by the development of many other cation-selective electrodes containing naturally occurring or rationally designed biomimetic ionophores. Specifically, other antibiotics

**Table 1.** Selectivity Coefficients for Several Antibiotics Used as Ionophores in ISE Membranes

| Antibiotic Ionophores | Primary Ion | Selectivity Coefficients ($K_{ij}$) | | | | | | | | | |
|---|---|---|---|---|---|---|---|---|---|---|---|
| | | $H^+$ | $Li^+$ | $Na^+$ | $Mg^{2+}$ | $K^+$ | $Ca^{2+}$ | $Rb^+$ | $Sr^{2+}$ | $Cs^+$ | $NH_4^+$ |
| Monensin [ref 11] | $Na^+$ | 15.7 | $7.7x10^{-2}$ | 1 | $1.1x10^{-3}$ | $6.7x10^{-2}$ | $2.6x10^{-2}$ | $2.1x10^{-2}$ | $3.3x10^{-3}$ | $2.3x10^{-2}$ | $2.2x10^{-2}$ |
| Valinomycin [ref 8] | $K^+$ | $5.6x10^{-5}$ | $2.1x10^{-4}$ | $2.6x10^{-4}$ | | 1 | | 1.9 | | $3.8x10^{-1}$ | $1.2x10^{-2}$ |
| Antibiotic A23187 [ref 9] | $Ca^{2+}$ | $2.5x10^{4}$ | | $2.3x10^{-1}$ | $5.1x10^{-2}$ | 16 | 1 | | 97 | | |
| Antibiotic 3016 [ref 10] | $Mg^{2+}$ | | $7x10^{-2}$ | $1x10^{-2}$ | 1 | $1x10^{-1}$ | 20 | $5x10^{-1}$ | 2 | 4 | $9x10^{-2}$ |
| Nonactin (72%) Monactin (28%) [ref 8] | $NH_4^+$ | $1.6x10^{-2}$ | $4.2x10^{-3}$ | $2.0x10^{-3}$ | | $1.2x10^{-1}$ | $1.7x10^{-4}$ | $4.3x10^{-2}$ | | $4.8x10^{-3}$ | 1 |

have been successfully mimicked or incorporated into polymer membranes with varying degrees of success resulting in the development of ammonium,[8] calcium,[9,10] and sodium-selective sensors[11] [Table 1]. The Nicolsky equation, Eq. 1, is a modified form of the Nernst equation and enables one to calculate $K_{ij}$, the selectivity coefficient with respect to a given interfering species. In this equation, K is a cell constant, R is the gas constant, T the temperature in K, $z_i/z_j$ the charge (including the sign) on the analyte ion (*i*) and interferent ion (*j*), F is Faraday's constant, and $a_i$ the activity of the analyte ion *i*. Various crown ethers have been synthesized in an attempt to mimic the macrocyclic ether properties of valinomycin. Several of these crown ether receptors have proven valuable in the development of ISEs for cations.[12-14]

$$E_{cell} = K + \frac{RT}{z_i F} \ln \left(a_i + K_{ij} a_j^{z_i/z_j}\right) \qquad \text{Eq. 1}$$

Since the 1960's the development of cation-selective electrodes has been extensive, as evidenced by the routine use of such electrodes in many applications. However, despite the appearance of a few commercially available anion-selective electrodes, selective electrodes for many important anions have not yet been developed. There are many factors that hinder the development of selective-anion recognition elements, such as the size of anions compared with cations (anions are larger and more difficult to encapsulate) and the natural occurrence of anionic sites in the polymeric membranes, which results in permselectivity towards cations in these membranes.[15] Thus, the use of a biomimetic approach toward the design of selective ionophores for anions should assist in the development of anion-selective electrodes.

## PORPHYRINS AND CORRINS AS BIOMIMETIC IONOPHORES FOR ANIONS

A few successful biomimetically designed ionophores that are selective for anions have been reported. Lipophilic cobyrinates are vitamin $B_{12}$ derivatives that lack the nucleotide part of the vitamin. Figure 3A shows two cobyrinates, differing in the "R" group, that have been used in ISE membranes. Through appropriate selection of the R group, the cobyrinates have been made sufficiently hydrophobic to prevent them from leaching out of the polymer membranes. These ISEs have demonstrated a high degree of selectivity towards the anions thiocyanate, salicylate, and nitrite.[16-20] Compared to ionophores that have no selective interaction with anions (i.e., ionophores that attract anions simply based on electrostatics), the order of response by the ISEs to the anions tested has been altered using vitamin $B_{12}$ derivatives. It should be mentioned that the order of "non-selective" response to anions, termed the Hofmeister series,[21] mirrors the lipophilicity of the anions with the most lipophilic anions preferred: organic anions > $ClO_4^-$ > $SCN^-$ > salicylate > $I^-$ > $Br^-$, $NO_2^-$ > $HSO_3^-$ > $CH_3COO^-$, $H_2PO_4^-$. This non-selective response is typically observed with quaternary ammonium and phosphonium ionophores.

Vitamin $B_{12}$ derivatives that have a proximal base (e.g., the natural 5,6-dimethylbenzimidazole of $B_{12}$ or just an imidazole ring), are known as cobalamins.[22] Figure 3B shows a lipophilic cobalamin that was tested as an ionophore in an ISE membrane and was found to possess selectivity properties that greatly differ from the selectivity of the cobyrinates, demonstrating a highly selective response towards iodide.[23] Porphyrins, which structurally mimic natural oxygen carriers such as myoglobin and hemocyanin, have been used to develop selective electrodes for thiocyanate,[24-28] salicylate,[29,30] nitrite,[31,32] and hydroxide.[33]

## OTHER BIOMIMETIC ANION CARRIERS

Schiff-base complexes, which mimic porphyrins, have been successfully employed in the development of an iodide-selective electrode.[34] A selective response to the polyanion heparin has been achieved by mimicking protamines, a naturally occurring class of proteins

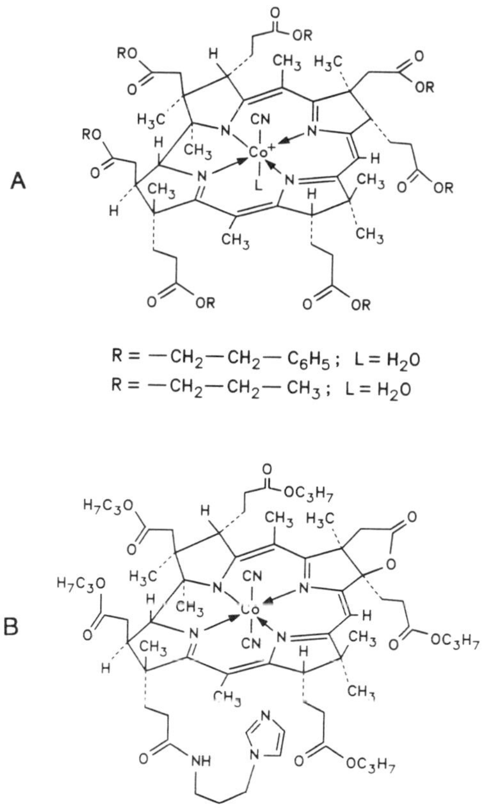


**Figure 3.** Types of vitamin $B_{12}$ derivatives that have been tested in ISE membranes and have demonstrated unique selectivity properties: (A) cobyrinate, (B) cobalamin.

that bind tightly to heparin's polyanionic structure to halt its anticoagulation activity.[35] By incorporating the appropriate quaternary ammonium salt into polymer membranes the interaction between protamine and heparin was mimicked and a heparin-selective electrode developed.[36] Finally, ionophores that mimic the base-pairing of nucleotides have been developed that contain a cytosine derivative attached to a hydrophobic tail. These ionophores, one of which is shown hydrogen-bound to a GMP derivative in Figure 4, have proven successful in achieving selective electrodes for guanine-containing compounds.[1]

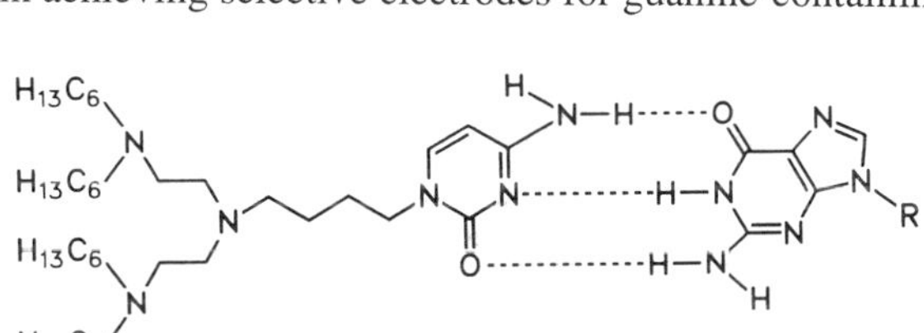

**Figure 4.** Hydrophobic ionophore (on the left) containing a cytosine moiety based-paired with the guanine group on GMP (on the right).

## SYNTHETIC RECEPTORS FOR OXOANIONS

Despite these successes in using biomimetically designed ionophores to selectively complex anions, there are many important anions for which no selective ionophore has been developed. The selective complexation of oxoanions, including phosphate, nitrate, hydrogen sulfite, carboxylate-containing compounds, and others, despite their importance, has generally not been achieved. Most oxoanions are hydrophilic in nature, which makes them more difficult to complex into lipophilic membranes from aqueous solutions due to the high energy of hydration.[37] With the goal of developing ionophores that selectively complex oxoanions, the manner by which proteins complex these anions was investigated. Because of the importance of the phosphate oxoanion, many studies have been performed that describe the key components found in naturally occurring phosphate binding proteins.[38,39] One study investigated 34 different proteins that bind either the phosphate or sulfate moiety, focusing on the amino acids that participated in the binding of these oxoanions.[40] They reported that arginine was assisting in the complexation of these anions with a frequency more than twice that of any other amino acid. Indeed, the guanidinium moiety of arginine has been shown to have a strong affinity for oxoanion species, and several crystal structures have been obtained of such complexes. Figure 5 shows the guanidinium group hydrogen-bound with the oxoanion nitrate.[41] In addition, crystal

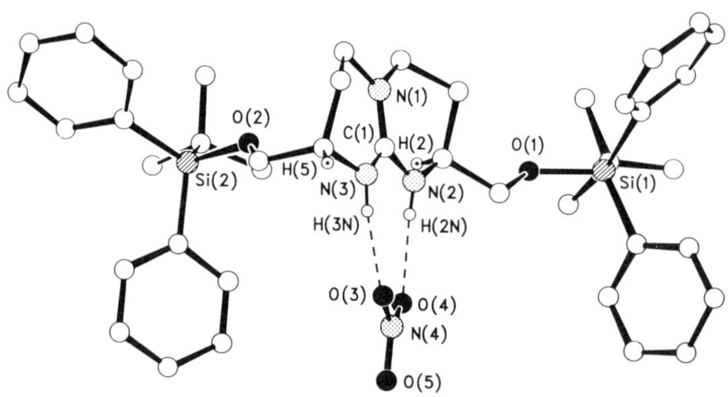


**Figure 5.** Crystal structure of a rigid guanidinium compound complexed to the oxoanion nitrate [reproduced with permission from reference 41].

structures with acetate,[42] sulfate,[39] and phosphate[38] have also been reported. In an attempt to mimic the coordination of oxoanions by the guanidinium group, many compounds have been synthesized that contain this arginine functional unit. Binding studies with these compounds demonstrate that, by incorporating the guanidinium unit into a restricted or rigid site, the strength of the interaction increased, and the selectivity toward the targeted oxoanions improved.[37,43] While much work had been done by others investigating the binding properties of compounds containing the guanidinium moiety, the use of such compounds in the development of sensors had not been attempted.

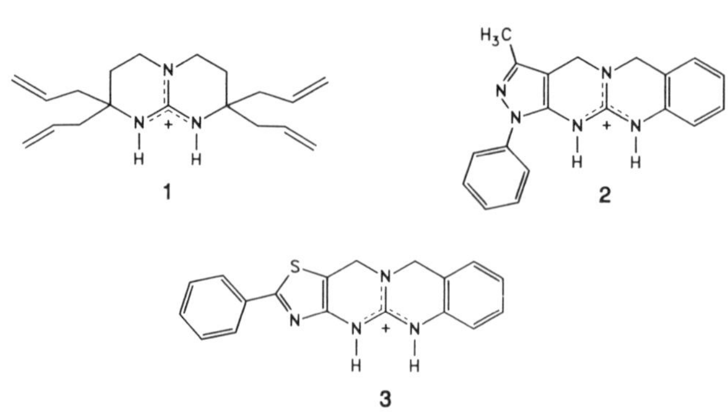

**Figure 6.** Structures of three guanidinium ionophores that have been tested in poly(vinyl chloride) membrane ISEs.

The bicyclic guanidinium **1**, seen in Figure 6, was synthesized according to a modified version of Schmidtchen's protocol,[44] and prepared for initial testing as an ionophore in ISE membranes. The selectivity of this compound towards various anions was observed to resemble closely that of the Hofmeister series. Perchlorate is the preferred ion, with significant response also observed towards the oxoanions salicylate and thiocyanate. The only difference in selectivity observed between ionophore **1** and that of the Hofmeister selectivity pattern is that **1** showed an improved response toward the oxoanion salicylate. While the response to salicylate was not analytically useful, it did provide evidence that the guanidinium compounds exhibit strong interactions with oxoanions. The second guanidinium ionophore tested, **2** (Figure 6), contained a phenyl side arm located near the oxoanion docking site.[45] This compound was prepared to investigate the effect that an

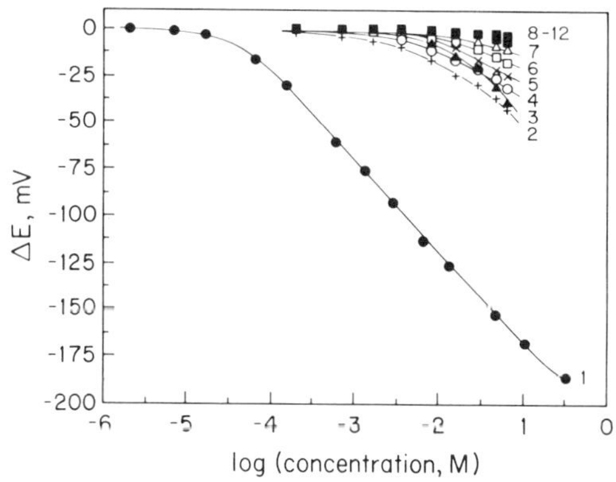


**Figure 7.** Anion selectivity of an ISE based on **2**·$BF_4$ in 0.500 M bis-tris/HCl, pH 6.00 buffer. The tested anions were prepared using the same buffer and the calibration curves shown represent the average response of the electrode to six or more trials. The anions tested were: hydrogen sulfite (1), perchlorate (2), salicylate (3), benzoate (4), thiocyanate (5), iodide (6), acetate (7), nitrite (8), thiosulfate (9), nitrate (10), sulfate (11), and phosphate (12) [reproduced with permission from reference 46].

aromatic pendent group near the guanidinium moiety might have on the selectivity. Figure 7 demonstrates the remarkable affinity and selective response that ISEs based on ionophore **2** have for hydrogen sulfite.[46] Despite being a very hydrophilic anion, hydrogen sulfite is preferred by several orders of magnitude over the more lipophilic anions perchlorate, thiocyanate, nitrate, etc. The stoichiometry of the complex formed between **2** and $HSO_3^-$ was investigated via a titration experiment using $^1$H-NMR spectroscopy. The initial location of the chemical shifts (ppm) of **2** were observed, and subsequently the change in the chemical shift (Δppm) was recorded as varying amounts of $HSO_3^-$ were added. The change in the chemical shift observed (Figure 8) at various ionophore-to-$HSO_3^-$ mole ratios is evidence of the formation of a complex with a 1:1 stoichiometry.

By incorporating the guanidinium salt **3** (Figure 6) into polymer membranes we were able to investigate the importance of the phenyl side-arm location with respect to the guanidinium moiety on the selectivity of the ISEs. The phenyl ring is located near the guanidinium group in **2** and strong anti-Hofmeister selectivity was observed for the oxoanion hydrogen sulfite. With **3**, which has the phenyl ring located further from the oxoanion binding site, a near-Hofmeister selectivity pattern was observed, with perchlorate being the preferred anion. These studies are important in understanding the structural requirements that must be in place to design oxoanion-selective ionophores. Additional binding sites that are able to participate in cooperative binding of targeted oxoanions must be included. This requirement also exists in protein binding of oxoanions, where typically more than one functional group assists in the complexation of these anions.[40] This biomimetic approach involving guanidinium compounds has also proved successful in the

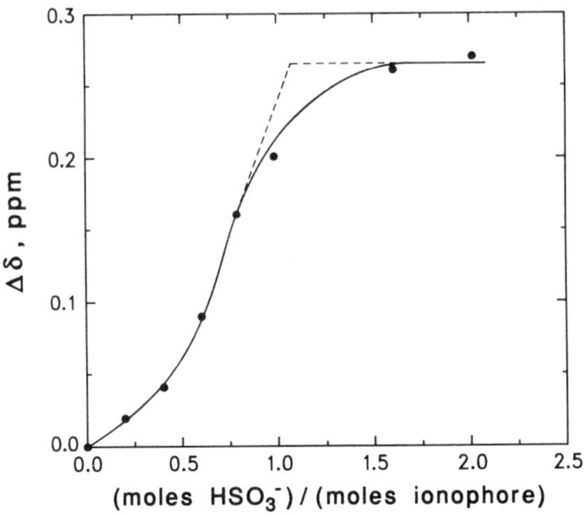


**Figure 8.** Changes in the $^{1}$H-NMR chemical shift of the four $CH_2$ protons that are neighbors to the unprotonated nitrogen in the guanidinium compound **2**·$BF_4$ observed as the mole ratio of sodium hydrogen sulfite to ionophore increases. The ionophore complexes in a 1:1 fashion with hydrogen sulfite.

development of two ionophores selective for the oxoanion salicylate,[47] and has shown promise in the development of selective ionophores for both acetate and phosphate (work in progress).

## SENSOR ARRAYS

The potential applications in which ISEs may be used have grown through the development of chemical sensing arrays. An array of sensors, coupled with an appropriate data acquisition system, extends the limits of sensitivity and selectivity, enhances the confidence level of the data obtained, and makes possible the determination of many components in a complex sample matrix. Earlier mention was made of a successful attempt to mimic some of the functions of olfactory membranes.[4] Another example involves an array of voltammetric gas sensors that was used to develop an integrated microelectronic device that can mimic the human sense of smell.[48] These "electronic noses" have been shown to discriminate effectively between very similar odorants. Potentiometric sensor arrays have also been successfully used to analyze more than one component simultaneously in a sample, and shown to work as effectively as individual ISEs.[49,50] The fabrication of potentiometric microsensing arrays, such as those in Figure 9 for *in vivo* cardiovascular applications, has been facilitated by the use of screen-printing[50] and photolithographic techniques.[51] The ability to mass-produce microchemical sensing arrays with various combinations of ion- and biosensors will greatly expand the use and need for biomimetically designed sensors.

## CONCLUSIONS

It has been demonstrated that by observing and mimicking the way in which nature selectively complexes anions, ionophores can be synthetically designed that are highly selective for these species. This has been demonstrated in the case of vitamin $B_{12}$, porphyrins, protamines, nucleotides/nucleosides, and the guanidinium moiety of arginine. This biomimetic approach toward the development of selective receptors, that can be used to design biofunctional membranes, can be extended to target other anions, in addition to cations, neutral molecules, and proteins. Indeed, there is truth to the familiar phrase: "mother nature knows best."

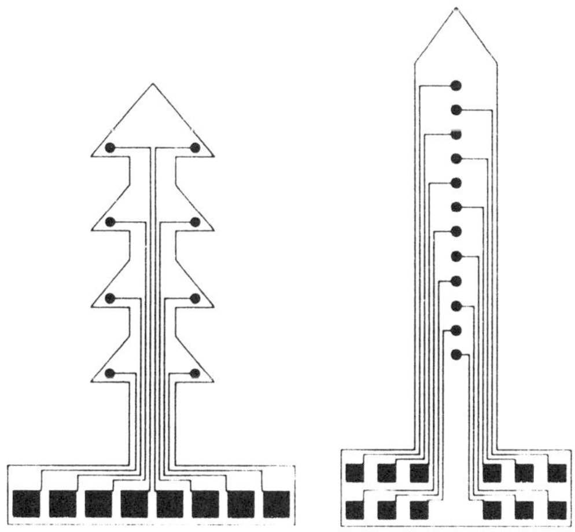

**Figure 9.** Two Kapton®-based microelectrode array designs developed for *in vivo* use [reproduced with permission from reference 51].

## ACKNOWLEDGMENTS

The research described from the authors' laboratory was funded by the National Science Foundation, the Kentucky Space Grant Consortium, and the National Aeronautics and Space Administration. The authors would also like to thank Sylvia Daunert, Antonio Florido, Preeti Bansal, Shelly Wallace Young, and Allan Witkowski for their contributions to this work, which is described in references 19, 20, 23, 27, and 47. We also acknowledge Monsangu Shabangi and Li Sun for the preliminary testing of several guanidinium ionophores.

## REFERENCES

1) Tohda, K.; Naganawa, R.; Lin, X. M.; Tange, M.; Umezawa, K.; Odashima, K.; Umezawa, Y.; Furuta, H.; Sessler, J. L. *Sens. Actuators B* **1993**, *13-14*, 669-672.
2) Siswanta, D.; Hisamoto, H.; Tohma, H.; Yamamoto, N.; Suzuki, K. *Chem. Lett.* **1994**, 945-948.
3) Cygan, A.; Luboch, E.; Biernat, J. F. *J. Incl. Phenom.* **1988**, *6*, 215-220.
4) Krull, U. J.; Thompson, M. *Trends Anal. Chem.* **1985**, *4*, 90-96.
5) Maeda, M.; Nakano, K.; Takagi, M. *ACS Symp. Ser.* **1994**, *556*, 238-251.
6) Rechnitz, G. A. *Chem. Eng. News* **1988**, *September 5*, 24-36.
7) Simon, W. *Swiss Pat.* **1969**, 479870.
8) Simon, W. *Pure Appl. Chem.* **1971**, *25*, 811-823.
9) Covington, A. K.; Kumar, N. *Anal. Chim. Acta* **1976**, *85*, 175-178.
10) Erne, D.; Stojanac, N.; Ammann, D.; Hofstetter, P.; Pretsch, E.; Simon, W. *Helv. Chim. Acta* **1980**, *63*, 2271-2279.
11) Schindler, J. G.; Schindler, M. M. *Fresenius' Z. Anal. Chem.* **1985**, *320*, 258-260.
12) Lockhart, J. C. In *Inclusion Compounds, Vol. 5*; Atwood, J. L., Davies, J. E. D., MacNicol, D. D., Eds.; Oxford University Press: New York, **1991**; pp 345-363.
13) Allen, J. R.; Cynkowski, T.; Desai, J.; Bachas, L. G. *Electroanalysis* **1992**, *4*, 533-537.
14) Ohki, A.; Lu, J. P.; Huang, X.; Bartsch, R. A. *Anal. Chem.* **1994**, *66*, 4332-4336.
15) Lindner, E.; Gráf, E.; Niegreisz, Z.; Tóth, K.; Pungor, E.; Buck, R. P. *Anal. Chem.* **1988**, *60*, 295-301.
16) Schulthess, P.; Ammann, D.; Simon, W.; Caderas, C.; Stepánek, R.; Kräutler, B. *Helv. Chim. Acta* **1984**, *67*, 1026-1032.
17) Schulthess, P.; Ammann, D.; Kräutler, B.; Caderas, C.; Stepánek, R.; Simon, W. *Anal. Chem.* **1985**, *57*, 1397-1401.

18) Stepánek, R.; Kräutler, B.; Schulthess, P.; Lindemann, B.; Ammann, D.; Simon, W. *Anal. Chim. Acta* **1986**, *182*, 83-90.
19) Daunert, S.; Witkowski, A.; Bachas, L. G. *Prog. Clin. Biol. Res.* **1989**, *292*, 215-225.
20) Florido, A.; Daunert, S.; Bachas, L. G. *Electroanalysis* **1991**, *3*, 177-182.
21) Hofmeister, F. *Arch. Exp. Pathol. Pharmakol.* **1888**, *24*, 247-260.
22) Pratt, J. M. *Inorganic Chemistry of Vitamin $B_{12}$;* Academic Press: New York, **1972**; pp 14-29.
23) Daunert, S.; Bachas, L. G. *Anal. Chem.* **1989**, *61*, 499-503.
24) Ammann, D.; Huser, M.; Krautler, B.; Rusterholz, B.; Schulthess, P.; Lindemann, B.; Halder, E.; Simon, W. *Helv. Chim. Acta.* **1986**, *69*, 849-854.
25) Hodinár, A.; Jyo, A. *Chem. Lett.* **1988**, 993-996.
26) Hodinár, A.; Jyo, A. *Anal. Chem.* **1989**, *61*, 1169-1171.
27) Daunert, S.; Wallace, S.; Florido, A.; Bachas, L. G. *Anal. Chem.* **1991**, *63*, 1676-1679.
28) Brown, D.; Chaniotakis, N.; Lee, I.; Ma, S.; Park, S.; Meyerhoff, M. E. *Electroanalysis* **1989**, *1*, 477-484.
29) Chaniotakis, N. A.; Park, S. B.; Meyerhoff, M. E. *Anal. Chem.* **1989**, *61*, 566-570.
30) Park, S. B.; Matuszewski, W.; Meyerhoff, M. E.; Liu, Y. H.; Kadish, K. M. *Electroanalysis* **1991**, *3*, 909-916.
31) Li, X; Harrison, D. J. *Anal. Chem.* **1991**, *63*, 2168-2174.
32) Malinowska, E.; Meyerhoff, M. E. *Anal. Chim. Acta* **1995**, *300*, 33-43.
33) Abe, H.; Kokufuta, E. *Bull. Chem. Soc. Jpn.* **1990**, *63*, 1360-1364.
34) Yuan, R.; Chai, Y. Q.; Lin, D.; Gao, D.; Li, J. Z.; Yu, R. Q. *Anal. Chem.* **1993**, *65*, 2572-2575.
35) Yim, H. S.; Kibbey, C. E.; Ma, S. C.; Kliza, D. M.; Liu, D.; Park, S. B.; Torre, C. E.; Meyerhoff, M. E. *Biosens. Bioelectron.* **1993**, *8*, 1-38.
36) Ma, S. C.; Yang, V. C.; Fu, B.; Meyerhoff, M. E. *Anal. Chem.* **1993**, *65*, 2078-2084.
37) Schmidtchen, F. P.; Gleich, A.; Schummer, A. *Pure Appl. Chem.* **1989**, *61*, 1535-1546.
38) Luecke, H.; Quiocho, F. A. *Nature* **1990**, *27*, 402-406.
39) Herzberg, O.; Reddy, P.; Sutrina, S.; Saier, M. H. Jr.; Reizer, J.; Kapadia, G. *Proc. Natl. Acad. Sci. USA* **1992**, *89*, 2499-2503.
40) Chakrabarti, P. *J. Mol. Biol.* **1993**, *234*, 463-482.
41) Gleich, A.; Schmidtchen, F. P.; Mikulcik, P.; Muller, G. *J. Chem. Soc., Chem. Commun.* **1990**, 55-57.
42) Muller, G.; Riede, J.; Schmidtchen, F. P. *Angew. Chem. Int. Ed. Engl.* **1988**, 27, 1516-1518.
43) Echavarren, A.; Galán, A.; Lehn, J. M.; de Mendoza, J. In *Inclusion Phenom. Mol. Recognit., (Proc. Int. Symp.), 5th*; Atwood, J. L., Ed.; Plenum: New York, NY, 1990; pp 119-124.
44) Schmidtchen, F. P. *Chem. Ber.* **1980**, *113*, 2175-2182.
45) Molina, P.; Alajarín, M.; Vidal, A. *J. Org. Chem.* **1993**, 58, 1687-1695.
46) Hutchins, R. S.; Molina, P.; Alajarín, M.; Vidal, A.; Bachas, L. G. *Anal. Chem.* **1994**, *66*, 3188-3192.
47) Hutchins, R. S.; Bansal, P.; Molina, P.; Alajarín, M.; Vidal, A.; Bachas, L. G. manuscript in preparation.
48) Shurmer, H. V.; Gardner, J. W. *Sens. Actuators, B* **1992**, *B8*, 1-11.
49) Otto, M.; Thomas, J. D. R. *Anal. Chem.* **1985**, *57*, 2647-2651.
50) Goldberg, H. D.; Brown, R. B.; Liu, D. P.; Meyerhoff, M. E. *Sens. Actuators, B* **1994**, *B21*, 171-183.
51) Lindner, E.; Cosofret, V. V.; Ufer, S.; Buck, R. P.; Kusy, R. P.; Ash, R. B.; Nagle, H. T. *J. Chem. Soc., Faraday Trans.* **1993**, *89*, 361-367.

# CLINICALLY USEFUL BIOSENSOR MEMBRANE DEVELOPMENT

Robert R. McCaffrey, Paul A. D'Orazio, Richard W. Mason, Thomas C. Maley and Peter G. Edelman*

Ciba Corning Diagnostics, 63 North Street, Medfield, MA 02052

## Introduction

This paper will begin with a conventional definition of a biosensor and the author's definition of clinically useful. To help define clinically useful, there is a brief discussion of performance specifications and system compatibility. Next there is a description of the function of biosensor membranes and membrane types. Membrane attributes as they pertain to sensor non-linearity is explained in some detail. The complete sensor construction process is described.

This paper describes a specific biosensor which was released for sale in November of 1994. This is the Ciba Corning Diagnostics 800 Series™ Glucose Sensor. Sensor construction is explained. The chemistry, swelling and transport characteristics of the outer membrane are explained in detail. Finally, sensor performance is presented.

## Biosensor Definition

For the purpose of this paper, a biosensor is defined as a combination of biocomponent and transducer for analysis. Biocomponents include biomolecules such as enzymes, antibodies, cells and DNA. Transduction can be by any physical means, such as electrochemical, optical (fluorescence, absorbance, index of refraction change, reflectivity change, surface plasmon resonance), gravimetry, surface acoustic wave or calorimetry. Analysis is defined as measurement of change in concentration or some environmental condition in a quantitative manner. The measurement media can be *in vivo*, *in vitro*, *ex vivo*, or for environmental monitoring.

## Clinically Useful

Clinically useful means meeting performance specifications which are designed to meet customer requirements. This needs to occur after 6 months to 1 year shelf life, for the required use life, after a minimum equilibration time. Six months to 1 year shelf life is desirable to allow worldwide distribution taking into consideration transportation time and

*Biofunctional Membranes*, Edited by
D.A. Butterfield, Plenum Press, New York, 1996

inventory. Clinically useful also implies system compatibility, manufacturability, ease of use, competitive cost per test, etc.

**Performance Specifications**

Table 1 lists a summary of the Ciba Corning Diagnostics 800 Series™ Glucose Sensor performance specifications. These are internal development specifications. While the customer is primarily concerned only with a total error and sample throughput rate, these development specifications separate the sources of error to individual components. These are useful and important during the design stage of the project.

The specifications are based on several sources. The specifications must meet National Committee for Clinical Laboratory Standards (NCCLS) requirements. In addition there are the Clinical Laboratory Improvement Act (CLIA) and the Health Care Financing Administration (HFCA) requirements[1]. There are also practical considerations regarding the level of accuracy needed as a function of the glucose concentration with respect to the different treatments mandated by the results[2]. The performance must also be competitive with existing products in the field. The only other system at this time which measures whole blood glucose simultaneously with blood gases and electrolytes is sold by Nova.

**Table 1: Specifications Summary**

| Parameter | Specification |
|---|---|
| Range of Operations | 10-999 mg/dl |
| Bias from Reference Method, 10-200 mg/dl | ≤ ± 15% |
| Bias from Reference Method, 500 mg/dl | ≤ ± 20% |
| Bias from Reference Method, 999 mg/dl | ≤ ± 30% |
| Imprecision | ≤ 3.3 mg/dL or 4% whichever is greater |
| CarryOver | ≤ 1.0 % |
| Correlation | To the CDC Hexokinase Method |
| End Point Time | ≤ 90 s for 180 cal, 60 s for samples |
| Interference | < 6 mg/dL glucose error @ 2 mg/dl acetaminophen |
| Equilibration time | Three 2 point cals, 1/2 hr, 2 successful 2 point cals in ≤ 4 hours. |
| Slope Drift | ≤ 3.5% in 30 minutes |
| Sample Induced Drift | ≤ 3.5% (or 3 mg/dl, whichever is greater) total drift between calibrations when sampled at maximum sample rate. |
| First Sample Effect | ≤ 4% (or 3.5 mg/dl, whichever is greater) difference from first to second sample after cal. |
| Recovery Imprecision | ± 6.4% at 2 SD |
| Range of Glucose Slope in Active Electrode | 0.04 < slope< 0.40 nA/mg/dl |
| Range of Acetaminophen Slope in Inactive Electrode | 0.25 < slope< 25 nA/mg/dl |
| Use Life | > 30 Days or 1000 Samples |

In addition to error contributions from sources such as carry over and drift, it is also necessary to evaluate error from interferents which are endogenous in blood or are added as therapeutic drugs or drugs of abuse. The list shown in Table 2 indicates some of the interfering substances which have been evaluated as part of the development of the 800 Series™ Glucose Sensor.

The summary table (Table 1: Specifications Summary) breaks down all the individual sources of error that contribute to total error. The total error is really the bottom line in terms of allowable error because it is what the customer sees. Our specification for total error is depicted graphically in Figure 1. As you can see from this figure, the allowable total error is small in the range near critical decision points around 80 and 200 mg/dl, but at high glucose (≥ 300 mg/dl) there is more allowable error, as the physician's decision is not as critically influenced by a few mg/dl.

**System Compatibility**

On a stand alone glucose analyzer, the system can be relatively simple and optimized for the single analyte. The sensor described in this paper however resides on a much more complicated system. In addition to glucose, this system does simultaneous measurement on CO-ox fractions, blood electrolytes, ($Na^+$, $K^+$, $Ca^{++}$, $Cl^-$, Ag/AgCl reference), and blood gases (pH, $pCO_2$, and $pO_2$). A lactate sensor is scheduled to be added to the system in 1995. Because of the number of parameters measured, development of a glucose sensor to be integrated into this system was a formidable challenge. This system integration was the most significant and critical part of sensor development.
The system controls data acquisition, fluid control, sample introduction, calibrations, washes between samples, etc. The system consists of hardware, pumps, tubing and electronics. Software is developed for data processing, reporting and storage of results and software for the user interface and for error detection.

**Table 2: Interferent Testing**

| INTERFERENT* | Therapeutic or Normal Concentration | NCCLS High Test Level | 860 Test Concentration |
|---|---|---|---|
| Drugs | | | |
| acetaminophen | 1.0-2.0 mg/dL | 20 mg/dL | 20 mg/dL |
| salicylic acid | 15-30 mg/dL | 50 mg/dL | 50 mg/dL |
| ethanol | — | 350 mg/dL | 350 mg/dL |
| chlorpromazine | 0.1 - 0.5 mg/dl | 5 mg/dl | 5 mg/dL |
| dopamine | 0.015 mg/dl | 13 mg/dl | 12 mg/dL |
| Endogenous Substances | | | |
| L-ascorbic acid | 2.0 mg/dL | 3.0 mg/dL | 6 mg/dL |
| uric acid | 10 g/dL | 20 mg/dL | 10 mg/dL ** |
| lactate | 20 mg/dL | 100 mg/dL | 100 mg/dL |
| acetoacetate | 1.0 mg/dl | 20 mg/dL | 40 mg/dL |
| hydroxybutyrate | Insignificant | 200 mg/dL | 200 mg/dL |
| urea | 38 mg/dl | 500 mg/dL | 500 mg/dL |
| creatinine | 1.5 mg/dl | 30 mg/dl | 30 mg/dL |
| bilirubin (total) | 1.2 mg/dl | — | 34 mg/dL |
| bilirubin (direct) | 12 mg/dl | 20 mg/dl | 30 mg/dL |
| Anticoagulants | | | |
| heparin | 4000 IU/dL | 8000 IU/dL | 8000 IU/dL |
| fluoride | 600-1000 mg/dL | 1000 mg/dL | 1000 mg/dL |
| citrate | 500 mg/dL | 1000 mg/dL | 1000 mg/dL |
| oxalate | 400 mg/dL | 800 mg/dL | 800 mg/dL |

* Interference Testing in Clinical Chemistry. NCCLS Document EP7-P Vol. 6 No. 13 .
* *Highest soluble concentration.

The system also includes reagents (calibrants, wash and 2 cleaning solutions). The reagents include calibrants, wash and cleaning solutions. There are 2 gas calibrators for $pO_2$ and $pCO_2$, 2 electrolyte calibrators containing $Na^+$, $K^+$, $Ca^{++}$, $Cl^-$, and 2 metabolite calibrators containing glucose and lactate.

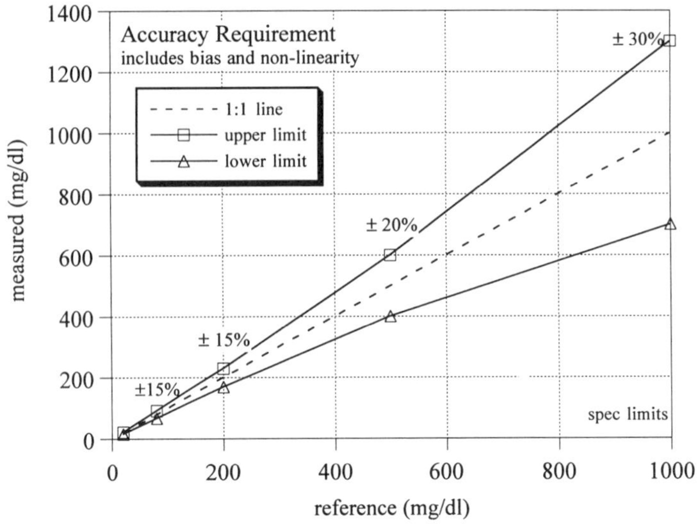


**Figure 1: Total Error. The upper and lower lines represent the upper and lower total error limit. Reference values are from the CDC Hexokinase Reference Method. Measured is from the Ciba Corning 860 Analyzer.**

Quality Control test fluids are a product sold separately which can also be considered a part of the total system. The QC's contain controlled levels of all analytes to be measured. In the case of the multianalyte 800 Series of which this glucose sensor is a part, the QC used is Certain GL™. Certain GL™ is an ampouled product which contains controlled, stable levels of $Na^+$, $K^+$, $Ca^{++}$, $Cl^-$, $pO_2$, $pCO_2$, pH, glucose and lactate. There is also a dye based QC (Certain Elite™) which, in addition to the gases and electrolytes, contains dyes to give 4 positive hemoglobin fractions for the 800 Series™ instruments which have the CO-oximeter module added. The QC's are formulated to work with the sensors, and the sensors may have to be modified to work with the QC's.

## Biosensor Membrane Function

Why are membranes used on biosensors? There are several reasons. Membranes can be used for attachment or immobilization of the biocomponent to or on the transducer. Or the membrane can be part of the transducer, such as usage for attachment of chromophores or fluorophores. The properly chosen membrane can help stabilize the biocomponent, usually a protein which needs to maintain its native conformation to be active. A membrane can help extend the use-life of the device by protecting the *transducer* from fouling. Use-life extension can also be achieved by prevention of fouling of the *biocomponent*. For some applications, such as for *in vivo* sensors, non-thrombogenic characteristics may be critical attributes supplied by the membrane. Membranes can extend the useful range of the sensor by providing transport limitation. Finally, membranes can impart some permselectivity which can minimize interferent concentration or maximize the concentration of a necessary cofactor, such as oxygen in the case of sensors based on oxidase enzymes. A good discussion of membranes for sensors can also be found in Reference 3.

## Membrane Types

There are several membrane types that are used for biosensors. These include solution/diffusion type, track etch, composite and tortuous path membranes. Figure 2 shows these different types. Good examples of the solution/diffusion type membranes are hydrogels. There are many types of hydrogels that have been applied to sensors, such as poly(hydroxyethylmethacrylate) and crosslinked poly(vinyl alcohol). Track etch membranes are unique. They are made by exposing a film to neutron bombardment, followed by development in aqueous base. The neutron bombardment causes molecular level damage. This molecular damage is magnified during the base etch, resulting in pores that are narrow straight channels through the film. These track etch pores can be as small as a few nanometers and be as large as few microns. Track-etch membranes have conventionally been made from polycarbonate, but recently other types of polymer films have also been used, such as cellulose nitrate, cellulose triacetate, polyethyleneterephthalate, polyimide, polypropylene, and polyvinylidene fluoride[4,5]. Tortuous path or asymmetric membranes can also be used on sensors. These are made at the free film stage by using a solvent/non-solvent film casting approach[6]. They are then applied to the sensors after film formation.

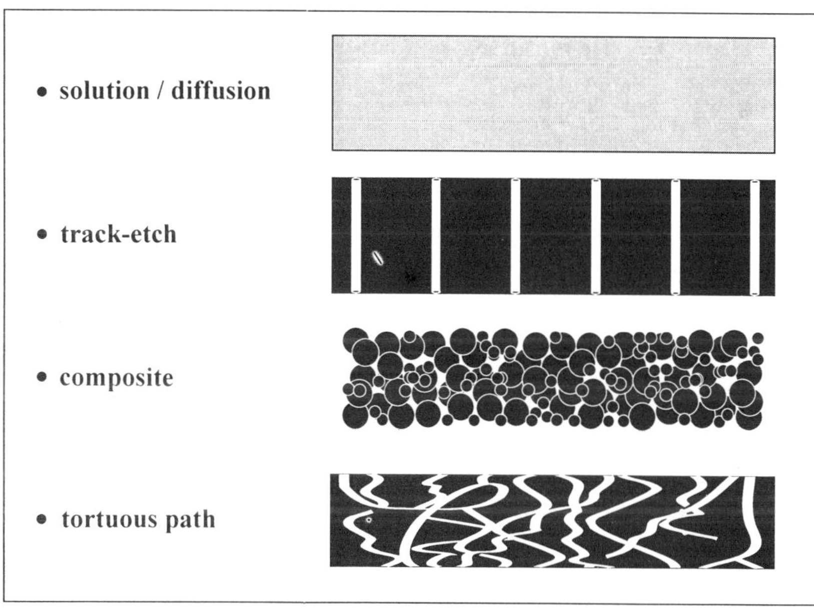


**Figure 2: Membrane Types. See Text for explanation.**

Early glucose biosensors utilized a three layer membrane held in place over an anode and counter/reference electrode. The three layer membrane consists of glucose oxidase crosslinked with glutaraldehyde sandwiched between cellulose acetate and track etch polycarbonate (Nucleopore®). A diagram is shown in Figure 3.

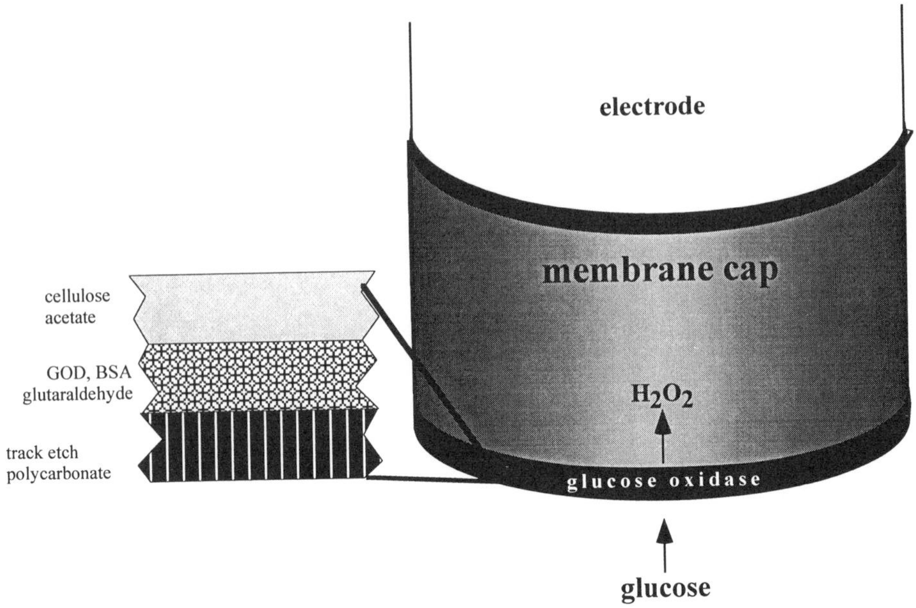


**Figure 3: Early glucose sensor design with three layer membrane system. The sensor is based on glucose oxidase sandwiched between cellulose acetate and track-etch polycarbonate. The enzyme containing film converts glucose to $H_2O_2$. The cap system holds the membrane over the amperometric electrode.**

This membrane system is still in use by Yellow Springs Instruments and Nova for glucose and lactate. The Nucleopore® membrane serves to exclude large protein and limits transport of substrate. The cellulose acetate acts as a size exclusion membrane which selectively transports the peroxide generated from the bioreaction while limiting the transport of electrochemical interferents including drugs such as acetaminophen (paracetamol, Tylenol®, 4-methyl phenyl acetamide) and also limiting the transport of endogenous electrochemical interferents such as ascorbic and uric acid.

The detection scheme most commonly utilized by glucose oxidase containing electrochemical sensors is depicted in Figure 4. β-D-Glucose diffuses into the enzyme environment, where it is converted to δ-gluconolactone. The enzyme is converted from oxidized to reduced. On reconversion of the enzyme from reduced back to oxidized, $O_2$ is converted to $H_2O_2$. The reaction at the electrode is oxidation of the $H_2O_2$. The current generated is proportional to the glucose concentration.

As can be discerned in Figure 4, there are other possible ways to detect glucose. There is generation of gluconolactone, which is automatically converted to gluconic acid resulting in a pH change which can be measured by techniques common for pH measurement. There is also the consumption of $O_2$. Therefore oxygen electrodes with bound GOD can also be used for measurement of glucose concentration[7]. In fact, some of the first biosensors were of the $O_2$ measurement type.[8,9,10,11]

A schematic of the electrode reaction is seen in Figure 5. The $H_2O_2$ is oxidized at the electrode to yield electrons (which gives rise to the measurement current), protons and $O_2$. As can be seen in this schematic, the electrochemical oxidation of $H_2O_2$ also gives rise to a pH and $O_2$ concentration change.

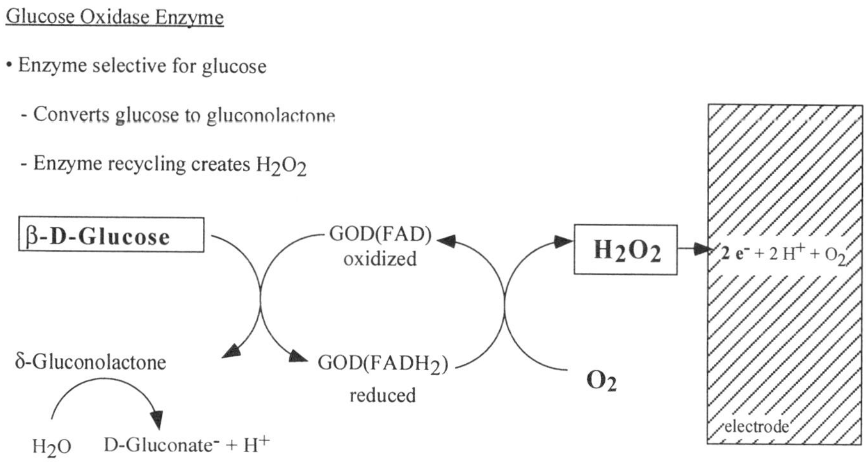


**Figure 4: Glucose Reaction at Electrode**

## Sensor Non-Linearity

In a biosensor of the design described in this paper, the membrane is very important in minimizing response non-linearity. Non-linearity in and of itself may not be a problem, with the help of non-linear correction algorithms in the system software. But if the non-linearity changes over time, then a two point calibration, as is conventionally used, will not be sufficient to measure high concentrations. A typical non-linear response is shown in Figure 6.

For the enzyme electrode based on peroxide oxidation, (using an oxido-reductase enzyme), there are several sources of non-linearity. The membrane can address several of these sources. The non-linearity can be due to $O_2$ limitation, enzyme limitation, electrochemical area limitation or byproduct inhibition. Refer back to Figure 4 to see these contributions to nonlinearity. By performing controlled experiments, the specific contribution(s) can be identified. $O_2$ is used by the enzyme to regenerate the active form (oxidation of $FADH_2$ to FAD). If there is insufficient $O_2$ (at high glucose concentration), the production of $H_2O_2$ will fall off ($O_2$ limitation). High glucose concentrations will at some point saturate the amount of enzyme (enzyme limitation). Membranes can simultaneously help correct both the problem of $O_2$ limitation and enzyme limitation. By increasing the transport ratio, $O_2$ / glucose, a membrane can improve sensor linearity. By the same token, a substrate limiting membrane will improve linearity if the non-linearity is due to electrochemical area limitation.

Glucose Electrode Reaction

$$H_2O_2 \longrightarrow \boxed{2\,e^-} + 2\,H^+ + O_2$$

- Oxidation of hydrogen peroxide occurs
- Electrons are generated (current is generated)
- The amount of current depends on the amount of hydrogen peroxide
- The amount of hydrogen peroxide depends on the amount of glucose

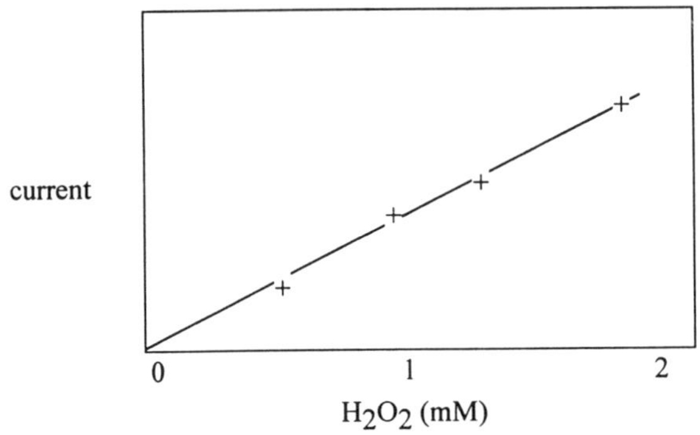


**Figure 5: Peroxide Oxidation Reaction**

To develop a membrane to help with byproduct inhibition is an interesting challenge. Refer back to Figure 4. In the case of our glucose sensor, the byproducts that could affect sensor output are gluconolactone (gluconate and proton). If it is the pH drop which inhibits the reaction, the rate limitation can be exacerbated by protons from the electrochemical reaction. One can perform experiments to look at each of these affects individually. Based on the results, a membrane can be designed that has different permselectivities to allow rapid transport of byproducts away from the generation source while still limiting transport of reactant into the electrode.

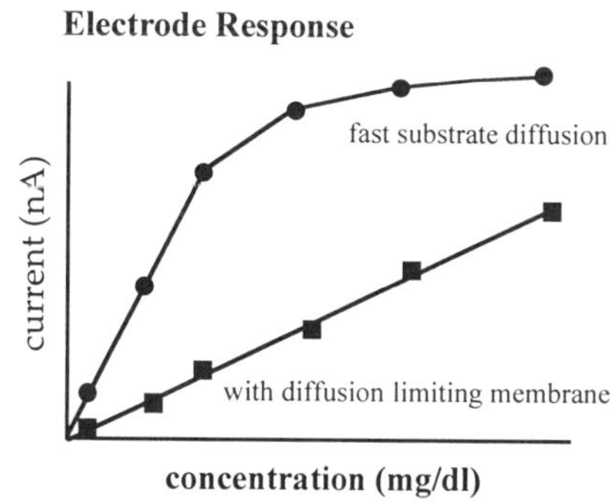


**Figure 6: Sensor Non-Linearity. Non-linearity may be due to 1) oxygen limitation; 2) enzyme limitation; 3) electrochemical limitation; and/or 4) byproduct inhibition.**

## Sensor Construction Process

The sensors are constructed using screen printing and spin coating techniques. Both of these are fabrication methods borrowed from the microelectronics industry. The electrochemical transducer is screen printed, including the layer containing the enzyme component in platinum activated carbon (PAC). Use of platinum activated carbon brings two advantages. The first is that the applied potential for peroxide oxidation can be lower. The surface is electrocatalytic for peroxide oxidation. This lessens the contribution due to other oxidizable interferences. The second advantage is seen in the large, macro-electrode with the PAC bound to carbon paper in a Teflon™ matrix. There is less $O_2$ dependence for these electrodes when using oxidase enzymes that need $O_2$ as a cofactor for regeneration of active form[12].

Application of the PAC layers is followed by spin coating of the cover membrane. To the point just prior to sensor packaging into individual plexiglass bodies, fabrication is done at the wafer level. To complete assembly, the wafers are singulated to individual sensors and ultrasonically welded into a plexiglass body which provides electrical contacts to the instrument as well as a flow channel to the sensor area. The last step is to bag, box and label for shipping. Storage conditions are at 4°-8°C.

Screen printing is done to make the basic electrochemical transducer. Starting from a bare ceramic blank, the layers are printed as indicated in Figure 7.

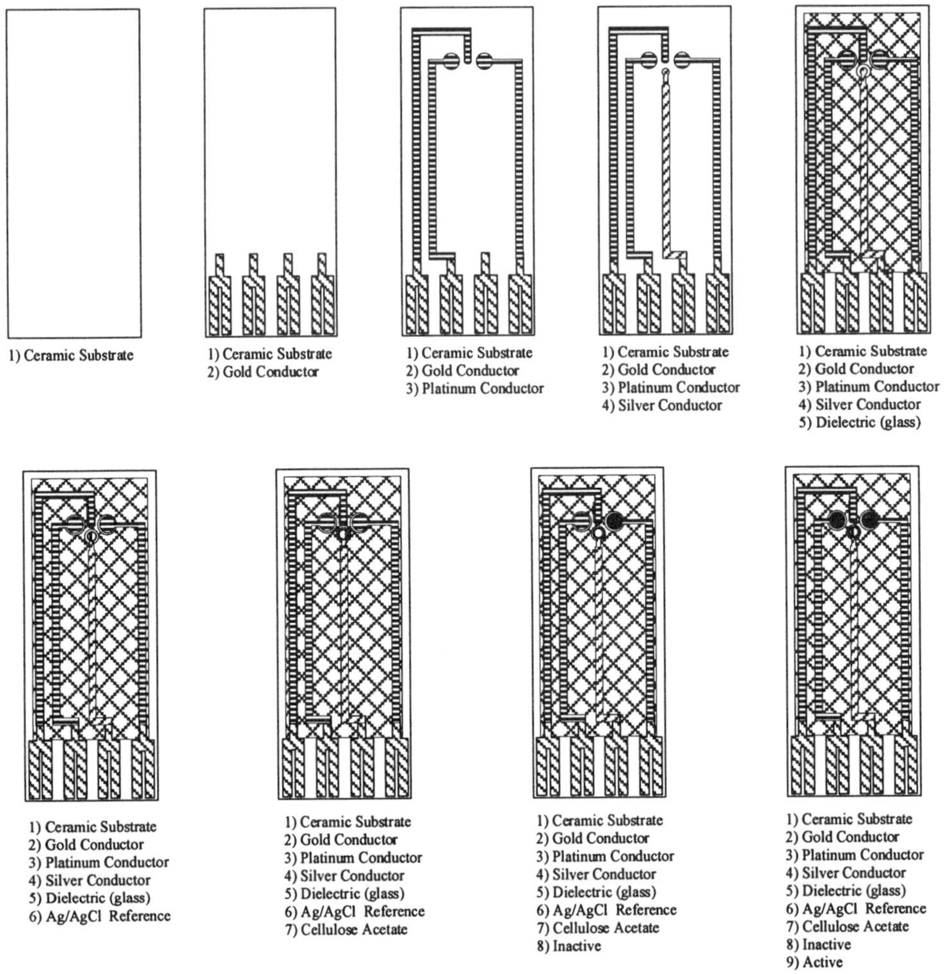


**Figure 7: Thick Film Process**

Over the working electrodes are printed a layer of PAC formulated into a printing ink which also contains bovine serum albumin and enzyme. This sensor has a correcting electrode which is a working electrode that contains no enzyme. Use of this type of dual working electrode system has been described earlier[13,14]. The signal due to interferents, such as acetaminophen, ascorbate or urate are electronically subtracted out. Our sensor is the first example of a commercial sensor that uses this principle. One individual chip is shown in Figure 8. Here can be seen the two working electrodes.

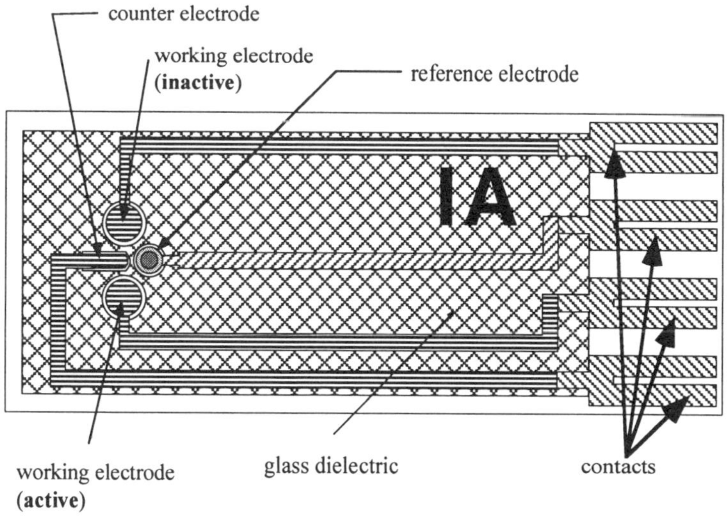


**Figure 8: Single Sensor Chip**

The process we use allows us to fabricate 40 sensors at a time on a 40-up wafer. An example of a full wafer is shown in Figure 9. After the PAC layers are applied, the cover membrane is applied by spin coating. This is a water based silicone emulsion.

After fabrication is complete, the individual sensor chips are assembled into the individual flow cells which fit into the sample block of the instrument. A diagram of the packaging assembly is shown in Figure 10.

## Spin Cast Water Based Silicone Emulsion for Cast -In-Place Membrane

A practical requirement for all planar sensors is that the membrane be physically attached to the electrode. This needs to be done in such a way that no air bubbles can be

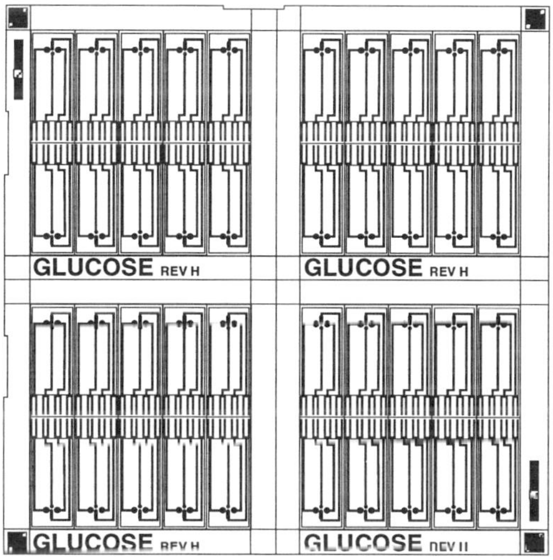


**Figure 9: Full 40 Up Wafer, 2" x 2"**

trapped under the membrane. A methodology that allows for this is to cast the membrane in place on the planar electrode. There are several ways to do this that are known in the art. These include dip coating, spot casting, spray coating and spin coating. Spin coating can be followed by some photopatterning. For the sensors which are the subject of this paper, photopatterning was not necessary.

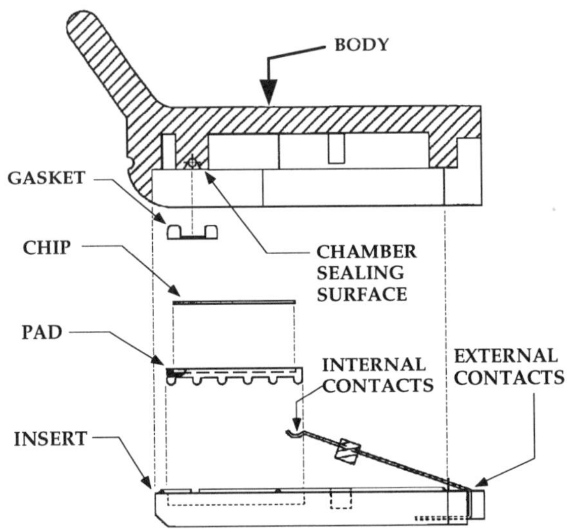


**Figure 10: Chip Packaging in Flow Cell**

The emulsion is formulated from several ingredients. Each play a role in determining the final transport characteristics. A list of the ingredients are shown in Figure 11. The silicone polymer serves to provide structural integrity, while providing high gas permeability. High gas permeability is desirable to allow $O_2$ transport needed for the bioreaction (see Figure 4: Glucose Reaction at Electrode). This type of membrane has been used in larger "macro" electrodes.[15]

The characteristics of this material has been described in some detail. [16,17] The amorphous silica acts as a filler and crosslink site. The silanol OH groups of the PDMS oligomer react to the silanol OH groups of the silica as well. Since there are many silanol groups on each silica particle, the crosslinking can be extensive. There are two low molecular weight amines present. These evaporate during the film curing stage. As they evaporate, the pH drops. Lowering of the pH accelerates the silanol condensation reaction. The copolymer of acrylic acid and ethyl acrylate serves as a pH sensitive rheology modifier. It also imparts some anion rejection characteristics to the membrane which helps screen out negatively charged interferences, such as ascorbate and urate. The dodecylbenzenesulfonic acid is an acid catalyst which catalyzes silanol condensation. Finally there is some tin catalyst present which remains from the reaction to form the siloxane oligomers from cyclic siloxanes.

The cure reaction and film formation reactions are shown in Figure 12 & Figure 13. While the film curing and physical chemical characteristics have been described,[16,17] a detailed characterization of the transport properties has not been presented elsewhere.

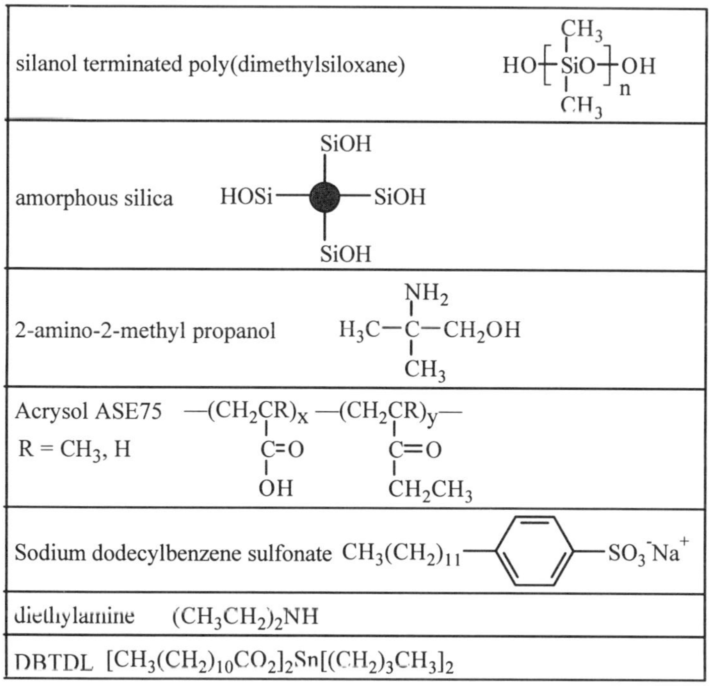


**Figure 11: Ingredients in Water Based Silicone Emulsion**

The precured emulsion depicted in Figure 12 is the material that is spin cast on the wafer. After the film formation reaction, the film is an insoluble elastomer. Consideration of the illustration in Figure 13 suggests the locus of transport through this film. In the precured emulsion, there exists micellular emulsion particles with reactive groups on the surface. After film formation, these particles are bound together. All the hydrophilic material may phase separate into domains between the hydrophobic silicone latex spheres. The transport of glucose must necessarily be through this hydrophilic domain.

Spin cast film thickness can be controlled by controlling the viscosity of the solution and the spin speed. This is well know from the micro-electronics industry. In practice, we control the spin speed and hold the viscosity relatively constant. Figure 14 shows a spin speed versus thickness calibration curve. These curves always follow power law relationship. Here the relationship is: thickness $= 5.1\times10^{5}$ (spin speed)$^{-1.24}$. Thickness is in microns and spin speed is in rpm.

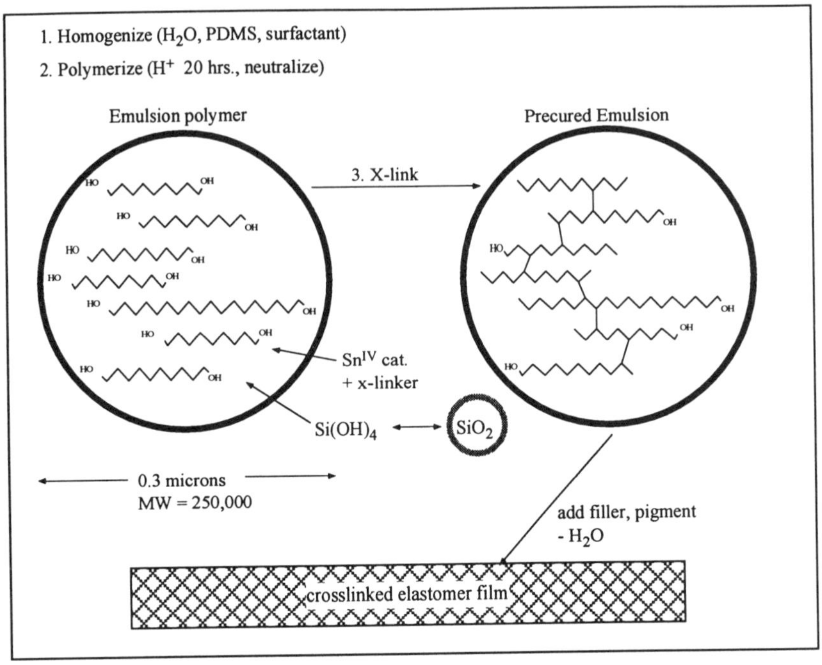


**Figure 12: WBSE Cure Reaction**

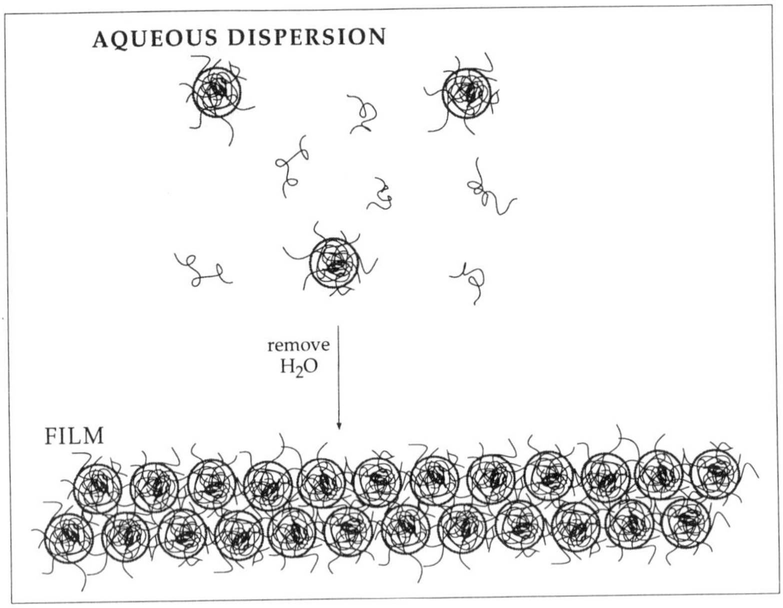


**Figure 13: WBSE Film Formation**

## WBSE Film Swelling Characteristics

The cover membrane material is mostly silicone. However, due to the hydrophilic components, it has some characteristic swelling properties in aqueous media. The behaviors exhibited are characteristic of an ion containing elastic matrix.

Charged polyelectrolyte behavior was observed when we characterized the water uptake difference between DI water and electrolyte solution. This electrolyte solution contains near physiological concentrations of $Na^+$, $K^+$, $Ca^{++}$, and $Cl^-$. The results are shown

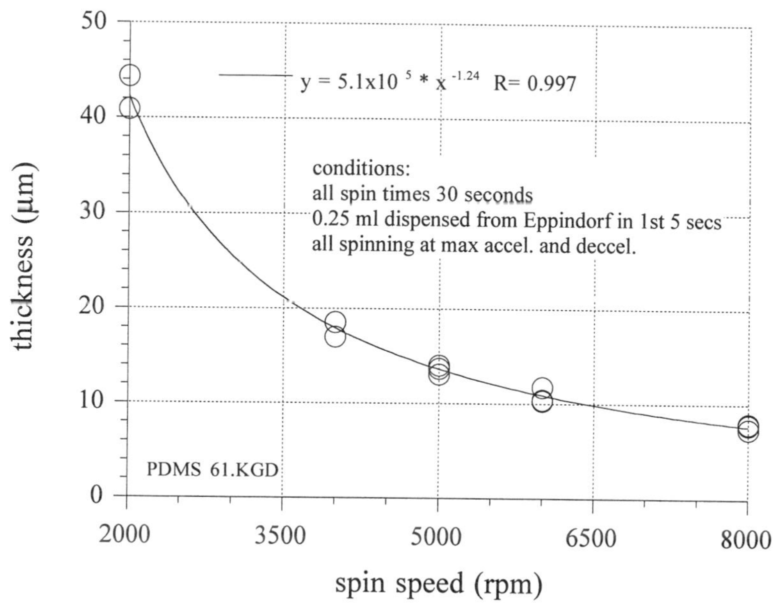


**Figure 14: Spin Speed vs. Thickness Calibration Curve**

graphically in Figure 15. In this graph the percent swelling is calculated as: % swelling = [100*(wet weight-dry weight)/wet weight]

Six films were knife cast onto teflon sheets After drying overnight in a vacuum at room temperature the films were weighed. Three were placed in DI water, three were placed in electrolyte solution. All swelling was done at 37°C, with gentle rocking. After 24 hours, the films were blotted dry with paper towel and weighed. Next the three films that were in DI water were switched in the electrolyte solution and the films that were in electrolyte solution were switched into DI water. Again after 24 hours the films were blotted dry and weighed.

The films that swelled the most were the ones first placed into DI water. Due to the low degree of shielding in DI water, swelling is increased for charged polyelectrolytes, as compared to neutral polymers. This is analogous to the increase in viscosity seen for linear polyelectrolytes going to solutions of lower ionic strength[18]. This has also been observed with crosslinked polyelectrolyte gels[19]. This increased swelling in low ionic strength media results because like charges have minimal shielding from other like charges so the charge repulsion is more pronounced. Our system is neither linear polyelectrolyte nor polyelectrolyte gel. It can be described as linear polyelectrolyte encapsulated, trapped or immobilized into a crosslinked hydrophobic gel matrix. By definition this is a semi-interpenetrating polymer network.

When this film was placed into electrolyte solution, the percent swelling decreased from 19.1% to 7.5%. This is to be expected as now there are many ions in solution which shield the like charges from each other so the charge repulsion is less.

When films first go into the electrolyte solution (soak1: 4,5&6) the percent swelling is greater than for films that first went onto DI water, then were transferred to the electrolyte solution (soak2: 1,2&3). This can be explained by loss of some of the water-soluble polyelectrolyte during the DI soak for films 1,2&3. There are charged, ionic components which dissolve out of the film. This is indicated by the decreased degree of swelling for films 4,5&6 after transfer to DI water after 24 hours in electrolyte. The swelling is less than for films 1,2&3 which start out in the DI water.

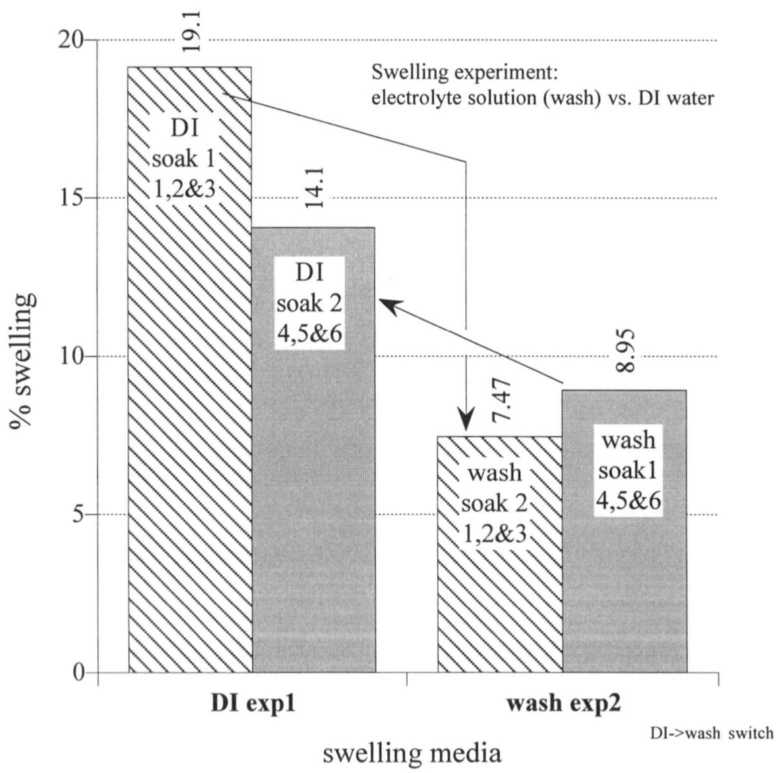


**Figure 15: Effect of Ionic Strength on WBSE Film Swelling**

The charged component of these membranes has been further characterized in an evaluation of the pH sensitivity of swelling. A sheet of cast WBSE was cut into 24 pieces. Eight solutions were prepared at pH increments of about 0.5 pH units, spanning the range from pH 4 to 7. After overnight drying in a vacuum oven at room temperature, the films were weighed and three were placed at each pH and kept in a gently agitating bath at 37°C. After 24 hours the films were blotted dry and weighed. The results are shown in Figure 16. There is an equivalence point at about pH 6.3.

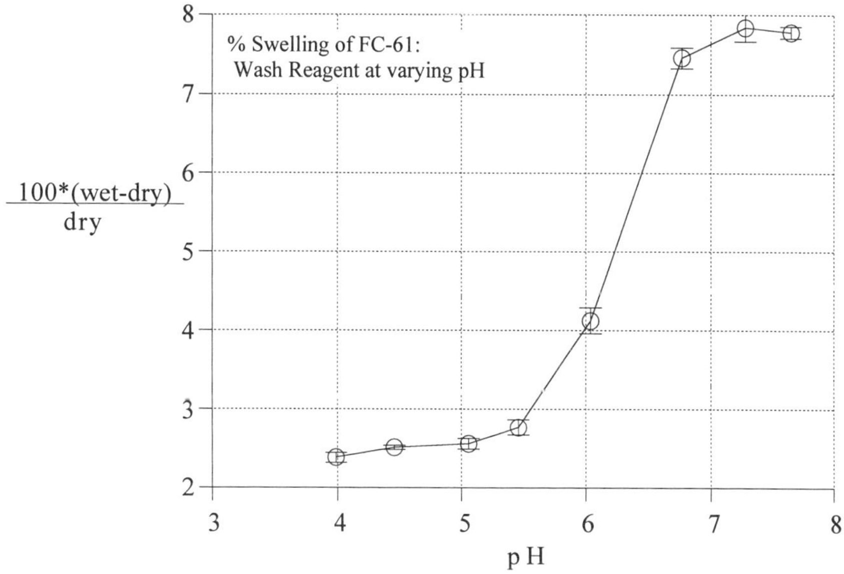


**Figure 16: Effect of pH on WBSE Swelling**

This equivalence point in a swelling curve is typical for acrylic acid containing copolymers[20]. We have determined that the formulation contains a poly(acrylic acid-co-ethyl acrylate) (by FTIR spectral matching). Therefore it can be concluded that pH dependent swelling behavior is due to this material. Poly(acrylic acid-co-ethyl acrylate) is added as a rheology modifier to the WBSE formulation.

In order to further characterize films made from this WBSE material, a permeability measurement device was fabricated. This device is shown in Figure 17. The device consists of a plexiglass membrane-holding fixture with a chamber on either side of the membrane. To hold WBSE films in place between two solutions with out mechanical deformation, WBSE films were cast onto ceramic filter supports (Anapore™). One chamber is for the feed solution, the other is for the permeate. The feed solution contains the permeant, and is 100 ml. The permeate solution carries whatever diffuses across the membrane past the detector. For this apparatus, the detectors used are electrochemical sensors, since the permeates of interest are electrochemically active. The permeate volume is 1/20th of the feed volume (5 ml), to maximize the concentration of the diffusing species.

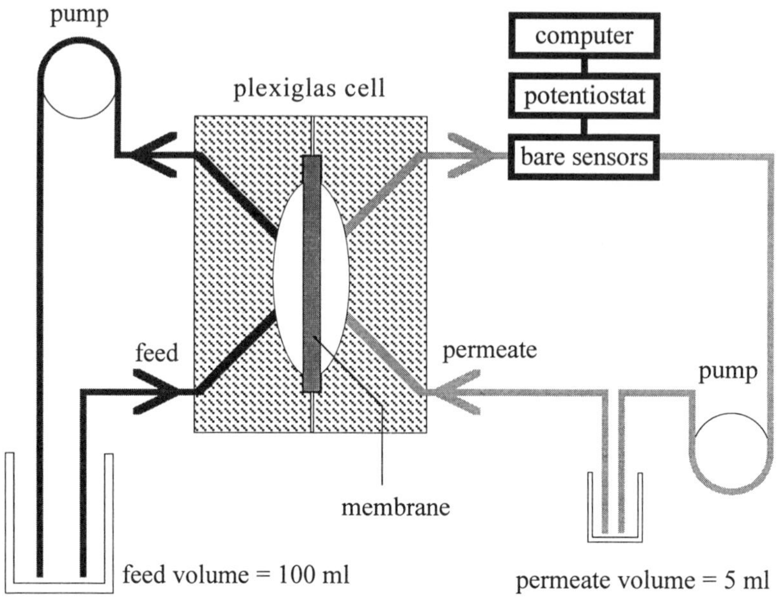


**Figure 17: Membrane Diffusion Set-up**

This apparatus was used to evaluate dependence of glucose transport on pH. This experiment would be confounded using our standard sensor because of the pH effect on enzyme activity. This would have to be decoupled from differences in transport.

Two runs of 4 replicates were done at pH 6.8 and two runs of 4 replicates were done at pH 5.5. At pH 6.8 the membrane is in its swollen state, and at pH 5.5 it is less swollen. These pH's bracket the swelling transition (see Figure 16). As expected the transport is significantly greater at pH 6.8, as seen in Figure 18.

To characterize membrane transport decoupled from enzyme activity, we characterize the transport of three electroactive molecules. Acetaminophen (4-acetamidophenol) is of interest because it is used extensively in hospitals for pain therapy. It is very electrochemically active, as are other phenolic molecules. Ascorbic acid is another molecule of interest. It is present in blood at an endogenous level of 2 mg/dl, so it can cause interference. Hydrogen peroxide is the species that is actually oxidized as the product of the bioreaction. It is very useful to characterize the peroxide response, to characterize membrane transport to small molecules and to determine the basic electrochemical activity of the electrode. The structures of these molecules are shown in Figure 19.

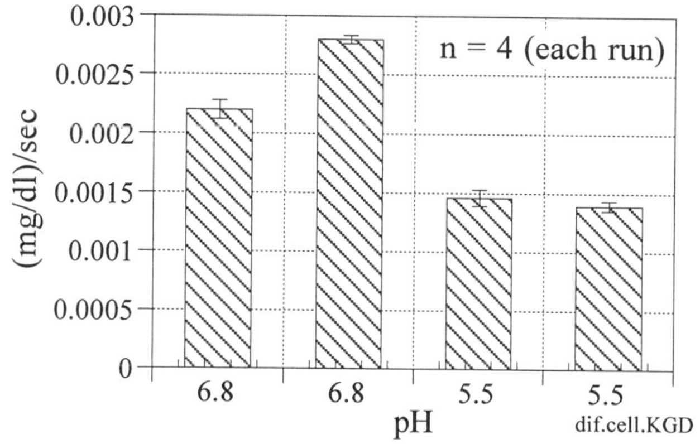


**Figure 18: pH Effect on Glucose Diffusion**

Membranes that are anionic, as the membrane discussed in this paper, are permselective for neutral and cationic molecules over anionic molecules. Anion rejection can be a good property for a membrane that needs to transport a neutral molecule, such as glucose, while rejecting negatively charged interferents, such as ascorbic and uric acids.

To evaluate the anion rejection properties of this membrane, the diffusion cell can be used, or actual sensors can also be used. An additional parameter that we investigated was how surfactants affect anion rejection. In this experiment, we compared two non-ionic surfactants, Brij 35 and Brij 700. These molecules have similar structures, $CH_3(CH_2)_x$-$O(CH_2CH_2O)_yH$, where x and y are 11 and 25 for Brij 35. For Brij 700, x and y are 17 and approximately 100, respectively.

The results are shown in Figure 20. On the y-axis is plotted the ratio of ascorbate slope to acetaminophen slope. The slopes are the electrochemically measured oxidation responses with the electrode poised at 400 mV (Ag/AgCl ref.). In the presence of Brij 35, the ratio of ascorbate transport to acetaminophen transport is less than with Brij 700 in solution. This means the anion rejection is greater with Brij 35, while Brij 700 decreases the permselectivity.

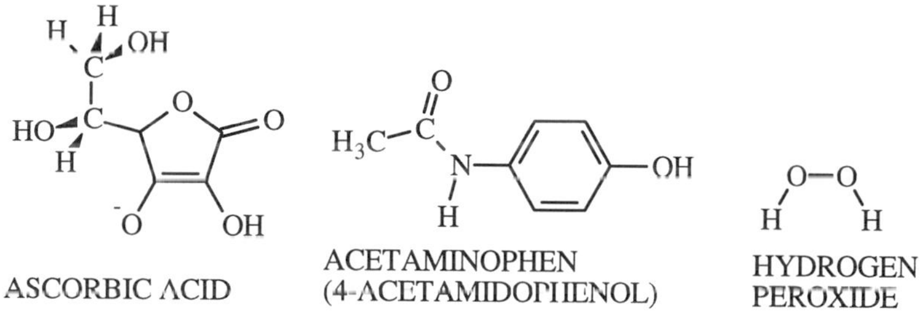


**Figure 19: Membrane Probes**

A possible explanation for this lies in the differences in the structures of the two surfactants. To affect the permselectivity of a neutral permeant compared to a negatively charged permeant, the surfactant must affect the degree to which the membrane's own anionic character is shielded. Brij 700 has 4 times the length of poly(ethylene oxide) segment as compared to Brij 35. This is the most distinguishing difference. It is known that polyethylene oxide can hydrogen bond with poly(acrylic acid), even to the point of forming insoluble complexes[21]. This phenomenon has also been observed with Brij-type surfactants and poly(methacrylic acid)[22]. Therefore it is proposed that the reason to explain why Brij 700 allows higher anionic transport than Brij 35 is because of the greater degree to which the PEO segment can complex with the polyacrylic acid copolymer, which decreases the shielding. However, in order for PEO to hydrogen bond to acid groups, the pH must be low enough for the acid groups to be protonated[23]. If the acid groups are protonated, they are neutral. If the acid groups are neutral, then they will not affect the transport of anionic molecules. Because we have always observed some degree of anion rejection, there must always be some dissociated acrylate present.

**Sensor Performance**

Part of the definition of Clinically Useful includes manufacturability. Manufacturability means reproducibility within-lot and lot-to-lot. Not only is the process described in this paper robust, (there have been no bad lots made since we started production in November of 1994), but also the reproducibility of all lots has been excellent. A graph showing start-up slope variability over 9 lots is shown in Figure 21.

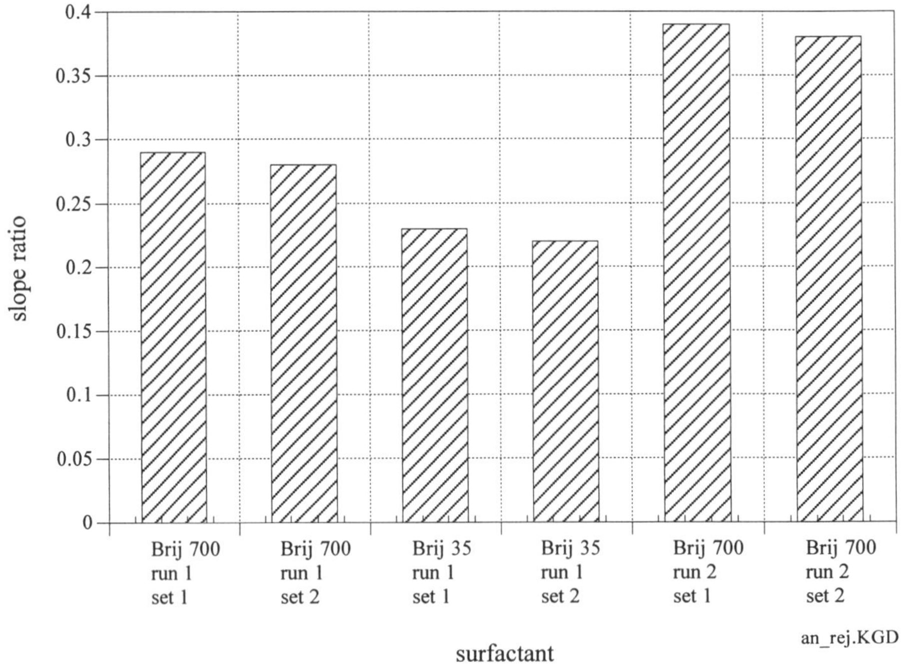


**Figure 20: Surfactant Effect on Anion Rejection**

Sensor uselife is defined as the time that a sensor meets *every* performance requirement. To the customer, it is time that a sensor meets the Total Error Requirement. A measure of sensor output over life is only a measure of enzyme activity. Since the sensors are calibrated in use at least every two hours, slope, in and of itself, is not an indicator of performance.

Yet the slope change over life is can be correlated to changes in sensor performance. For this sensor configuration, what we typically see is shown in Figure 22. For the first third of sensor life the slope increases. Then for the last two thirds of sensor life the slope declines. The slope cut-off point has been designated as 0.05 nA/mg/dl. This is based on system electronics, as well as the fact that low slope sensors generally have increased response times and worse carryover.

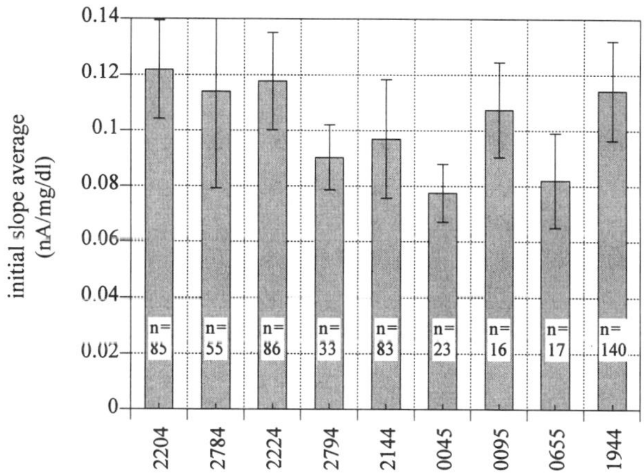


**Figure 21: Slope Reproducibility Lot to Lot**

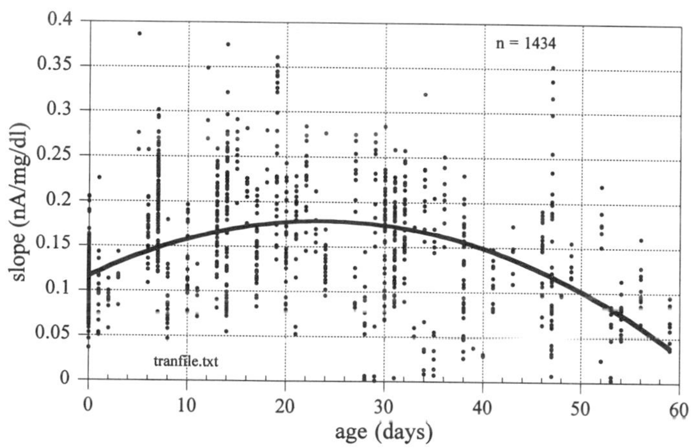


**Figure 22: Change in Slope over Sensor Life**

The best measure of sensor life is how the sensor reads whole blood as compared to a reference method. The NCCLS has specified that the reference method for clinical glucose analyzers is the Center for Disease Control (CDC) Hexokinase Method[1]. The recovery values for whole blood (normally at 80 mg/dl) over time for a population of sensors is shown in Figure 23. The normal range for serum glucose in non-diabetic, fasting adults is

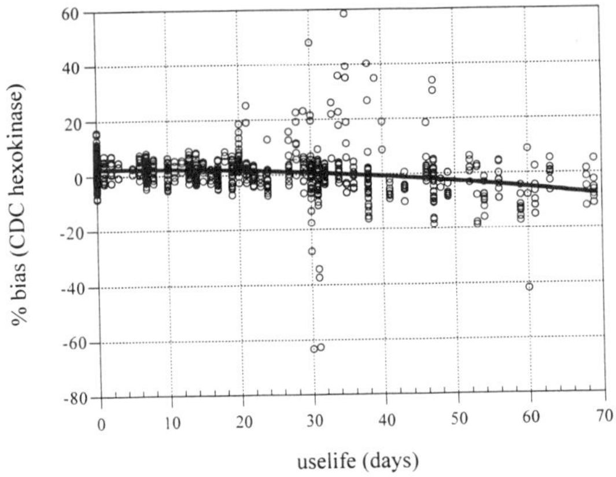


**Figure 23: Percent Bias at 80 mg/dl**

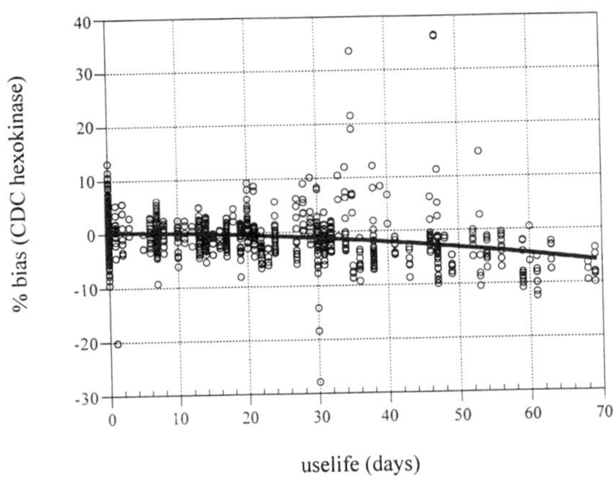


**Figure 24: Percent Bias at 200 mg/dl**

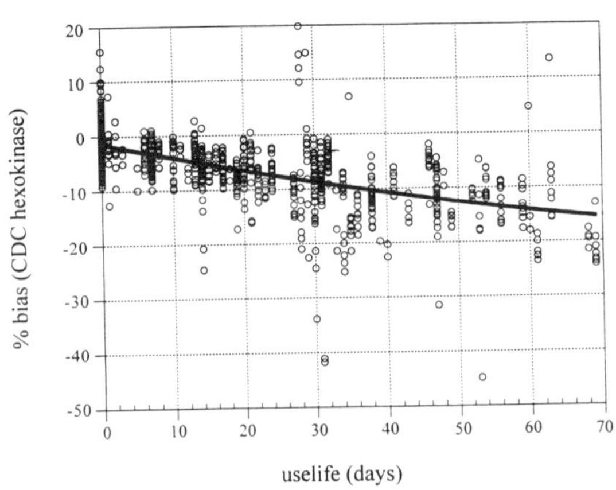


**Figure 25: Percent Bias at 500 mg/dl**

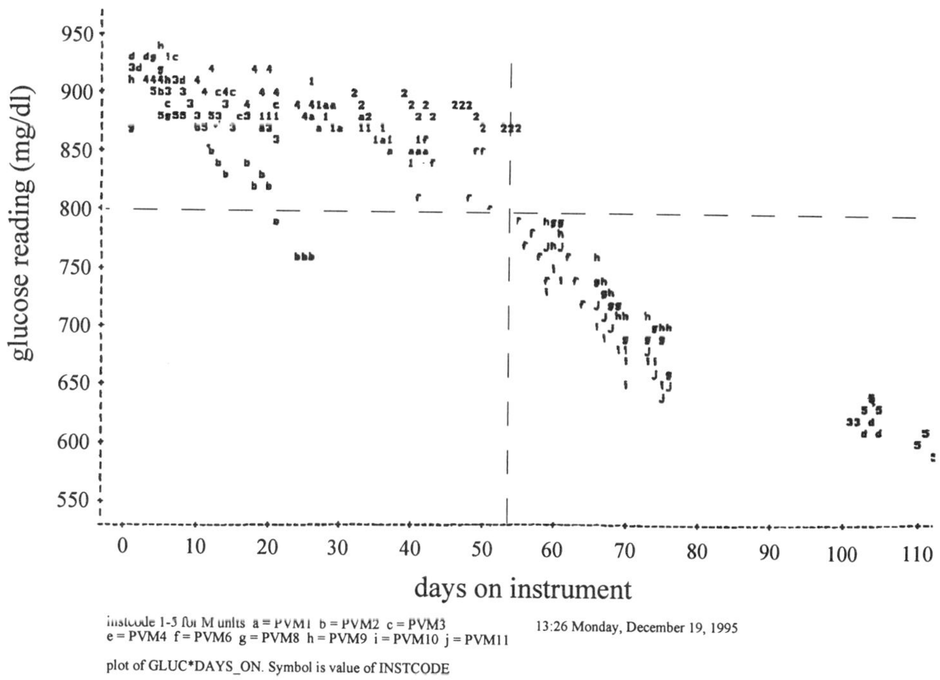


**Figure 26: Bias Drift at 950 mg/dl**

70-105 mg/dl[24]. If this were the only range we needed to measure in, the life of the sensor would be on the order of 70 days.

At 200 mg/dl, the performance is about the same as at 80 mg/dl, see Figure 24. At 500 mg/dl, there is a much more noticeable fall-off in recovery, as seen in Figure 25. And finally, at 950 mg/dl glucose, (rarely observed clinically), the sensors fall out of the required usability range at about 50 days, see Figure 26. The level of performance that is required at different glucose levels depends on how the information is to be used. The level of performance also depends on where the location of the critical decision points for the physician making a diagnosis[2].

## Conclusion

This paper has documented part of what it takes to develop a clinically useful biosensor. The glucose sensor which resulted from this development effort measures glucose from 10 mg/dl to 1000 mg/dl from as short as 30 minutes after installation up to 45 to 60 days. This is done on whole blood on a system which simultaneously measures $pO_2$, $pCO_2$, pH, $Na^+$, $K^+$, $Ca^{++}$, $Cl^-$,total hemoglobin, hemoglobin fractions and lactate. The time to result for all measurements is usually about 45 to 60 seconds.

The membrane is a key part of the sensor design. The membrane must impart specific transport characteristics for the analyte, as well as for interferents and cofactors. What gets lost from most papers which discuss sensor design and fabrication is the manufacturability aspects. It is of little value to create a sensor design that is not capable of being constructed by manufacturing personnel, in quantities sufficient to supply demand. What has been presented here is a design that can be manufactured by personnel using batch fabrication processes.

Another aspect that gets lost from most papers that present sensor research is that the final sensor must be integrated into some kind of system. This is part of the clinically useful description as well. System integration can be the most difficult stage of sensor development.

The membrane developed for this process is a spin cast water based silicone emulsion. The emulsion was adapted for sensor use. The spin coating is done over an electrode substrate consisting of conventional and non-conventional screen printed layers. Spin coating is an ideal method for this process due to its conformal nature. It is a relatively reproducible process, as long as the viscosity of the solution is kept constant.

The resultant membrane structure is heterogeneous. There are hydrophobic domains of silicone elastomer which impart high gas permeability, and domains of hydrophilic material. The hydrophobic silicone phase gives a high level of gas transport which improves the non-linearity due to $O_2$ limitation. It is the nature of the hydrophilic domain that controls the transport of the analyte and interferences. As water soluble material washes out of the membrane over time, there is a change in performance. This change in performance is within a range which can be compensated for by system calibration.

## ACKNOWLEDGMENTS

Special thanks goes to Mr. Robert Bergquist for rheological characterizations, Dr. Mark Boden for various chemical characterization, Mr. Kal Gyemesi and Mr. Barry Heller for specification development and use life testing and to Mr. Brian Holman for system integration.

## REFERENCES

1. J.O. Westgard, HFCA/CLIA Proficiency Testing for Analytical Quality, Clin. Chem. 38/7, 1226-1233, (1992).
2. W.L. Clarke, D. Cox, L.A. Gonder-Frederick, W. Carter and S.L. Pohl, Evaluating Clinical Accuracy of Systems for Self-Monitoring of Blood Glucose, Diabetes Care, 10/5, 622-628, (1987).
3 . P.H. Treloar, I.M. Christie and P M. Vadgama, "Engineering the Right Membranes for Electrodes at the Biological Interface; Solvent Cast and Electropolymerized", Biosens. & Bioelec., **10**, 195-201, (1995).
4. H.B. Lück, B. Heinrich and G. Wedell, "Particle Track Technology for Custom-made Membranes", Polym. Mat., Sci. and Eng. Preprints, **70**, 327, (1994).
5. D. Pfeiffer, K. Setz, T. Schulmeister, F.W. Scheller, H.B. Lück and D. Pfeiffer, "Development and Characterization of an Enzyme-based Lactate Probe for Undiluted Media", Biosens. and Bioelec., 7, 661-671, (1992).
6. L.S. Kuhn and S.G. Weber, "Novel Solvent Systems for the Preparation of Phase-Inversion Cellulose Acetate Size Exclusion Membranes: Voltammetric Investigations", Electroanalysis, **3**, 941-948, (1991).
7 . H. Suzuli, A. Sugama, N. Kojima, F. Takei and K. Ikegami, "A Miniature Clark-type Oxygen Electrode Using a Polyelectrolyte and its Application as a Glucose Sensor", Biosensors and Bioelectronics, 6, 395-400, (1991).
8. L.C. Clark, Jr., L.K. Noyes, R.B. Spokane R. Sudan and M.L. Miller, "Long-term Implantation of Voltammetric Oxidase/peroxide Glucose Sensors in the Rat Peritoneum," in K. Mosbach (ed), Methods in Enzymology, Vol. 137, Academic Press, New York, 1988, pp68-89.
9. L.C. Clark, "Electrochemical Device for Chemical Analysis", US Patent 2,913,386, Nov. 17, 1959.
10. L.C. Clark, Jr., Trans. Am. Soc. Artif. Organs, 2, 41, (1956).
11. P. Atanasov and E. Wilkins, "Glucose Biosensor Baed on Oxygen Electrode Part II: Long-term Operational Stability of the Rechargeable Glucose Biosensor", Anal. Lett., **26**(10), 2079-2094, (1993).
12. K.R. Greenough, A.W. Skillen and C.J. McNeil, "Potential Glucose Sensor for Perioperative Blood Glucose Control in Diabetes Mellitus", Biosens. & Bioelec., **9**, pp23-28, (1994).
13. G. Urban, G. Jobst, F. Keplinger, E. Aschauer, O. Tilado, R. Fasching, and F. Kohl, "Miniaturized Multi-enzyme Biosensors Integrated with pH Sensor on Flexible Polymer Carriers for *in vivo* Applications", Biosens. and Bioelec., 7, 733-739, (1992).

14. G. Urban, G. Jobst, F. Kohl, A. Jachimowicz, f. Olcaytug, O. Tilado, P. Goiser, G. Nuer, F. Pittner, T. Schalkhammer & E. Mann-Buxbaum, "Miniaturized Thin-Film Biosensors using Covalently Immobilized Glucose Oxidase, Biosens. & Bioelectronics, **6**, pp555-562, (1991).
15 . M.A. Genshaw, "Enzyme Electrode for Determining Glucose in Whole-blood, Clin. Chem., 34 1717-1719, (1988).
16 . D.T. Liles and H.V. Lefler III, "Silicone Rubber Latex Technology", Proc. 18th Water-Borne, Higher-Solids and Powder Coatings Symp., R.F. Sotrey and S.F. Thames, Eds., Feb. 6-8, 1991, 161-173.
17 . J.C. Saam, D. Graiver and M. Baile, "Room-Temperature-Cured Polydimethylsiloxane Elastomers from Aqueous Dispersions", Rubb. Chem. Tech., 54, 976-87, (1981).
18 . P.J. Flory, Principles of Polymer Chemistry, Cornell University Press, ©1953, p. 635.
19 . I. Ohmine and T. Tanaka, "Salt Effects on the Phase Transition of Ionic Gels." J. Chem Phys., 77(11), (1982) 5725-5729.
20. T. Sawai, S. Yamazaki, Y. Ikariyama and M. Aizawa, "pH-Responsive Swelling of the Ultrafine Microsphere", Macromolecules (1991), 24, 2117-2118.
21. Kinetics of the Cooperative Complex Formation and Dissociation of Poly(acrylic acid) and Poly(oxyethylene)". B. Bednar, H. Morawetz and J.A., Shafer, Macromolecules 1984, 17, 1634-1636.
22 . V. Yu. Baranovsky, S. Shenkov and G. Borisov, "Complexation of Poly(methacrylic Acid) with Poly(ethylene Glycol) Nonionic Surfactants in Aqueous Solutions, European Polymer Journal, 29/8, 1137-1142, (1993).
23. S. Nishi and T. Kotaka, "Complex-Forming Oxyethylene:Poly(acrylic acid) Interpenetrating Polymer Networks III. Swelling and Mechanochemical Behavior", Polymer Journal, 21/5, 393-402, (1989).
24. Teitz Textbook of Clinical Chemistry, 2nd Edition, D.A. Burtis and E.R. Ashwood, © 1986, W.B. Saunders Company.

# SECTION III.

# SEPARATIONS AND RECOVERY

# AFFINITY CHROMATOGRAPHY FOR RECOMBINANT PROTEINS

Nathaniel G. Hentz, Vesna Vukasinovic, Sergi Díez[1], Manuel Valiente[1] and Sylvia Daunert*

Department of Chemistry and Center of Membrane Sciences
University of Kentucky
Lexington, KY 40506-0055

[1]Departament de Química
Unitat de Química Analítica
Universitat Autònoma de Barcelona
08193-Bellaterra (Barcelona), Spain

## INTRODUCTION

The emergence of biotechnology has provided unique opportunities for the design and high level production of a variety of biomolecules, including recombinant proteins and peptides. Recombinant proteins/peptides with potential therapeutic value are being rapidly developed by the pharmaceutical industry. The purification of these therapeutic proteins must be performed under the most stringent conditions in order to achieve the homogeneity, purity and potency required by Federal regulations. Thus, comes the need for expanding the repertoire of conventional bioseparation techniques with novel methods of purification.

Ion-exchange chromatography is a widely used separation technique for biomolecules where one can choose from a variety of anion- or cation-exchange separation media. In this technique, proteins are separated on the basis of their charge characteristics. It is possible to alter the interaction between the protein and the ion-exchange medium by adjusting the pH or ionic strength of the elution buffer, thereby changing the selectivity of the separation. In some cases, it may be advantageous to add charged groups to biomolecules to enhance the ion-exchange ability of the target molecule.[1] Incorporation of longer spacer arms between the support matrix and the ion-exchange functional group leads to the development of tentacle ion-exchange chromatography.[2] This method is reported to have higher loading capacities and efficiencies than conventional ion-exchange resins.[3-5]

Another property of proteins that may be exploited for purification is hydrophobicity. The basis of this hydrophobic-interaction separation technique is that proteins have hydrophobic regions, usually characterized by tryptophan, phenylalanine, tyrosine, histidine, isoleucine, etc. By starting the mobile phase with a high concentration of salt the proteins are effectively "forced" to interact with the hydrophobic functional group. The proteins are released by decreasing the salt concentration. A similar method, reversed phase chromatography, is often used in conjunction with hydrophobic-interaction chromatography, where the hydrophobic properties of a protein are used in the separation process. Reversed-phase chromatography differs in that the mobile phase usually goes from polar to less polar with a concomitant extraction of the protein from the nonpolar stationary phase.

Yet another characteristic that can be utilized in protein purification is the size and, to some extent, the shape of the protein. This size-exclusion chromatography is often used as a crude separation step to isolate the target protein from the bulk of the preparation before other separation techniques can be used.

A relatively new and increasingly diverse area of chromatography is that of affinity chromatography. Affinity chromatography is a technique that takes advantage of the specific binding affinity of the target protein toward an immobilized ligand. For affinity chromatography to work, binding pairs must exist, where the two components comprising the binding pair have a highly specific affinity toward one another. While the other techniques usually require more than one separation step to purify a protein to homogeneity, a high degree of purification is often obtained with affinity chromatography. Other categories within affinity chromatography are those that use biomimetic approaches that mimic a ligand-protein interaction; for example, dyes such as Ramazol Yellow GGL[6] and triazine derivatives[7] have been used in this manner. Finally, immobilized metal affinity chromatography, takes advantage of the coordination sites of an immobilized metal to purify the target protein.[8]

Since about the mid-1980's we have seen the growth of a new area in the separation of biomolecules: that of membrane chromatography.[9,10] This membrane technology has been applied to affinity,[11-15] ion-exchange,[16-21] hydrophobic-interaction[18-20] and reversed-phase[22-24] chromatography. Membranes offer several advantages over conventional packed-bed chromatography supports such as small bed heights, large cross-sectional surface area and negligible intraparticle diffusion. These characteristics lead to low back-pressures, short residence times and high volumetric throughput.

Biofunctional membranes incorporate biochemical or biomimetic interactions as part of a membrane system. As such, they combine the advantages that are common to membrane chromatography with the inherent selectivity found in biological systems. Although biofunctional membranes have been used previously in membrane chromatography[9], there are still a wide variety of potentially useful separation methods that are yet to be developed which could take advantage of the unique chemical recognition properties of proteins and other biomolecules. With that aim in mind, our group has worked on strategies that should result in novel membrane affinity separation systems for recombinant proteins, as well as amino acids. Conventional chromatography media were used initially to test the feasibility of the designed separation systems. Following the successful demonstration of feasibility, membrane affinity chromatography systems are going to be developed based on these strategies. In this article we will review our progress toward achieving these goals.

## AFFINITY CHROMATOGRAPHY

As mentioned previously, affinity chromatography has as a prerequisite the existence of a binding pair, one component of which is the target protein. Examples of binding pairs include binding protein/ligand, antibody/antigen, lectin/carbohydrate, enzyme/substrate, receptor/hormone, etc. The basis for affinity chromatography is relatively simple. One component of the binding pair is attached to an inert support matrix, which can be either an appropriately derivatized particle or a membrane. Then, the sample containing the protein of interest is loaded onto the affinity matrix. Extraneous proteins do not bind and can thus be washed away, while the target protein remains specifically bound to the affinity matrix. The target protein can be released from the affinity matrix by changing the elution conditions (i.e., pH, ionic strength, etc.) or by adding a high concentration of competing compound. Affinity chromatography allows 1000- to 10000- fold purification of the target protein.

Before choosing a matrix for affinity chromatography, several considerations must be explored. First, the matrix should be easily derivatized (i.e., functional groups should be available for the covalent attachment of one of the components of the binding pair). The matrix should also be mechanically and chemically stable. The matrix should possess a surface, such that the immobilized component is accessible to binding by the target protein. Finally, the matrix should possess good flow characteristics to allow fast elution.

Despite the advantages of affinity chromatography, the technique has some limitations. If an affinity pair complimentary to the target protein does not exist, then affinity chromatography offers no advantage over other separation techniques. One approach to get around this limitation is to attach an affinity tail to the target protein, thereby introducing the desirable affinity properties. This affinity-tail approach will be discussed in the next section.

## AFFINITY TAILS

Affinity tails have been introduced to circumvent the limitation of conventional affinity chromatography described in the previous section and allowing the facile purification of biomolecules. With the advent of molecular biology techniques, affinity tails can be attached to the target protein by genetically fusing the "tail" to the N- or C-terminal amino acid of the target protein. In this technique, a plasmid is designed that incorporates an oligonucleotide that encodes for the affinity tail fused in the same reading frame to the gene that encodes the production of the target protein. The plasmid is then inserted into a host (bacteria, yeast, etc.) where a fusion protein with the affinity tail attached to the target protein is expressed (Figure 1).

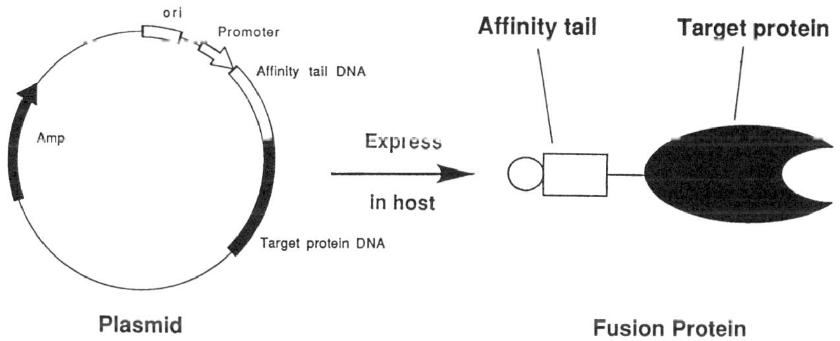


**Figure 1.** General schematic diagram of the genetic fusion of the affinity tail to the target protein.

After expression of the fusion protein, the cells are lysed and centrifuged. The cell lysate supernatant, containing the target protein, is then loaded onto the affinity column (or affinity membrane) where the unwanted proteins are eluted while the fusion protein remains attached through the affinity tail portion. Then, the elution conditions are changed to alter the interaction between the affinity tail and the column, thus releasing the fusion protein which is collected (see Figure 2). It is important to note that the native target protein does not need to possess any specific binding abilities. It is because of the genetical fusion of an affinity tail to the target protein that the target protein acquires specific binding toward an affinity matrix. To isolate the target protein from the fusion protein, a site-specific endopeptidase cleavage site must be introduced between the target protein and the affinity tail during the design of the plasmid. The appropriate endopeptidase can then be introduced either directly to the column, where the fusion protein is still attached, or after removal of the fusion protein. This will allow the affinity tail portion to be cleaved from the target protein.

A number of fusion tails have been genetically attached to target proteins in the development of affinity chromatography systems (for review see reference 25). By genetically fusing a variety of polyamino acid tails to target proteins, affinity purification by ion-exchange chromatography,[1,26-29] immobilized metal ion affinity chromatography[30-33] or hydrophobic-interaction chromatography[34] is possible. Recombinant proteins can also be purified by attaching biotin to a fusion tail through posttranslational modification in which the (strept)avidin binding interaction can be taken advantage of.[35] Affinity columns containing immobilized polypeptides[36,37] or carbohydrates[38,39] have been used to purify target molecules in which binding proteins are attached as fusion tails. Finally, by genetically fusing an enzyme or antigen to the target protein, columns of immobilized substrate/inhibitor[40,41] or monoclonal antibody,[42-44] respectively, have been used to purify

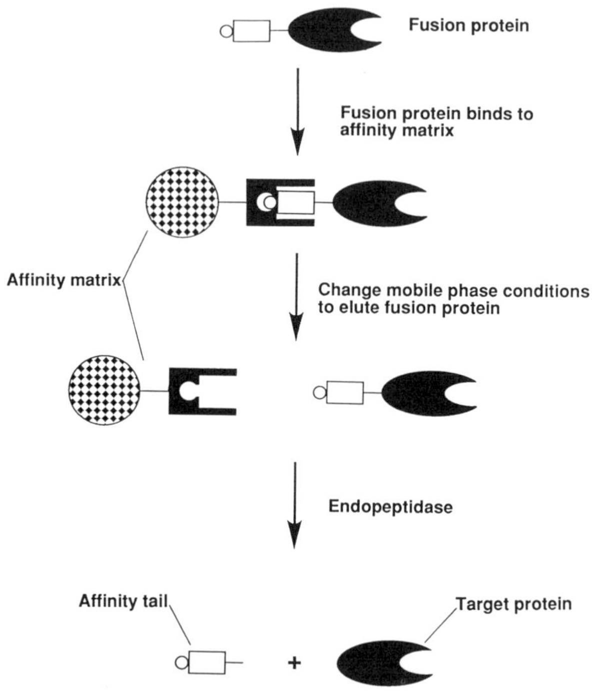


**Figure 2.** Affinity separation of recombinant proteins using the fusion tail approach.

the target protein. Thus, it may now be possible to use affinity chromatography for the purification of recombinant proteins that otherwise possess no specific binding properties.

Our strategy in the development of an affinity tail system took advantage of the $Ca^{2+}$-dependent conformational change of calmodulin.[45] Calmodulin is a regulatory protein with four $Ca^{2+}$-binding sites. In the presence of $Ca^{2+}$, the conformation of calmodulin changes such that a hydrophobic pocket is exposed on each end of the dumbbell-shaped protein. It is when these hydrophobic regions are exposed that calmodulin exhibits binding toward a number of peptides[46,47] and hydrophobic compounds,[48] such as the drug phenothiazine.[49] This hydrophobic recognition site on calmodulin is no longer available upon removal of the $Ca^{2+}$, allowing complete dissociation of the calmodulin:ligand complex.

In this study, we chose phenothiazine as the ligand to be immobilized on the stationary phase and calmodulin as the affinity tail to be genetically fused to the target protein. Protein A was used as a model protein. After the calmodulin-protein A fusion protein was expressed in either *Escherichia coli* or *Saccharomyces cerevisiae*, the crude cell lysate was loaded onto the phenothiazine affinity column in the presence of a $Ca^{2+}$-containing buffer. To elute the fusion protein, an EGTA-containing buffer was then passed through the column (Figure 3). This method produced very pure calmodulin-protein A as confirmed by SDS-PAGE and Western blotting.

This affinity chromatography method can be easily extended to design purification systems for other proteins using appropriately designed calmodulin-protein fusions. In addition, a selective endopeptidase cleavage site can be incorporated between the calmodulin and the target protein. After the fusion protein is expressed and purified via the affinity chromatography step, the target protein can be isolated by treatment with the appropriate endopeptidase. Work in this direction is currently in progress in our laboratory.

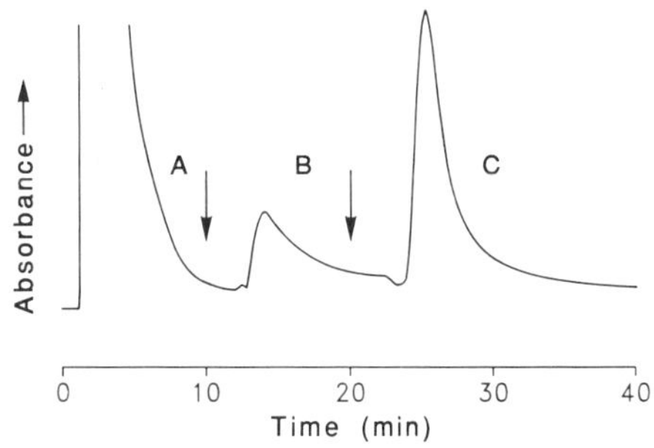


**Figure 3.** A typical affinity chromatogram of calmodulin-protein A that was expressed in *S. cerevisiae*. Buffer A (10 mM HEPES, pH 7.0 containing 0.5 mM $CaCl_2$), buffer B (10 mM HEPES, pH 7.0 containing 0.5 mM $CaCl_2$ and 0.25 M NaCl) and buffer C (10 mM Tris, pH 8.0 containing 5 mM EGTA) were used to elute the calmodulin-protein A. The buffer changes are indicated by the arrows.

## IMMOBILIZED METAL ION AFFINITY CHROMATOGRAPHY

Immobilized metal ion affinity chromatography (IMAC) was first introduced in 1975 by Porath et al.[9] However, it was not until the mid-1980's that IMAC gained wide recognition. IMAC is based on the interaction between biomolecules in solution and immobilized metal ions, usually coordinated to hydrophilic cross-linked polymers[50] or membranes.[51] In addition to protein purification, IMAC has also been used for the separation of smaller biomolecules such as peptides[52,53] and amino acids.[54]

There are several features that make IMAC an attractive alternative to conventional chromatographic methods for the separation of biomolecules. Different metal ions can be immobilized on the stationary phase relatively easily. By changing the metal ion, one can add another optimization parameter for the separation of the target biomolecule. Also, the metal ions can be removed easily by eluting the column with a strong chelator or strong electron-donating buffer additive, yielding a near quantitative recovery of the target biomolecule. The elution conditions can be fairly mild, decreasing the chance of protein denaturation. Since the column strongly binds heavy metals, bacterial growth is inhibited. Finally, it has been shown that the IMAC columns can be regenerated hundreds of times.[50]

The hard-soft acid-base theory of Pearson[55] provides guidelines for the selection of metals for IMAC systems. Metal ions are classified as hard, intermediate and soft,[55] where the soft metals are more easily polarizable. Almost all IMAC work has been performed with either hard metals or intermediate-type metals. Given the different properties of soft metals, it appears worthwhile to investigate the potential use of soft metals, such as $Pd^{2+}$, $Pt^{2+}$, $Hg^{2+}$, $Ag^{+}$, etc., as coordination centers in IMAC systems. Besides work performed in our laboratories,[56] there has been only one additional report on IMAC using a soft metal. In particular, Miles and Garcia have recently demonstrated the usefulness of Pt-based IMAC to separate biotinylated proteins from nonbiotinylated proteins.[57]

Our group has explored the use of $Pd^{2+}$-based IMAC, where the metal was immobilized on Spheron Oxine 1000 resin through an 8-hydroxyquinoline functional group (Figure 4). The system was optimized so that $Pd^{2+}$ formed a 1:1 complex with the 8-hydroxyquinoline on the resin. Therefore, in the immobilized complex, 8-hydroxyquinoline occupies two of the four coordination sites of $Pd^{2+}$, the other two being occupied by two chloride ions.

The IMAC system was characterized using glycine as a model amino acid.[56] Glycine partitions onto the resin and displaces chloride from the $Pd^{2+}$:8-hydroxyquinoline complex. Glycine can be eluted by using either HCl or a buffer containing imidazole at pH 6.5 (mild conditions). Since the model system using glycine showed promise for the separation of amino acids, IMAC studies using phenylalanine (Figure 5), tryptophan (Figure 6) and tyrosine (Figure 7) were also performed. Figures 5-7 show the amount of amino acid adsorbed per gram of $Pd^{2+}$-loaded resin versus the initial amount of amino acid added.

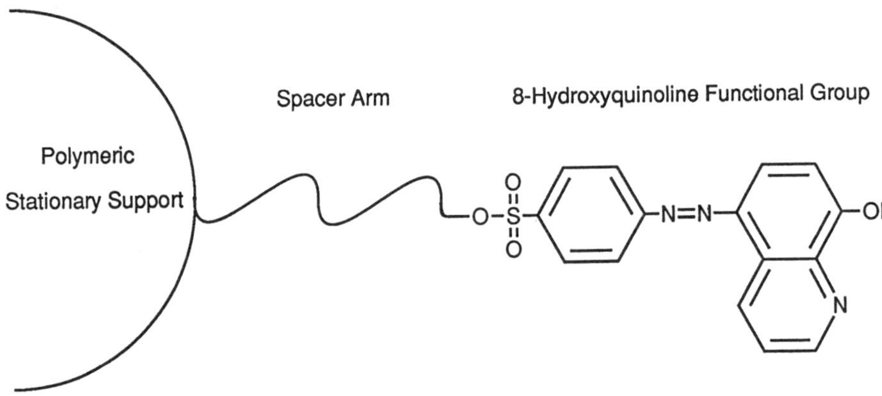


**Figure 4.** Structure of the 8-hydroxyquinoline functional group that is used to immobilize $Pd^{2+}$.

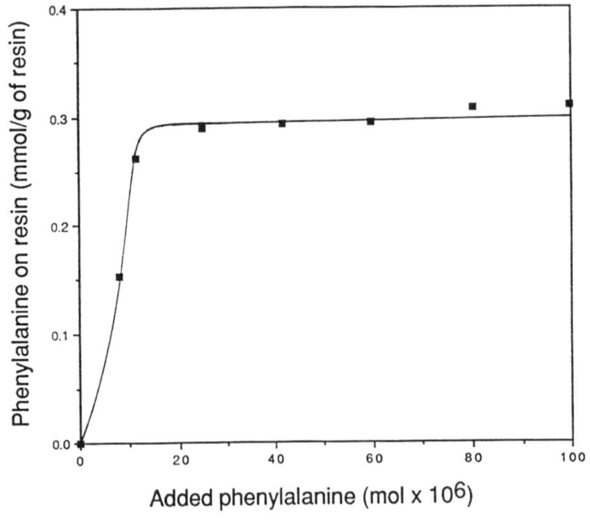


**Figure 5.** Phenylalanine partition in Spheron Oxine 1000 resin loaded with Pd(II).

The results showed that at saturation 0.24 mmol tryptophan, 0.31 mmol phenylalanine, 0.25 mmol tyrosine and 0.31 mmol glycine could be adsorbed per gram of resin. The amino acids could be eluted under acidic conditions with HCl or $H_2SO_4$. In the case of glycine, the maximum eluting efficiencies were found to be 98% and 34% for HCl and $H_2SO_4$, respectively. The higher eluting efficiency of HCl was due not only to the protonation of the amino acids, but also to the competition for the $Pd^{2+}$ coordination sites by the $Cl^-$. It should be noted that elution of the amino acids is also possible under milder conditions (pH 6.5) by adding imidazole to the elution buffer. In this case, imidazole competes with the amino acids for the coordination sites available on the immobilized $Pd^{2+}$.

This work demonstrated that a $Pd^{2+}$-based IMAC system can be used for the separation of amino acids, a step that is necessary before extending this system to the separation of larger biomolecules, such as proteins or peptides. It was also shown that the elution buffer needs to be optimized in terms of the different types of complexes that can form during the partition and elution processes.

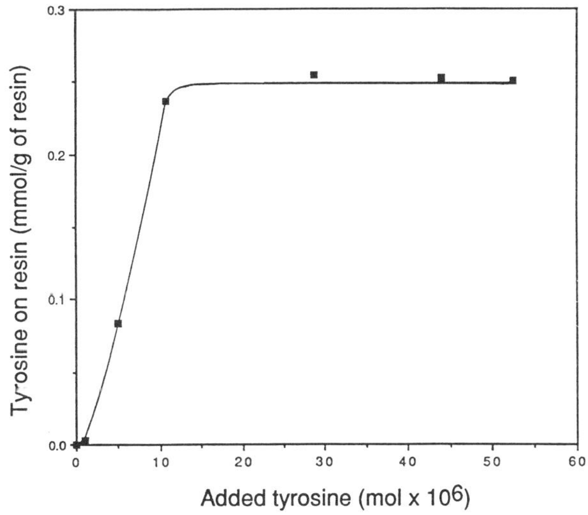


**Figure 6.** Tryptophan partition in Spheron Oxine 1000 resin loaded with Pd(II).

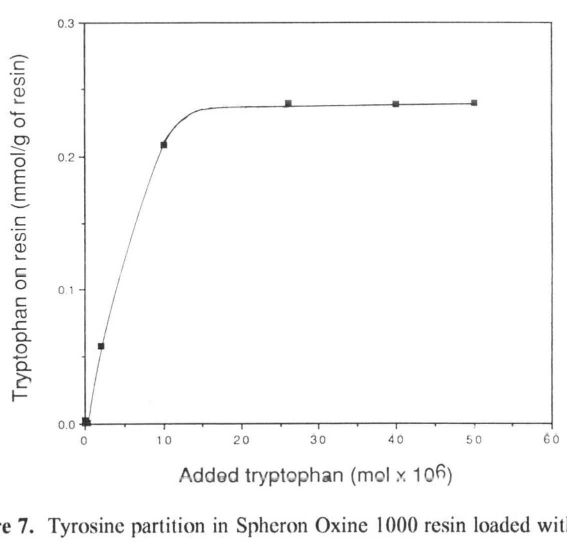


**Figure 7.** Tyrosine partition in Spheron Oxine 1000 resin loaded with Pd(II).

## CONCLUSIONS

It has been demonstrated that there are many techniques to choose from when developing a separation protocol for recombinant proteins. One can take advantage of a variety of characteristics of the protein such as its charge characteristics, hydrophobicity, size/shape, or affinity toward other molecules. In addition, through the use of genetic engineering, a protein can be altered to induce or enhance one or more of these characteristics.


## ACKNOWLEDGMENTS

This work was supported by the National Aeronautics and Space Administration (NCCW-60). Manuel Valiente acknowledges support from the Spanish Commision of Research and Development and from CYCIT (project MAT93-0621-C03-01).


## REFERENCES


1) Heng, M.H.; Glatz, C.E. *J. Chromatogr.* **1995**, *689*, 227-234.
2) Tsuneda, S.; Saito, K.; Furusaki, S.; Sugo, T. *J. Chromatogr.* **1995**, *689*, 211-218.
3) Ditz, R.; Hauke, G.; Müller, W. *Austr. J. Biotechnol.* **1991**, *5*, 101-102.
4) Müller, W. *J. Chromatogr.* **1990**, *510*, 133-140.
5) Kopaciewicz, W.; Rounds, A.; Fausnaugh, J.; Regnier, F.E. *J. Chromatogr.* **1983**, *266*, 3-21.
6) Regnault, V.; Rivat, C.; Vallar, L.; Geschier, C.; Stoltz, J.F. *J. Chromatogr. B* **1992**, *584*, 93-99.
7) Dean, P.D.G.; Watson, D.H. *J. Chromatogr.* **1979**, *165*, 301-319.
8) Porath, J.; Carlsson, J.; Olsson, I.; Belfrage, G. *Nature* **1975**, *258*, 598-599.
9) Klein, E. In *Affinity Membranes*, Wiley: New York, NY, **1992**.
10) Roper, D.K.; Lightfoot, E.N. *J. Chromatogr.* **1995**, *702*, 3-26.
11) Chui, W.K.; Wainer, I.W. *Anal. Biochem.* **1992**, *201*, 237-245.
12) Gou, W.; Shang, Z.; Yu, Y.; Guan, Y.; Zhou, L. *Biomed. Chromatogr.* **1992**, *6*, 95-98.
13) Nachman, M. *J. Chromatogr.* **1992**, *597*, 167-172.
14) Nachman, M.; Azad, A.R.; Bailon, P. *J. Chromatogr.* **1992**, *597*, 155-166.
15) Josic, D.; Bal, F.; Schwinn, H. *J. Chromatogr.* **1993**, *632*, 1-10.
16) Wenisch, E.; Vorauer, K.; Jungbauer, A.; Katinger, H.; Righetti, P.G. *Electrophoresis* **1994**, *15*, 647-653.
17) Lütkemeyer, D.; Bretschneider, M.; Büntemeyer, H.; Lehmann, J. *J. Chromatogr.* **1993**, *639*, 57-66.
18) Josic, D.; Reusch, J.; Loster, K.; Baum, O.; Reutter, W. *J. Chromatogr.* **1992**, *590*, 59-76.
19) Luksa, J.; Menart, V.; Milicic, S.; Kus, B.; Gaberc-Porekar, V.; Josic, D. *J. Chromatogr.* **1994**, *661*, 161-168.
20) Josic, D.; Lim, Y.P.; Strancar, A.; Reutter, W. *J. Chromatogr. B* **1994**, *662*, 217-226.
21) van Huynh, N.; Motte, J.C.; Pilette, J.F.; Decleire, M.; Colson, C. *Anal. Biochem.* **1993**, *211*, 61-65.
22) Wang, Q.C.; Svec, F.; Fréchet, J.M. *Anal. Chem.* **1995**, *67*, 670-674.
23) Wang, Q.C.; Svec, F.; Fréchet, J.M. *J. Chromatogr.* **1994**, *669*, 230-235.
24) Wang, Q.C.; Svec, F.; Fréchet, J.M. *Anal. Chem.* **1993**, *65*, 2243-2248.
25) Ford, C.F.; Souminen, I.; Glatz, C.E. *Prot. Expres. Purif.* **1991**, *2*, 95-107.
26) Smith, J.C.; Derbyshire, R.B.; Cook, F.; Dunthorne, L.; Viney, J.; Brewer, S.J.; Sassenfeld, H.M.; Bell, L.D. *Gene* **1984**, *32*, 321-327.
27) Sassenfeld, H.M.; Brewer, S.J. *Bio/Technology* **1984**, *2*, 76-81.
28) Brewer, S.J.; Sassenfeld, H.M. *Trends Biotechnol.* **1985**, *3*, 119-122.
29) Heng, M.H.; Glatz, C.E. *J. Chromatogr.* **1995**, *689*, 227-234.
30) Hutchens, T.W.; Yip, T.T. *Anal. Biochem.* **1990**, *191*, 160-168.

31) Hochuli, E.; Bannwarth, W.; Döbeli, H.; Gentz, R.; Stüber, D. *Bio/Technology* **1988**, *6*, 1321-1325.
32) Todd, R.J.; Johnson, R.D.; Arnold, F.H. *J. Chromatogr.* **1994**, *662*, 13-26.
33) Kasher, M.S.; Wakulchik, M.; Cook, J.A.; Smith, M.C. *Biotechniques* **1993**, *14*, 630-641.
34) Persson, M.; Bergstrand, M.G.; Bülow, L.; Mosbach, K. *Anal. Biochem.* **1988**, *172*, 330-337.
35) Cronan, J.E. *J. Biol. Chem.* **1990**, *265*, 10327-10333.
36) Moks, T.; Abrahmsén, L.; Holmgren, E.; Bilich, M.; Olsson, A.; Uhlén, M.; Pohl, G.; Sterky, C.; Hultberg, H.; Josephson, S.; Holmgren, A.; Jörnvall, H.; Nilsson, B. *Biochemistry* **1987**, *26*, 5239-5244.
37) Schmidt, T.G.M.; Skerra, A. *J. Chromatogr.* **1994**, *676*, 337-345.
38) Guan, C.; Li, P.; Riggs, P.D. *Gene* **1988**, *67*, 21-30.
39) Blondel, A.; Bedouelle, H. *Eur. J. Biochem.* **1990**, *193*, 325-330.
40) Ullmann, A. *Gene* **1984**, *29*, 27-31.
41) Guan, K.; Dixon, J.E. *Anal. Biochem.* **1991**, *192*, 262-267.
42) Field, J.; Nikawa, J.I.; Broek, D.; MacDonald, B.; Rodgers, L.; Wilson, I.A.; Lerner, R.A.; Wigler, M. *Mol. Cell. Biol.* **1988**, *8*, 2159-2165.
43) Hopp, T.P.; Prickett, K.S.; Price, V.L.; Libby, R.T.; March, C.J.; Cerretti, D.P., Urdal, D.L.; Conlon, P.J. *Bio/Technology* **1988**, *6*, 1204-1210.
44) Knappik, A.; Pluckthun, A. *Biotechniques* **1994**, *17*, 754-761.
45) Hentz, N.G; Vukasinovic, V.; Daunert, S. in preparation.
46) Maulet, Y.; Cox, J.A. *Biochemistry* **1983**, *22*, 5680-5686.
47) Tanaka, T.; Ohmura, T.; Hidaka, H. *Mol. Pharmacol.* **1982**, *22*, 403-407.
48) Levin, R.M.; Weiss, B. *Mol. Pharmacol.* **1977**, *13*, 690-697.
49) Foster, W.S.; Jarrett, H.W. *J. Chromatogr.* **1987**, *403*, 99-107.
50) Porath, J. *Prot. Expr. Purif.* **1992**, *3*, 263-281.
51) Seratica, G.C.; Pimbley, J.; Belfort, G. *Biotechnol. Bioeng.* **1994**, *43*, 21-36.
52) Hansen, P.; Lindeberg, G. *J. Chromatogr.* **1995**, *690*, 155-159.
53) Hansen, P.; Lindeberg, G. *J. Chromatogr.* **1994**, *662*. 235-241.
54) Wirth, H.J.; Hearn, M.T. *J. Chromatogr.* **1993**, *646*, 143-151.
55) Pearson, R.; Pearson, R.G. (Ed) *Hard and Soft Acids and Bases*, Dowden, Hutchington & Ross, Stroudsburg, PA 1973.
56) Díez, S.; Zhou, P.; Daunert, S.; Valiente, M. *Anal. Chim. Acta* **1995**, in press.
57) Miles, D.; Garcia, A.A. *J. Chromatogr.* **1995**, *702*, 173-189.

# PROTEIN RECOVERY USING FIBROUS MATERIALS

Li Ang Chen[*], George A. Serad[†] and Ruben G. Carbonell[*]

* Department of Chemical Engineering - North Carolina State University
Box 7905 - Raleigh, NC 27695
† Hoechst Celanese Corporation
P. O. Box 32414 - Charlotte, NC 28232-6085

## INTRODUCTION

Tighter environmental regulations challenge industry to improve processes to minimize waste and conserve energy (Metzner, 1988). Traditionally, solid wastes resulting from effluent treatment systems have been disposed by incineration or by sending to landfills (Swinehart, 1990). These approaches are no longer appropriate, since incineration may lead to toxic gaseous emissions and air pollution, and landfills could result in contamination of water supplies.

Recovery of by products for reuse in the process or in other systems has been the solution in most cases. Solid and liquid wastes from the food processing, breweries and distillers, and the pharmaceutical industries present a significant amount of valuable biological materials, which could be recovered for use as animal feed. This procedure not only saves money but also reduces pollution and generates revenues (Crickenberger and Carawan, 1991). Beszedits and Netzer (1982) have identified the major sources of proteins that could be recovered from food processing waste waters and gave an overview of the most common techniques used for the recovery: coagulation and flocculation followed by dissolved air flotation (DAF), electroflocculation, ion exchange, ultrafiltration, reverse osmosis and single cell protein production.

By products to be used as feedstuffs should be free of toxins or other substances that may be harmful to animals (Crickenberger and Carawan, 1991). Coagulants, such as ferric chloride and alum, have been widely used for fat and protein recovery, but they result in a high metal content in the sludge which is undesirable in animal feed (Rusten *et al.*, 1993). Alternatively, a series of polyacrylamide-based polyelectrolytes is gaining a large share of the market as flocculants, but it does not have Generally Recognized as Safe (GRAS) approval for addition in animal feed formulation (Metzner, 1991; Powers, 1991). Lately, natural polymers, such as lignosulfonic acid (LSA) and chitosan, have been used as coagulants for recovery of by products. LSA is a cellulose derivative, generated in great amounts in the wood pulp and paper industry, and chitosan is recovered from crab and shrimp shells. On feeding to animals, both LSA and chitosan did not show adverse effects on growth rate or weight gain (Arai *et al.*, 1968; Hopwood, 1977).

These coagulants/flocculants have been used to increase recovery of proteins and fats during DAF (Dissolved Air Flotation) operations, one of the most common processes for waste water treatment in the food industries (Carawan and Valentine, 1991; Lunney, 1988). The skimming resulting from DAF systems has approximately 90% water, which is undesirable and could lead to high cost in the further dewatering processes, before rendering and adding to animal feedstuffs (Babbit, 1989).

In view of these technological limitations, the present project aims to study modified cellulose-based fibers that can be used as flocculants and filter aids for the recovery of proteins and other nutrients from waste water streams. They can also be incorporated into membranes for usage in depth filters. Charged cellulose (dyed cotton) fibers and charge-modified cellulose acetate fibrets (CAF) could adsorb proteins from solution by electrostatic interaction, whereas, unmodified CAF or cellulose triacetate fibrets (CTF) could lead to adsorption of proteins by hydrophobic interaction, at high ionic strengths. Cellulose, being a natural polymer and biodegradable, can achieve GRAS status for addition in animal feed formulations. Fibrets have high surface area (Smith, 1988), which could improve recovery of proteins from aqueous media, resulting in a drier product.

The model system chosen for the studies on recovery of soluble proteins from water solutions is bovine serum albumin (BSA) dissolved in 0.05M citrate buffer. Modified cellulose (dyed cotton) fibers and cellulose acetate fibrets were supplied by Hoechst Celanese Corporation (HCC). Studies on the recovery of the protein from solution using these fibrous materials were performed and compared with known systems, such as precipitation with polyelectrolytes (carboxymethyl cellulose (CMC)) and adsorption to cross-linked cellulose-based ion exchangers (SP/CM Sephadex C-50).

## MATERIALS AND METHODS

### Materials

Bovine Serum Albumin (BSA) of 99% purity, MW 67,000 and cation exchanger SP Sephadex C-50 were purchased from Sigma (St. Louis, MO). Carboxymethyl cellulose (CMC), average MW 250,000, with degree of substitution (DS) 0.7 was from Polysciences, Inc. (Warrington, PA), and cation exchanger CM Sephadex C-50 was obtained from Pharmacia Fine Chemicals (Piscataway, NJ). Citric acid, sodium citrate, ammonium sulfate and miscellaneous chemicals were all of analytical grade. The Hoechst Celanese Corporation (HCC) donated the following materials: cellulose acetate fibrets (CAF), DS 2.4, 20-200 µm length, 0.5-5 µm width, and 20 $m^2/g$ surface area; water soluble cellulose acetate (WSCA), white powder, <40 mesh, DS 0.6-0.8, low viscosity (P-38); cellulose triacetate fibrets (CTF), DS 3.0, oblong particles with 2 mm length, and 20 $m^2/g$ surface area; cotton fibers made charged by dyeing with 6% Schaeffer's acid (FB2C) and 6% R salt (FB4C); non dyed cotton slivers; and charge-modified CAF (CAF that were made charged by attaching CMC or bentonite on the surface).

### Precipitation of BSA using CMC

BSA solutions of 1 mg/mL were prepared by dissolving the lyophilized protein in 0.05 M citrate buffer at the appropriate pH. CMC stock solution of 1 mg/mL was prepared by dispersing the powdered polymer in deionized (DI) water and mixing for about 2 hours under slight heating until completely dissolved. Mixing was provided by a magnetic stirrer. Precipitations were performed in 14 mL batch solutions. CMC was added to BSA solutions at pre-determined dosages, under mixing conditions. Solutions were mixed for 3 minutes (magnetic stirrer) and incubated for 60 minutes in the shaker at room temperature. The final slurries were centrifuged at 20,000 *g* for 45 minutes. The supernatants were recovered for protein analysis by UV absorbance and the percentage protein removed from solution was calculated for each batch. The parameters varied in these experiments were solution pH and polymer dosage.

### Removal of BSA by adsorption to ion exchanger

A pH of 4.0 was chosen for these experiments, so that BSA is positively charged and two cation exchangers (negatively charged) were used: CM and SP Sephadex C-50. BSA solutions of 1 mg/mL at pH 4.0 were prepared as described previously. The cation exchangers were prepared according to manufacturer's instructions (Pharmacia, 1983): the exchanger in powder form was poured into 0.5 M citrate buffer at pH 4.0. The resulting gel was equilibrated at high ionic strength by washing 10 times with 0.5 M citrate buffer, then equilibrated again at low ionic strength, by washing 10 times with 0.05 M citrate buffer. The final gel concentration was 10 mg/mL. The gel was added to the BSA solutions at various dosages and the final slurries were incubated under shaking for one hour, then centrifuged at 5,200 *g* for 30 minutes. The supernatants were recovered for protein analysis by UV absorbance.

### Removal of BSA by adsorption to charged cotton fibers

Cotton fibers (FB2C, FB4C, and non-dyed) in sliver form were chopped to smaller sizes by hand (to about 5 to 10 mm in length), then equilibrated in buffer at various pHs, first at high ionic strength, then at low ionic strength. These fibers were added to BSA solutions at appropriate pHs and dosages, and adsorption took place under shaking for 2 hour at room temperature. Solutions of 0.3, 0.5 and 1.0 mg/mL BSA concentration were studied. Screening of the appropriate pH was performed, and pH 4.2 was chosen for the experiments at various fibers dosages. Slurries were filtered through 20 μm polypropylene frits (BioRad, Richmond, CA), and filtrates were recovered for protein analysis by UV absorbance or Bradford dye binding assay (Bradford, 1976). Non-dyed cotton slivers were used as controls for the UV absorbance readings.

### Adsorption of BSA using WSCA/CAF/CTF/Cotton Fiber in the presence of ammonium sulfate

**Protein removal vs. ammonium sulfate saturation.** Water-soluble cellulose acetate (WSCA) solutions were prepared by wetting and dispersing the particles in hot water, then adding cold buffer to dissolve the wet particles. Mixing was provided by a magnetic stirrer and kept for 2-3 hours with slight heating until a homogenous solution was obtained. Final concentrations were 20 mg/mL, in 0.05 M citrate buffer. BSA stock solutions were 5 mg/mL at pH 4.8. Cellulose acetate fibrets (CAF), cellulose triacetate fibrets (CTF), and cotton fibers were used without further treatment. Ammonium sulfate was added to the solutions to be precipitated at concentrations from 10 to 55% saturation (equivalent to 0.405 to 2.228 M). Precipitation batches were 5 mililiters in volume and the initial BSA concentration was kept at 1 mg/mL. All precipitations took place under shaking for 1 hour, then slurries were filtered through 20 μm frits. Filtrates were recovered for protein analysis by the Bradford assay.

**Adsorption isotherms of BSA on CTF.** BSA stock solutions of 10 and 15 mg/mL were prepared in 0.05 M pH 4.8 citrate buffer. 200 mg of CTF were added to each adsorption batch of 2 mililiters, with varying concentrations of BSA. Adsorption isotherms were obtained at 10, 15, 20, and 25% ammonium sulfate saturation and various temperatures (8, 28, 35, and 40 $^{o}$C). Adsorption was performed under shaking at 250 rpm for 2 hours. Solutions were then filtered through 20 μm polyethylene frits (Alltech, Deerfield, IL), and filtrates recovered for protein analysis by the Bradford assay.

### Removal of BSA using charge-modified CAF

Charged CAF were used as received, without further treatment: bentonite coated CAF and CMC coated CAF. BSA stock solutions were 5 mg/mL at pH 4.2. Precipitation batches were 2 mililiters. Initial protein concentrations were 0.25, 0.5 and 1 mg/mL. Modified CAF were added at various dosages, and precipitations were carried out under shaking for 1 hour at room temperature. Slurries were filtered through 20 μm frits, and filtrate recovered for protein analysis by the Bradford assay.

## RESULTS AND DISCUSSION

### Precipitation of BSA using CMC

The precipitation of proteins by polyelectrolytes is very efficient, requiring low dosages for high recovery. Figure 1 shows the results obtained in the precipitation of BSA by CMC. This polyelectrolyte is anionic and interacts with BSA by electrostatic interaction, as can be observed in the removal behavior at different pHs.

At pHs below 4.9, BSA molecules are positively charged in solution, and can be flocculated by negatively charged polymers by charge neutralization (Figure 1.(a)). At pHs 3.0 and 3.6, most of the CMC carboxyl groups are not normally ionized - the degree of ionization is around 0.1 at pH 3.0 and 0.3 at pH 3.6 (Chowdhury and Neale, 1963). There is some flocculation, but not the optimum level. At pH 4.2 the degree of ionization of CMC is about 0.5 (Chowdhury and Neale, 1963), which makes the negatively charged CMC molecules prone to electrostatic interaction with positively charged BSA molecules, resulting in the optimum precipitation level. At pHs 5.4 and 6.0, although the degree of ionization of CMC is even higher (0.8 and 1.0 respectively), the net charge on the protein is also negative and less flocculation is likely to occur. The effect of polymer dosage is evident in Figure 1(b). At low dosages, protein removal increased with increasing dosage, reaching a maximum removal level. Increasing the polymer dosage beyond this optimal level does not improve protein removal. For 98% removal, the optimum pH and dosage are 4.2 and 0.25 g/g BSA respectively.

### Removal of BSA by Adsorption to Ion Exchanger

The removal of BSA from an aqueous solution at pH 4.0 by adsorption to two different cation exchangers is presented in Figure 2. BSA has a net positive charge at this pH and both ion exchangers are negatively charged: CM Sephadex is weakly acidic and SP Sephadex is strongly acidic. This explains the behavior of the two adsorption curves: higher amounts of CM Sephadex are necessary for the same percentage removal of BSA, as compared to SP Sephadex. The protein molecules displace the counterions that balance the charged groups on the exchanger and occupy a site on the adsorbent. The amount of protein that can be adsorbed is a function of the ionic capacity of the exchanger and also of the number and size of the pores in the exchanger particle. The Sephadex C-50 ion exchangers have a good capacity for medium sized molecules with MW ranging from 10,000 to 100,000 (BSA has MW of 67,000) The amount of SP Sephadex necessary for 100% BSA removal at pH 4.0 is about 12 g / g BSA.

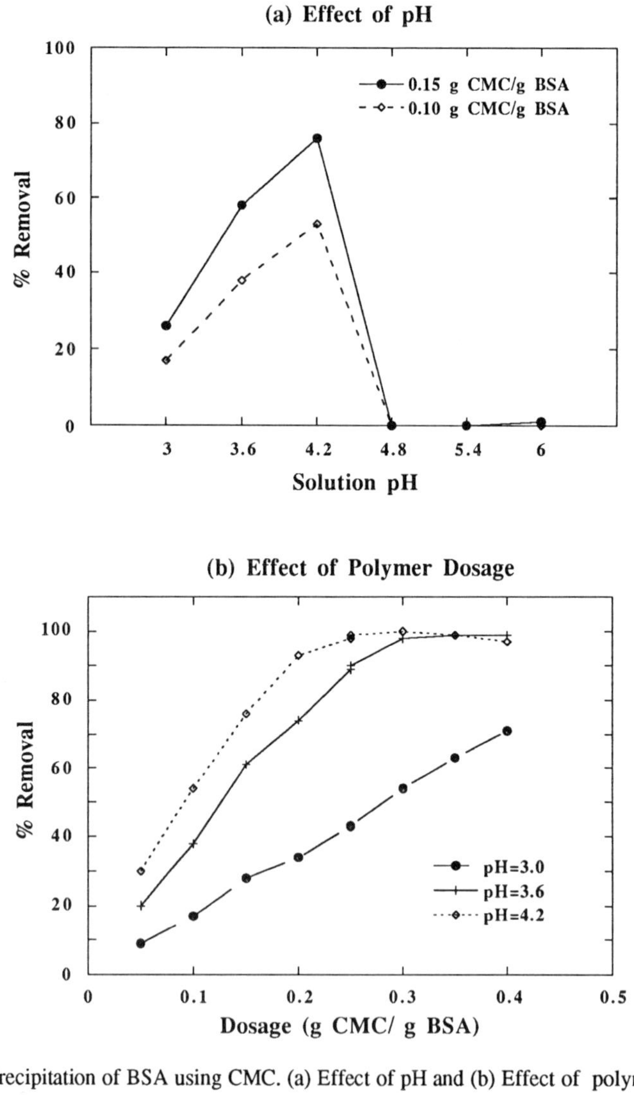


**Figure 1.** Precipitation of BSA using CMC. (a) Effect of pH and (b) Effect of polymer dosage.

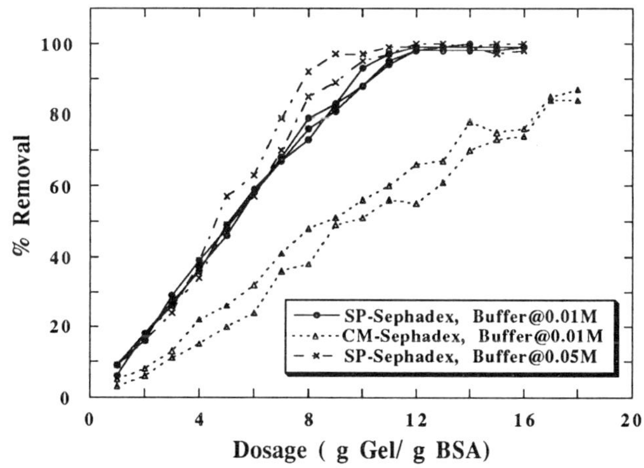


**Figure 2.** Removal of BSA by adsorption to cation exchangers, pH 4.0: SP Sephadex C-50 (strong exchanger) and CM Sephadex C-50 (weak exchanger).

## Removal of BSA by Adsorption to Charged Cotton Fibers

The ionizable groups on the charged cotton fibers are:

- FB2C (cotton dyed with 6% Schaeffer's acid): a sulfonic group (-$SO_3H$) and a hydroxyl group (-OH) ;
- FB4C (cotton dyed with 6% R salt): two sulfonic groups (-$SO_3H$) and a hydroxyl group (-OH).

The effect of fiber dosage and BSA initial concentration can be observed in Figure 3. Fibers 2C and 4C showed similar efficiencies in protein removal, at pH 4.2, though fiber 4C resulted in a slightly better performance. This could be explained by the presence of two sulfonic groups in fiber 4C, as compared to only one in fiber 2C. The sulfonic groups might be ionized at pH 4.2 (negatively charged) and interact with the positively charged protein molecules. A dosage of 100 g fiber/ g BSA is necessary for 98% protein removal.

## Adsorption of BSA to WSCA/CAF/CTF/Cotton Fiber in the presence of ammonium sulfate

**Protein removal as a function of ammonium sulfate saturation.** When BSA is precipitated by ammonium sulfate in the presence of water soluble cellulose acetate

(WSCA), cellulose acetate (CAF) and triacetate (CTF) fibrets, and cotton fibers, then filtered through 20 μm pore size frit, improvement in protein removal is obtained as compared to the precipitation with ammonium sulfate only. Figure 4 shows the result obtained with CAF, WSCA, and cotton fibers at increasing percentage saturation of ammonium sulfate, and Figure 5 compares the performances of CAF, WSCA, and CTF.

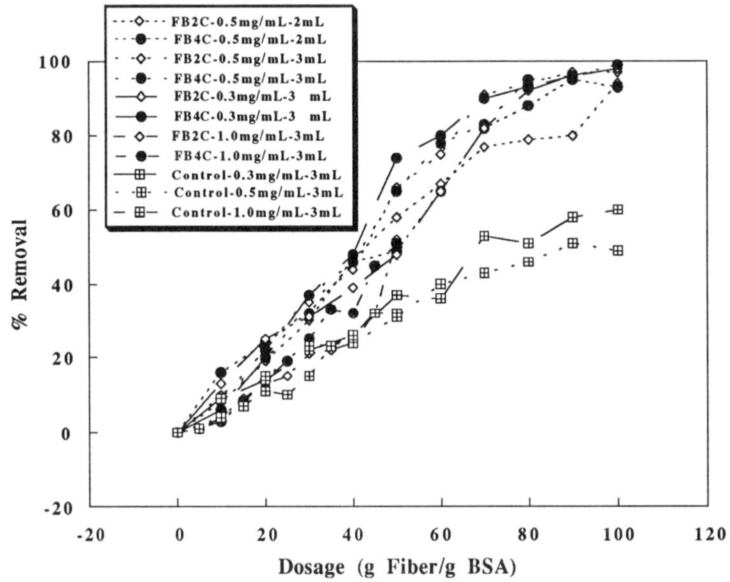


**Figure 3.** Removal of BSA from solution, pH 4.2, using charged cotton fibers. Effect of fiber dosage and type, and of BSA initial concentration.

Analyzing the results of BSA precipitation by ammonium sulfate and WSCA, followed by filtration, it can be observed that about 50% ammonium sulfate saturation is necessary for 100% protein removal from solution. When using ammonium sulfate only, the maximum protein removal reached only 20%, and an increase in the concentration of salt did not improve this value. One explanation for this behavior is the aggregate size. When precipitation is done only by addition of salt, the aggregates are normally smaller than 20 μm, the pore size of the filter used, and few particles are retained by filtration. On the other hand, precipitation in the presence of the polymer results in particles larger than 20 μm, which could be separated from solution by filtration through the 20 μm frits.

Three possible mechanisms have been envisaged for the improvement in protein recovery with the addition of polymer (WSCA). First, it may be that protein and polymer molecules interact with each other in solution, forming a complex which has its solubility decreased by salt addition. Second, the protein may adsorb to the polymer particles only after their formation. Third, protein aggregates and polymer aggregates may form independently, and on filtration, protein particles might be entrapped by the polymer aggregates.

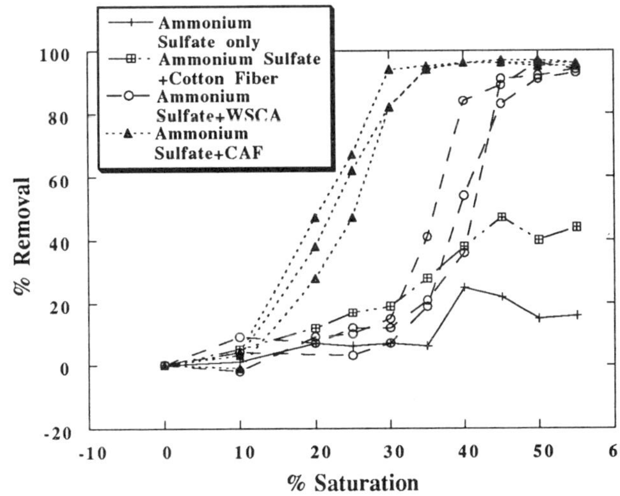


**Figure 4.** Removal of BSA from solution, pH 4.8, by ammonium sulfate + CAF/WSCA/ Cotton Fiber. Effect of ammonium sulfate concentration. Dosages of CAF/WSCA/Cotton are 24 g/g BSA.

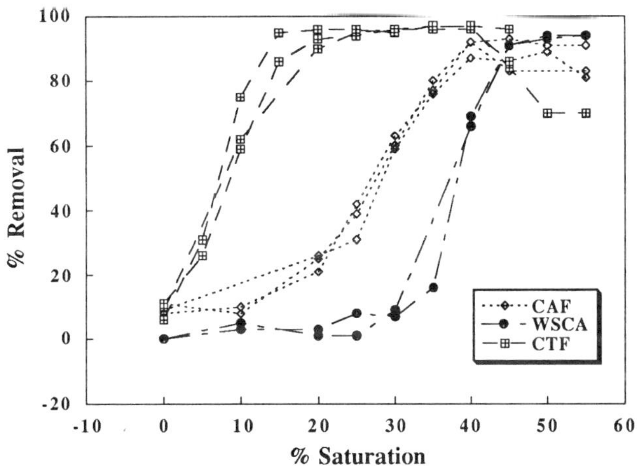


**Figure 5.** Removal of BSA from solution, pH 4.8, by ammonium sulfate + CTF/CAF/WSCA. Effect of ammonium sulfate concentration. Dosages are 30 g WSCA, 40 g CAF and 30 g CTF per g BSA.

Polymer precipitation starts to occur even at low salt concentration (at 10% ammonium sulfate saturation, WSCA aggregates were already observed in the precipitation samples), while significant removal of protein starts only at concentrations of 40% ammonium sulfate saturation. This behavior weakens the hypothesis that the formation of a protein-polymer complex may be the leading mechanism of protein recovery. Now, the two other hypothesis need to be analyzed.

From Figures 4 and 5, it is clear that addition of CAF, CTF and cotton fiber also improves protein recovery. These results show evidence that protein may be interacting with the polymer solid phase present in solution (after polymer aggregate formation), rather than forming a complex with the soluble polymer during precipitation. To verify if the cellulose acetate is acting as an adsorption matrix during protein precipitation or only as a filter aid that entraps protein aggregates formed by addition of salt, a set of experiments have been performed: in one batch, BSA was precipitated with ammonium sulfate, in the presence of CAF, then filtered through a 20 μm frit. In a second batch, BSA and CAF were treated separately with ammonium sulfate, and the sample with the CAF was filtered first through a 20 μm frit, resulting in a layer of fibret. Then, the salted out BSA sample was filtered through this CAF layer, to verify whether aggregates were retained by entrapment. No improvement in protein removal was observed when compared to filtration without the CAF layer. On the other hand, when CAF, protein and salt were added simultaneously during the precipitation process, 90% of the protein was removed at about 35% ammonium sulfate saturation. This demonstrated that the cellulose acetate in this case was not acting as a filter aid for entrapment of big protein aggregates, but rather as a matrix for protein adsorption.

The adsorption mechanism for the removal of protein by the cellulose acetate of various degrees of substitution (0.6-0.8 for WSCA, 2.4 for CAF and 3.0 for CTF) was postulated as being driven by hydrophobic interaction. This hypothesis is supported by the findings of Kiso (1986) and Kiso *et al.* (1989a), in which thorough studies of the interaction between organic solutes and cellulose acetate were made. They confirmed that the dominant effect on adsorption of organic compounds to cellulose acetate chromatographic columns was the hydrophobic interaction between cellulose acetate and the solute molecule (Kiso, 1986), and that this hydrophobic interaction was mainly through the acetyl groups of cellulose acetate and the alkyl chains of the solutes (Kiso *et al.*, 1989a). They used a liquid chromatographic technique for the studies (Kiso, 1985), comparing the performance of particulate (Kiso *et al.*, 1989b) and fibrous (Kiso *et al.*, 1986) cellulose acetates. Based on the results obtained in liquid chromatography, Kiso *et al.* (1992) studied the effects of molecular shape on the permeation of organic solute through cellulose acetate membranes.

Figure 5 compares the performance of salt-induced protein precipitation in the presence of WSCA, CAF and CTF, which have increasing acetyl contents. As the acetyl content of the cellulose acetate increases, less salt is necessary for the same amount of protein removed. Increasing the amount of salt in solution, increases the hydrophobic interactions among protein molecules and polymer surface. Kiso *et al.* (1989a) showed that retention times of organic solutes increased with increase of the DS values, indicating the relative hydrophobic property of the cellulose acetates: cellulose monoacetate< diacetate< triacetate (in our case, WSCA<CAF<CTF).

Another means of confirming adsorption by hydrophobic interaction, is to study the effect of temperature on protein recovery. Figure 6 shows the result of salt-induced precipitation in the presence of water soluble cellulose acetate (WSCA). At higher temperatures, lower concentrations of salt are necessary for the same amount of protein removed. Scopes (1987) stated that in the salting-out range, the solubility of proteins generally decreases with increasing temperature, which explains the results obtained here. In addition, through a detailed study of the adsorption and desorption of BSA, γ-globulin and fibrinogen on a series of polymer membranes, Dillman and Miller (1973) indicated that one of the mechanisms through which adsorption takes place increases with increasing temperature. They characterized this type of adsorption as being tightly bound, hydrophobic, and endothermic. In our case, a similar mechanism has been observed.

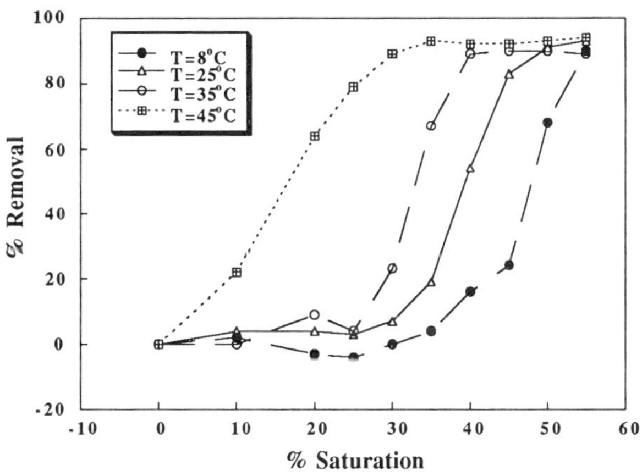


**Figure 6.** Effect of temperature on BSA precipitation by ammonium sulfate and water soluble cellulose acetate (WSCA).

## Adsorption isotherms of BSA on CTF

Following the elucidation of the adsorption mechanism as being hydrophobic and endothermic, an attempt to calculate the heat of adsorption was performed, by using a simple Langmuir isotherm model (Smith, 1981). Figure 7 shows the adsorption isotherms for CTF at 20% ammonium sulfate and temperatures of $8^o$, $28^o$, $35^o$, and $40^oC$ (adsorption isotherms at 10%, 15% and 25% are not shown here - see Chen (1994)). Using the Langmuir constants calculated from the model, the heats of adsorption were calculated for each salt concentration (Smith and van Ess, 1987):

| % Salt Saturation | $\Delta H_{ads}$ (cal/mol) |
|---|---|
| 10 | +3350 |
| 15 | +2700 |
| 20 | +2200 |
| 25 | +2250 |

Dillman and Miller (1973) found heats of adsorption of BSA, γ-globulin, and fibrinogen ranging from about 5 kcal/mol to about 20 kcal/mol, on various membranes such as cellophane and silicone rubber; while Borzin and Miller (1982) obtained enthalpies of adsorption of 1.5 kcal/mol for adsorption of albumin on xenograft. Although the values obtained for BSA on CTF are within the range of the ones obtained in the literature, it is worth noting that more recent studies have controversial views about the application of the Langmuir theory of adsorption for proteins at solid interfaces. Norde (1986) stated that in the case of protein adsorption, virtually none of the Langmuir premises is satisfied: adsorption is mostly irreversible and does not take place on fixed sites, the protein molecule usually changes upon adsorption, and lateral interaction may not be ignored. Norde and Lyklema (1978) used calorimetric determinations of the enthalpy of adsorption, instead of calculation from the temperature dependence from the adsorption isotherms according to Clausius-Clayperon, showing a discrepancy in the results obtained from both ways. On the other hand, Andrade (1985) stated that the classical Langmuir theory can be applied to protein adsorption from solution, if the solution is sufficiently dilute.

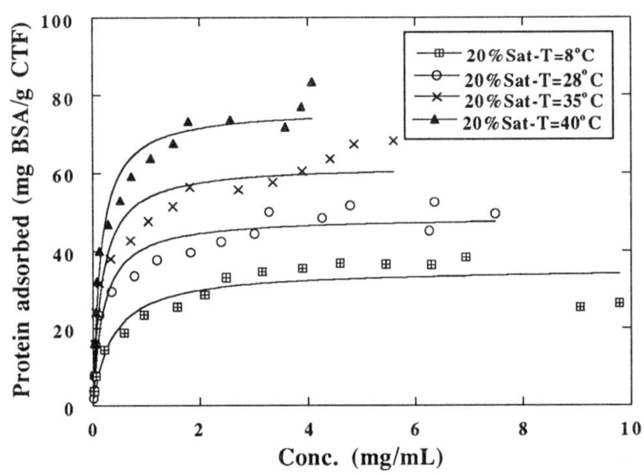


**Figure 7.** Adsorption Isotherm of BSA on CTF at 20% ammonium sulfate pH 4.8, at various temperatures: 8°, 28°, 35°, and 40°C. Experimental and calculated curves.

## Removal of BSA by Adsorption to Charge-Modified CAF

Fibrets modified with charged groups have been produced, by attachment of carboxymethyl cellulose (CMC) or bentonite on the surface, and tested in protein removal from solution. Figure 8 shows the result using CMC coated CAF, at pH 4.2. For 99% protein removal, a dosage of about 15 grams of fibret per gram of BSA is necessary. This is equivalent to about 0.15 grams of CMC per gram of BSA (the fibret is coated with 1% CMC), which is comparable to the dosage of 0.2 to 0.25 grams per gram of BSA when only CMC was used in the protein recovery by flocculation. If the CMC-CAF dosage is increased above the optimum level, protein recovery decreases, similarly to the case when only the polyelectrolyte is used.

Protein removal by bentonite coated CAF requires higher dosages than CMC-CAF:: for 85% removal, 200 grams of bentonite-CAF per gram of BSA were required at pH 3.6 (Chen, 1994).

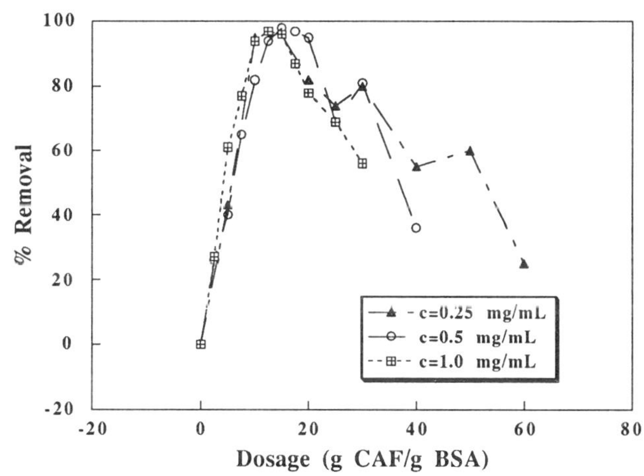


**Figure 8.** Removal of BSA from aqueous solution by adsorption to 1% CMC- CAF, pH 4.2. Effect of CMC-CAF dosage and initial BSA concentration.

## Comparison of the Dosages Used in Various Systems

Figure 9 presents a summary of the dosages necessary for protein removal from solution when polyelectrolyte (CMC), ion exchanger (SP Sephadex), charged cotton fibers (FB4C), cellulose acetate fibrets (CAF), cellulose triacetate fibrets (CTF), bentonite coated CAF, and CMC coated CAF were used. Polyelectrolytes have the highest efficiencies, requiring very low dosages. But, as has been mentioned before, for protein recovery from waste waters, the usage of polyelectrolytes resulted in precipitates with high water content which are difficult to handle, requiring high operation costs. CAF and CTF dosages used are comparable to ion exchanger dosages. CTF is slightly more efficient in terms of dosage, when compared to CAF, but, in terms of salt content necessary for protein removal, it is a more advantageous choice: for 93% protein removal, 15% ammonium sulfate saturation is necessary for CTF, as opposed to 40% saturation for CAF and 50% for WSCA.

Charged cotton fibers are not as efficient as CTF or CAF, due to lower surface area and low charge density for interaction with proteins. High dosages are required for the same amount of protein removal. The use of CMC coated fibrets resulted in dosages comparable to ion exchangers and CTF. But when compared to the usage of CMC alone, no greater advantages were found in terms of dosage and filterability of the final slurry. Bentonite coated CAF presented the lowest efficiency, requiring much higher dosages. This might be due to the low charge density available for protein interaction by electrostatic interaction.

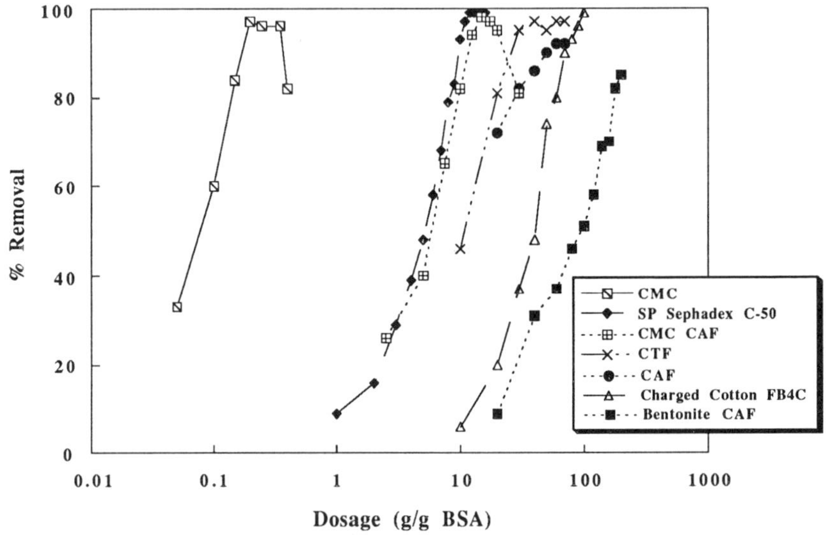


**Figure 9.** Comparison of polymer dosage necessary for recovery of BSA from aqueous solutions.

## CONCLUSIONS

Soluble protein recovery from aqueous solution using fibrous materials have been demonstrated by using a standard protein - bovine serum albumin. Cellulose acetate fibrets, cellulose triacetate fibrets, carboxymethyl cellulose coated fibrets, and charged cotton fibers showed good performance in the recovery of protein. In the case of CAF and CTF, the addition of salts is necessary, in order to achieve the desirable percentage removal of protein, though the required salt concentration in the case of CTF is relatively low. In some cases the normal amount of salt present in the food processing waste waters may allow good protein adsorption to CTF, without further salt addition.

In most of the cases, the resulting filter cake, containing the fibers/fibrets and the recovered protein, was dry and easy to handle. This is an advantage, if compared to traditional recovery methods, using chemical coagulants or polyelectrolytes. Although higher dosages are necessary when using the fibrous materials studied, the attainment of drier proteinated by-products that can be used directly as animal feed is a positive aspect worth consideration. It can lead to reduction in operating costs.

Future work using actual waste water samples from the food processing industries, breweries and distilleries, and pharmaceutical industries, is required, in order to verify the applicability of these fibrous materials to protein recovery from these effluent sources. It is also necessary to develop a process that would utilize these fibrous materials effectively.

## ACKNOWLEDGEMENTS

The authors wish to thank the Hoechst Celanese Corporation, the William R. Kenan Jr. Institute for Engineering, Technology and Science at NCSU, the Fulbright Commission and the American Chamber of Commerce in Brazil for the financial support during the course of these investigations.

## REFERENCES

Andrade, J. D. "Principles of Protein Adsorption." In Surface and Interfacial Aspects of Biomedical Polymers, ed. J. D. Andrade. 1. 2, Plenum Press, New York, 1985.

Arai, K., Kinumari, T., and Fujita, T. "On the Toxicity of Chitosan.", *Bull. Tokai Reg. Fish. Res. Lab.* , **56**, 89, 1968.

Babbit, S. S. "Thermally Enhanced Dewatering.", *Poultry Intern.* , April, 42-44, 1989.

Beszedits, S. and Netzer, A., Protein Recovery From Food Processing Waste waters., ed. B & L Information Services. Toronto, Canada, 1982.

Borzin, G. A. and Miller, I. F. "The Kinetics of Protein Adsorption on Synthetics and Modified Natural Surfaces.", *J. Colloid Interface Sci.* , **86**(2), 539-558, 1982.

Bradford, M. M. "A Rapid and Sensitive Method for the Quantitation of Microgram Quantities of Protein Utilizing the Principle of Protein-Dye Binding.", *Analyt. Biochem.* , **72**, 248-254, 1976.

Carawan, R. E. and Valentine, E. G. Dissolved Air Flotation (DAFs) for Bakeries. North Carolina Cooperative Extension Service, 1991. 12/91-3M-TAH-210576.

Chen, L. A. "Protein Precipitation Using Fibrous Materials." M.S. Thesis, North Carolina State University, Raleigh, 1994.

Chowdhury, F. H. and Neale, S. M. "Acid Behavior of Carboxylic Derivatives of Cellulose - Part I. Carboxymethyl cellulose.", *J. Polymer Sci.* , **A**(1), 2881-2891, 1963.

Crickenberger, R. G. and Carawan, R. E. Using Food Processing By-Products for Animal Feed. The North Carolina Cooperative Extension Service, 1991. 8/91-2M-TAH-210453.

Dillman, W. J. J. and Miller, I. F. "On the Adsorption of Serum Proteins on Polymer Membrane Surfaces.", *J. Colloid Interface Sci.* , **44**(2), 221-241, 1973.

Hopwood, D. "Effluent Treatment in Meat and Poultry Processing Industries.", *Proc. Biochem.* , **12**(8), 5-8 + 44 & 46, 1977.

Kiso, Y. "A HPLC Study of Organic Solutes of Cellulose Acetate.", *Bull. Chem. Soc. Jpn.* , **58**, 3637-3638, 1985.

Kiso, Y. "Factors Affecting Adsorption of Organic Solutes on Cellulose Acetate in an Aqueous Solution System.", *Chromatographia* , **22**(1), 55-58, 1986.

Kiso, Y., Jinno, K., and Nagoshi, T. "Liquid Chromatography in a Capillary Packed with Fibrous Cellulose Acetate.", *HRC & CC* , **9**, 763-764, 1986.

Kiso, Y., Kitao, T., Jinno, K., and Miyagi, M. "The effects of molecular width on permeation of organic membranes.", *J. Memb. Sci.* , **74**, 95-103, 1992.

Kiso, Y., Kitao, T., Sheng, G. Y., and Jinno, K. "Retention Characteristics of Aliphatic Compounds on Cellulose Acetates as a Stationary Phase with an Aqueous Mobile Phase.", *Chromatographia* , **28**(5), 279-284, 1989a.

Kiso, Y., Takayama, K., and Jinno, K. "Cellulose Acetate as Stationary Phase in Microcolumn LC.", *HRC* , **12**(3), 169-173, 1989b.

Lunney, R. "Focus on Flocculation.", *Food Processing* , December, 42-43, 1988.

Metzner, A. V."Protein Recovery Opportunities for Fibrets in the Treatment of Food Processing Waste water." Magma Resources, 1988.

Metzner, A. V. "By-Products: Facts and Considerations." Magma Resources, 1991.

Norde, W. "Adsorption of Proteins from Solution at the Solid-Liquid Interface.", *Advances in Colloid and Interface Science* , **25**, 267-340, 1986.

Norde, W. and Lyklema, J. "The Adsorption of Human Plasma Albumin and Bovine Pancreas Ribonuclease at Negatively Charged Polystyrene Surfaces.", *J. Colloid Inteface Sci.* , **66**(2), 295-302, 1978.

Pharmacia Fine Chemicals, "Ion Exchange chromatography - principles and methods.", Sweden , 1983.

Powers, E. J. Polyacrylamides in Animal Foods. Hoechst Celanese Corporation, 1991. EJP-91-034.

Rusten, B., Lundar, A., Eide, O., and Ødegaard, H. "Chemical Pretreatment of Dairy Waste water.", *Wat. Sci. Tech.* , **28**(2), 67-76, 1993.

Scopes, R., Protein Purification - Principles and Practice., 2 ed., Spring-Verlag, New York, 1987.

Smith, J. E. "Cellulose Acetate Fibrets: A Fibrillated Pulp with High Surface Area.", *Tappi J.* , **71**(12), 185-187, 1988.

Smith, J. M., Chemical Engineering Kinetics., 3rd. ed., McGraw-Hill chemical engineering series, ed. J. V. Brown. McGraw-Hill Book Co., Singapore, 1981.

Smith, J. M. and van Ess, H. C., Introduction to Chemical Engineering Thermodynamics., 4th. ed., McGraw-Hill chemical engineering series. McGraw-Hill Bood Co., Singapore, 1987.

Swinehart, C. "Waste Disposal Calls for Ingenuity, Creativity.", *The National Provisioner*, **203**(17), 10-11, 1990.

# SECTION IV.

# BIOMIMETICS AND ENZYME BIOREACTORS

# PREPARATION, CHARACTERIZATION, AND UTILIZATION OF BIOMIMETIC MEMBRANES

Janos H. Fendler

Department of Chemistry
Syracuse University
Syracuse, New York 13244-4100

## INTRODUCTION

Biological membranes define the very existence of the living cell and are intimately involved in *in vivo* syntheses, recognition, information transfer, and energy transduction.[1] Self-organization, predominantly bimolecular thickness, domain formation, temperature, media- and electrical signal-dependent fluidity, and permeability control are believed to be responsible for the effectiveness of the biological membrane in mediating these myriads of activities. The exploitation of biomembrane-dependent processes *in vitro* in relatively simple artificial biomimetic membranes for the compartmentalization of substrates, for acting as carriers, and for altering reaction rates, products, and stereochemistries has been the subject of intensive research activities.[2] Biomimetic membranes are defined by a utilitarian point of view as compartments which are able to accommodate selected substrates in desired microenvironments. Aqueous micelles, reversed micelles, monolayers, Langmuir-Blodgett (LB) films, bilayer lipid membranes (BLMs), freely suspended ultrathin films, surfactant vesicles (liposomes), cast multilayers, self-assembled films, and even such layered compounds as zeolites and pillared clays (or organoclay complexes) are considered to be biomimetic membranes in this broad definition. In contrast, the term biofunctional membrane is limited to an "entity in which biological molecules (or cells) are attached to polymeric supports cast in the form of porous membranes."[3]

Advances in biomimetic membrane research have been spectacular.[2,4] A large variety of new techniques, including surface force apparatus;[5,6] the different scanning force,[7] fluorescence,[8] and Brewster-angle[9,10] microscopies; and X-ray diffraction using synchrotron sources,[11] have been employed for the characterization of biomimetic membranes. New theories and improved molecular modeling have also contributed to

*Biofunctional Membranes*, Edited by
D.A. Butterfield, Plenum Press, New York, 1996

our understanding of the membrane-mimetic approach.[12] These vigorous activities have stimulated, in turn, the development of new constituents for biomimetic membranes and novel biomimetic membrane-based applications. Discussion here will be limited to monolayers, BLMs, multilayer films derived from surfactant vesicles, and self-assembled films. The biomimetic membranes discussed in the present survey are already biofunctional to some extent. Furthermore, they can be readily modified to become fully biofunctional in the sense given in Butterfield's definition.[3] Monolayers, BLMs, and surfactant vesicles were the initial models of biological membranes.[2] All of the biomimetic membranes that are surveyed in the present review have been fruitfully employed as templates in advanced materials synthesis.[4] Emphasis will be placed on recent advances in the preparation, characterization, and utilization of monolayers, BLMs, multilayer films derived from surfactant vesicles, and self-assembled films.

## MONOLAYERS

Formation of a monomolecularly thick film upon the spreading of a water insoluble surfactant (or a lipid) on the surface of an aqueous solution contained in a trough has been known for over a century.[13] Classical investigations of monolayers (sometimes referred to as Langmuir films), based primarily on surface pressure (Π) *vs.* surface area (A) isotherms, have established gaseous ⇋ liquid ⇋ solid phase transitions and molecular packing in the two dimensions (see Figure 1). Much research momentum was gained upon the optimization of conditions for the transfer of Langmuir films, monolayer-by-monolayer from the water surface, to solid substrates to form Langmuir-Blodgett (LB) films (Figure 2)[12] and by the recognition of the potential industrial and high-technological applications of LB films.[4,14,15]

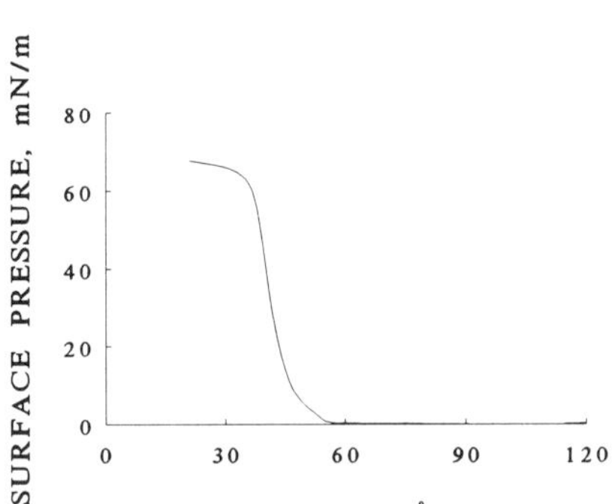


**Figure 1.** Surface-pressure/surface-area isotherm of a thiol surfactant

Monoparticulate thick layers of polystyrene microspheres,[16,17] silylated glass beads,[18] organoclay complexes,[19] proteins,[20] and size-quantized semiconductor (CdS,[21] $TiO_2$,[22] CdSe,[23] and CdS/CdSe[23] ), ferroelectric (piezoelectric barium titanate[23] and PZT[24]), and magnetic ($Fe_3O_4$)[25] nanoparticles have also been spread on aqueous solution surfaces. In all cases, particulate films were spread on water in a Langmuir trough by dropping solution aliquots from a Hamilton syringe. The characteristic surface pressure *vs.* surface area (Π *vs.* A) isotherms were then determined after permitting time for evaporation of the carrier solvent. The physical properties of the films were characterized both *ex-* and *in situ* on the water surface by using a range of physical techniques. The structures of the films on the water subphase were examined on the micron scale by Brewster-angle microscopy (BAM) and at higher magnifications by transmission electron microscopy (TEM). The mean and standard deviation of the particle-size distribution in each film was also evaluated by using TEM to analyze photographically enlarged micrographs. Absorption spectroscopy and steady-state fluorescence spectroscopy were applied where appropriate and reflectivity measurements permitted estimation of the film thicknesses on water. The films were also transferred to solid substrates by using the standard LB dipping technique. The particulate films were typically maintained at a constant surface pressure of approximately 30 mN/m, while a slide of either glass, silanized glass, or quartz was immersed and subsequently extracted from the subphase.

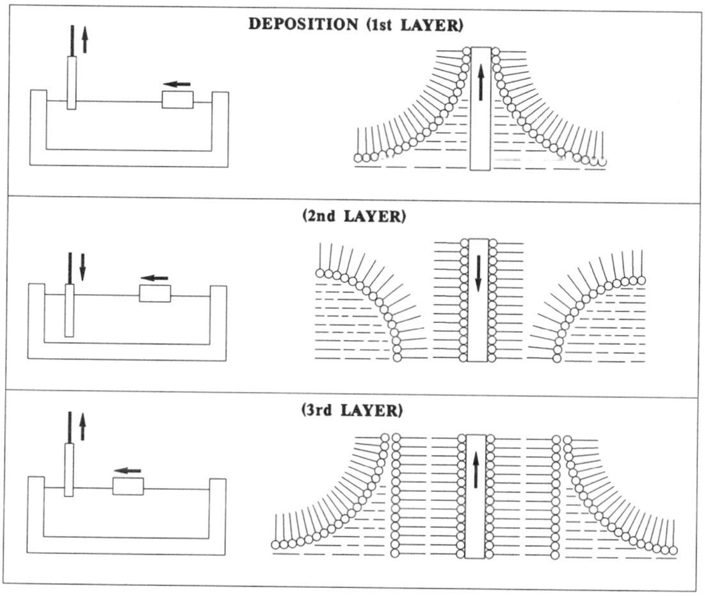


**Figure 2.** Schematics of Y-type Langmuir-Blodgett (LB) formation on a hydrophilic substrate.

Formation of a uniform monoparticulate film from hexadecyltrimethylammonium bromide-stabilized, 20±2 Å-diameter $TiO_2$ particles is given here as an example. The uniformity of the $TiO_2$ nanocrystalline film, during all stages of its compression, was demonstrated by Brewster-angle microscopy (Figure 3). The presence of the film in static pictures only became apparent when contrast was introduced by cracks in the images, caused by dust particles. An increase of the surface pressure to 20 mN/m brightened the image somewhat and closed some of the cracks (Figure 3b). The solid state of the $TiO_2$ nanocrystalline film at $\Pi = 50$ mN/m could be observed as a featureless bright field (Figure 3c), whereas after collapse and expansion the image revealed a series of stripes which corresponded to the disruption of the film by internal stress (Figure 3d).

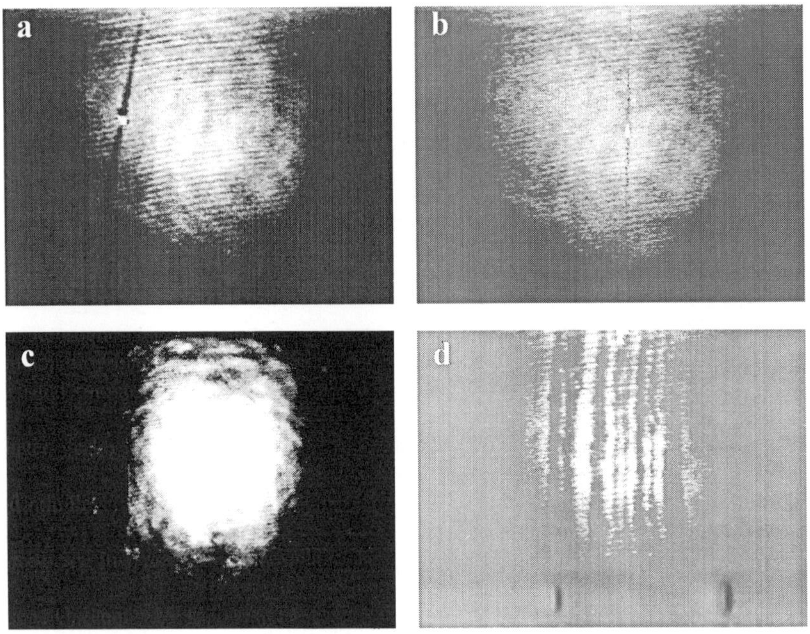


**Figure 3.** Brewster-angle microscopic images of a $TiO_2$ monoparticulate film at $\Pi = 5$ (a), 15 (b), 50 (c), and 0 mN/m; decompression after collapse (d).

Measurements of angle-dependent reflectance of p-polarized light have been fruitfully employed for the determination of the reflective index (n) and thickness (d) of monoparticulate films on water surfaces.[26] The measured reflectivity of the $TiO_2$ film was found to be significantly different from the bare water surface only at high surface pressures. The obtained mean thickness of 32±10 Å and reflective index of 2.2±0.2 are compatible with the presence of a monoparticulate thick nanocrystalline $TiO_2$ film on the water surface. This value should be compared to the diameter of the $TiO_2$ particles (22 Å), assessed from the absorption edge of the sol (thermal fluctuations and the presence of an organic surfactant shell are believed to be responsible for the larger diameters, observed by reflectivity, than those assessed from absorption spectrophotometry for the $TiO_2$ particles). Transmission electron microscopic images substantiated the formation of monoparticulate films.

Heat treatment of the $TiO_2$ sols dramatically decreased the interparticle distances in the spread monoparticulate layers. A $TiO_2$ particle prepared from sols in the absence of any prior heat treatment occupied a mean area of 1150 $Å^2$. The monoparticulate layers prepared from $TiO_2$ sols which had been heated for 90 minutes at 70°C, for 60 minutes at 90°C, and for 120 minutes at 90°C resulted in mean areas of 1000 $Å^2$, 650 $Å^2$, and 450 $Å^2$ occupied by a $TiO_2$ particle. Decrease of the interparticulate distances in the monoparticulate films is the consequence of thinning of the $TiO_2$ surface coating upon heating the sols. Heating alters the surface structure of the $TiO_2$ nanocrystallites which renders the binding of CTAB to be less favorable; heat treated $TiO_2$ nanocrystallites have thinner surfactant coats than untreated ones (Figure 4).

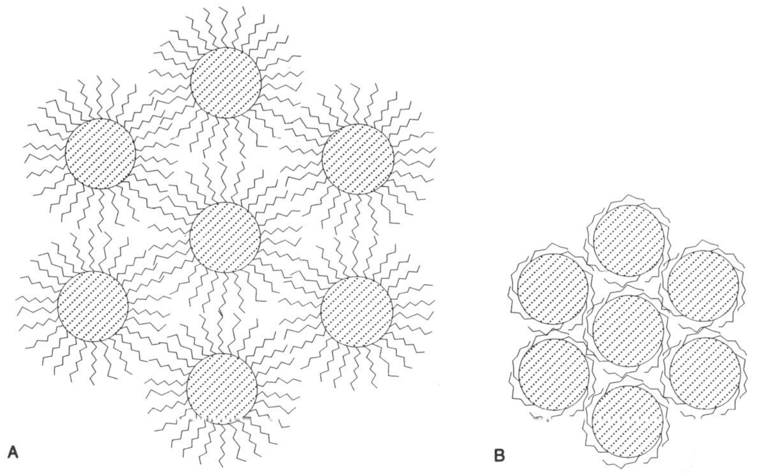


**Figure 4.** Schematics of $TiO_2$ monoparticulate films, viewed from above, with a longer (A) and with a shorter (B) interparticle distance.

There are numerous advantages to forming monoparticulate Langmuir films and constructing LB films from them. The particles are prepared prior to their incorporation into the films, thus permitting their dimensions, their physical properties, and the particle size distribution to be precisely controlled. Spreading the particles in a Langmuir trough enables the area of the subphase to be covered by the films and, thus, the interparticulate distances to be well-defined. Subsequent to film formation on water, the particulate films can be systematically transferred to a wide range of solid substrates; this contrasts with most deposition techniques where the quality of the film is highly dependent on the solid substrate itself. Thus, selective highly specific membranes and sensors can be constructed by transferring the monoparticulate films to porous supports and electrodes.

Growing proteins[27] and inorganic crystals (non-epitaxially and epitaxially) under monolayers[28] is inherently important for the rationalization of biomineralization[29,30] and for the development of innovative optical and electro-optical devices and sensors.

## BILAYER LIPID MEMBRANES (BLMs)

BLMs are produced on pinholes (diameters of 2.00 mm or smaller), drilled into a Teflon partition which separates two aqueous electrolyte solutions. Painting an organic solution of the surfactant (or lipid) across the pinhole results in the formation of a film which reflects the light with a gray color (Figure 5). Within a few minutes, the film thins and the reflected light exhibits interference colors which ultimately turn black.

At this point, the film is considered to be bimolecular and 5.0±1.0-nm thick. Separation of the aqueous solutions by the BLM allows electrical measurements by using macroscopic electrodes. Precise capacitance, conductance, and impedance measurements, in the absence and in the presence of ionophores, have greatly contributed to our understanding of impulse and ion transfer mechanisms.[31] Particularly significant has been the development of voltage clamping (*ie.*, holding the BLM at a predetermined potential and measuring the current flow) and single-channel recording.[32,33]

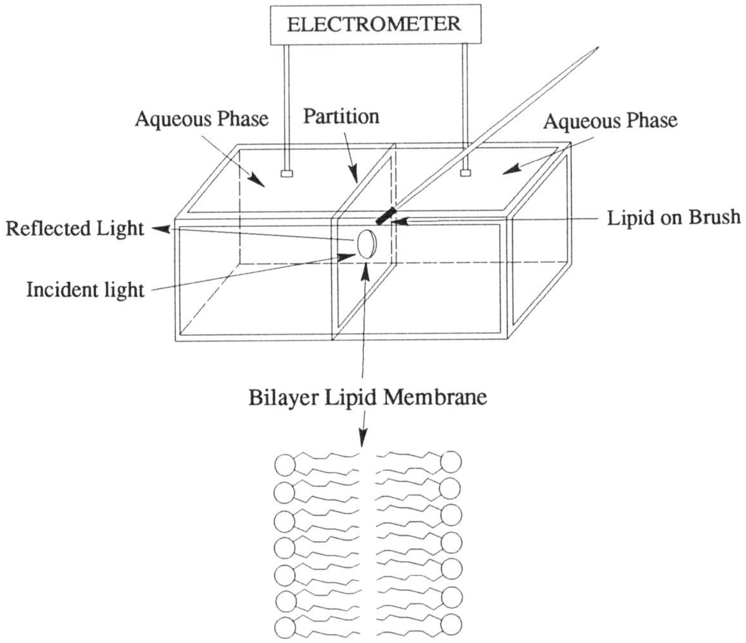


**Figure 5.** Top: formation of BLMs at a teflon pinhole which separates two compartments. Bottom: a schematic diagram of the molecular organization of the surfactant molecules in a bilayer lipid membrane (BLM).

Although BLMs are much more amenable to electrical measurements than are LB films, relatively little use has been made of them in application-oriented research. Lack of long-term stability of the early BLM preparations is believed to be the cause of this neglect. The use of polymerized BLMs,[34] hydrogel-supported BLMs,[35] and the incorporation of proteins,[31] semiconductor, metallic, and magnetic particles onto the BLM matrix[36,37] have overcome these stability problems and opened the door to potential applications.

The approach illustrated by the construction of semiconductor-semiconductor (SS') and semiconductor-insulator-semiconductor (SIS) junctions on BLM surfaces (see Figure 6).[38] Single-phase microcrystalline semiconductor particles incorporated onto only one side of the BLM represent the most straightforward system. The composition of the electrochemical cell used in investigating *System A* is explicitly shown in Equation 1.

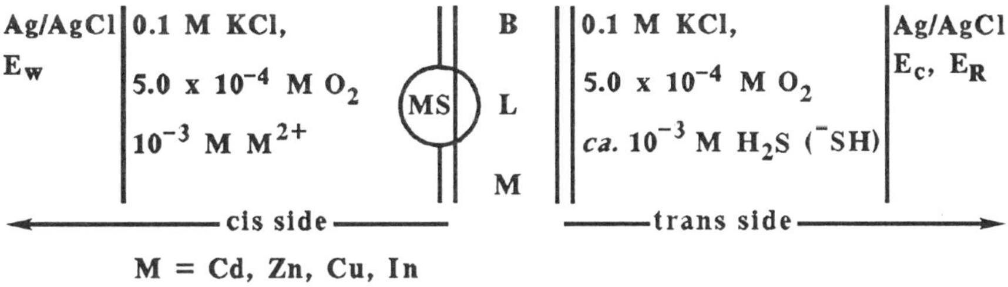


(1)

where $E_w$, $E_c$, and $E_R$ indicate the working, the counter, and the reference electrodes, respectively. Presumably, *Systems B* and *C* are also porous structures, explicitly shown in Equations 2 and 3, respectively.

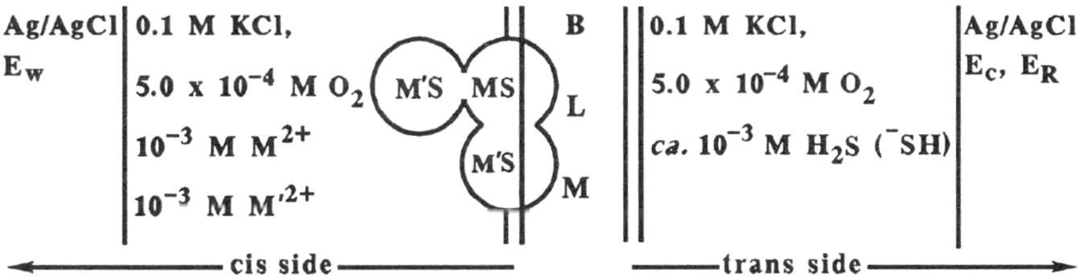


(2)

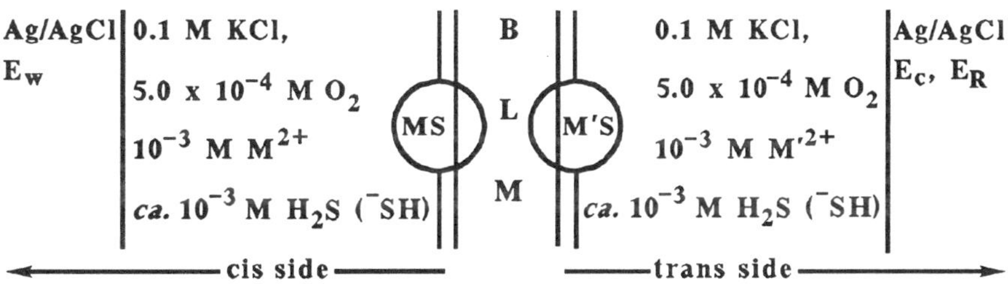


(3)

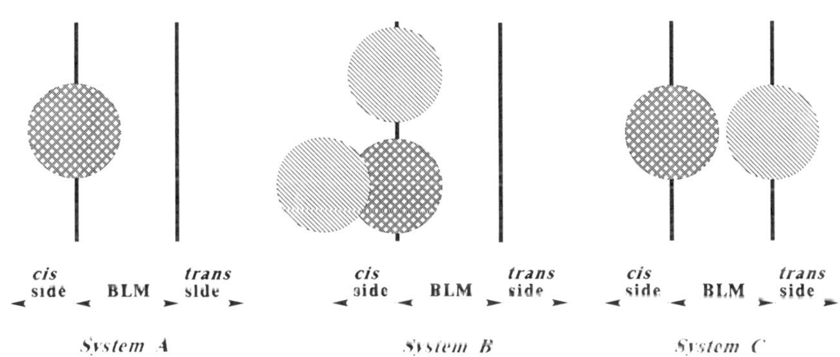


**Figure 6.** Schematic representation of the different semiconductor-coated BLMs. A single composition of particulate semiconductor deposited only on one side of the BLM constituted *System A*. Two different compositions of particulate semiconductors sequentially deposited on the same side of the BLM represent *System B*. Finally, two different compositions of particulates deposited on the opposite sides of the BLM made up *System C*.

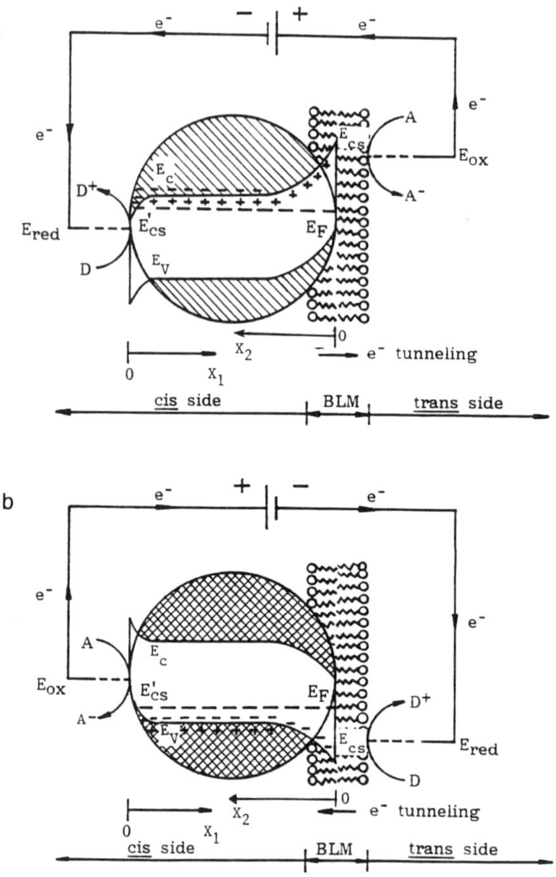


**Figure 7.** Band models for: an n-type (a) and a p-type (b) semiconductor-containing BLM (*System A*)-ES junction. $E_{cs}$ and $E'_{cs}$ represent the conduction-band edge at the semiconductor-BLM interface and at the semiconductor *cis* surface. $E_F$ is the Fermi energy level and $E_{red}$ and $E_{ox}$ are energy levels of the reducing and oxidizing agents, respectively. A and D stand for an electron acceptor and an electron donor, respectively.

Equations 1, 2, and 3, as well as Figure 6, are gross oversimplifications and should in no way be considered to convey any structural information beyond the location of the semiconductor particles and their penetration into the BLM. Band models of n- and p-type electrolyte semiconductor-containing BLM (*System A*), ES, junctions are drawn in Figure 7.[38] Charge injection into the conduction band of the n-type semiconductor by a sufficiently active surface donor or by an applied voltage (making the trans side positive relative to the cis side) results in an accumulation of the majority carriers at the space-charge region of the semiconductor particles (Figure 7a). Downward bending of the conduction-band energy ($E_{cs}$) provides for a favorable overlap with the energy level of the reducing agent ($E_{red}$) at the cis electrolyte surface; the semiconductor particle provides an ohmic contact with the cis electrolyte. A depletion layer is formed at the particle surface which is immersed in the BLM. Hence, the conduction band energy, $E_c$, is upwardly bent to the energy of the conduction band edge at the surface,

$E_{cs}$. Current flow is observed under cathodic (trans side positive with respect to the cis side) potential since $E_{red}$ is located in the band-gap region. Electrons are assumed to tunnel through the modified, very thin BLM (or to be transported by electrolytes) to the overlapping unoccupied energy levels of the oxidizing agent ($E_{ox}$). The exponential increase of cathodic current with applied voltage is the expected consequence of decreased band bending and increased surface electron density. The behavior of p-type semiconductors can be analogously rationalized.[38]

These studies have demonstrated semiconductor-coated BLMs to be viable candidates for advanced nanostructured electronic devices and sensors. Additionally, the reported direct and converse flexoelectric effects (curvature-induced electrical polarization and transmembrane voltage difference-induced curvature) in BLMs[39,40] have demonstrated the interconversion of mechanical and electrical energies in two-dimensional systems and opened the door to the development of nanometer-sized mechanical switches and actuators.

Related to BLMs are the freely suspended liquid crystalline multilayer films whose phases can be ordered and manipulated by temperature changes and/or application of an electric field.[41] Importantly, they have been transferred to solid substrates without any structural alteration.[41]

## MULTILAYER FILMS DERIVED FROM SURFACTANT VESICLES

Surfactant vesicles (or liposomes, if they are prepared from phospholipids) have been known since the late 1950s and are employed extensively in fundamental and applied research. Multi-lamellar vesicles (MLVs) are closed, onion-like, 1000-8000 Å-diameter, multibilayer aggregates which, upon sonication, form closed, 300-600 Å-diameter spherical unilamellar vesicles (SUVs). Publication of reports on the formation of well-characterized SUVs from long-chain dialkylammonium salts,[42,43] has prompted the syntheses of numerous single- and multiple-chained vesicle-forming surfactants.[44-49]

Once formed, SUVs, unlike aqueous micelles, do not break down upon dilution; there is no equivalent of critical micelle concentration (CMC) for SUVs. Additionally, depending on their chemical composition, vesicles remain stable for days to weeks. SUVs, like membranes, are osmotically active; addition of electrolytes shrinks the vesicles, while placing them in solutions more dilute than their internal electrolytic concentrations causes swelling. SUVs, like membranes, are destroyed (lysed) by the addition of detergents or alcohols.

Below their phase transition temperatures, surfactants in SUVs are arranged in tilted, one-dimensional lattices; they are in highly ordered "solid" states. Above the phase transition temperature, the surfactants are somewhat separated from each other and assume liquid-like configurations; they are in their "fluid" states. Temperature-induced phase transition and phase separation are important properties of vesicles, offering routes to permeability control and molecular recognition. Indeed, the phase behavior of over 9500 surfactants and lipids have been investigated and the obtained data has been compiled.[50] Phase separation and domain formation can be realized by constructing SUVs from surfactant mixtures.

Surfactant vesicles have been used as templates for the synthesis of ultrathin, spherical polymer membranes.[51] The approach involved the formation of SUVs from mixtures of diallylammonium dihexadecylphosphate and sodium dihexadecylphosphate, polymerization of the counterions coating the vesicles by ultraviolet radiation, removal of the surfactants (by treatment with HCl and dialysis), and redispersal of the remaining polymeric bags (termed "ghost vesicles"). Electron microscopy and ($^{14}$C)sucrose entrapments indicated the presence of 200- to 2000-Å-diameter, ghost vesicles which were closed spherical and relatively porous in character.[51]

A related system is that of the lipid-bilayer corked capsule membranes which are formed from ultrathin (ca. 1 μm thick), spongy, 2.0- to 2.5-mm-diameter, more-or-less spherical nylon bags in which multiple bilayers are immobilized.[52-54] They were considered to combine the advantages of mechanical and chemical stabilities of polymeric membranes with the controllable permeabilities of surfactant vesicles. Polymerization of the bilayers, in situ, alleviated the undesirable and premature release of surfactants from the matrices of the capsule membranes. The surfactants used contained polymerizable groups either in their tails, in their headgroups, or in their counterions.

Controlled evaporation of SUVs and MLVs on substrates has been shown to result in the formation of ultrathin films which retained the regular bilayer structure of vesicles.[47,55,56] These immobilized bilayers, termed as "cast bilayers", "cast multilayers", or "ordered cast (ultrathin) films", have provided an alternative to LB films.[55-75] ammonium surfactants with azobenzene and glutamate functionalities have been used, for example, in the preparation of cast-film-forming SUVs. X-ray diffraction measurements of cast films established the surfactant alignment to be perpendicular to the substrate surface with 45° and 65° tilts (Figure 8).[74]

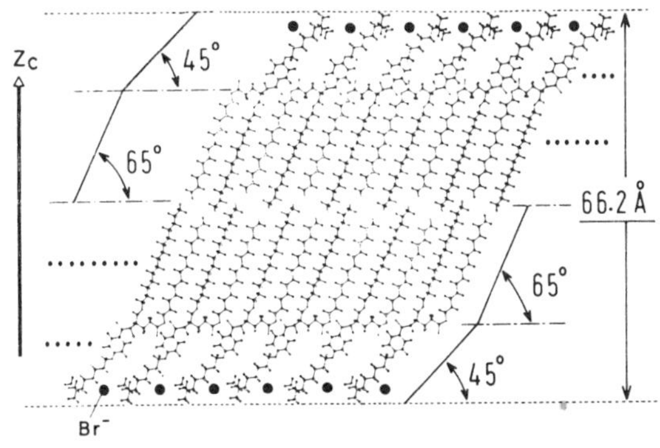


**Figure 8.** Plausible model of the molecular packing of the cast multibilayer of glutamate as produced by electron-density matching of the X-ray diffraction data. The plane of the benzene ring is arbitrarily set in the plane of the two alkyl chains. When the benzene plane is assumed to be in the perpendicular disposition, the tilt angles for the spacer and tail chains become slightly greater: 50° for the spacer and 75° for the tail.

Cast multibilayers prepared from an azobenzene-containing surfactant displayed only short-term stabilities and low solubilities in water. These disadvantages have been overcome by coating the bilayers with cellulose acetate and by using poly(vinyl alcohol) as a binder.[58]

Thin films have also been cast from fluorocarbon-surfactant-containing SUVs with bound poly(vinyl alcohol).[65,66] Wide angle X-ray diffraction measurements have been interpreted in terms of highly oriented bilayers and complete substrate coverage by the fluorocarbon surfactants, even at low surfactant-to-poly(vinyl alcohol) ratios (Figure 9).[66]

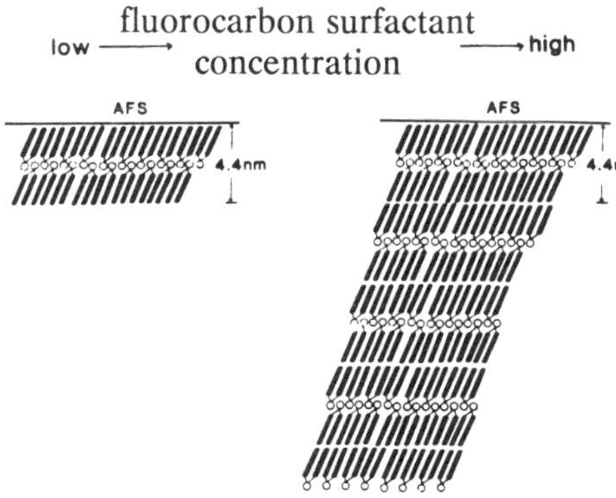


**Figure 9.** Schematic representation of the structure of the outermost layer of the composite thin film with small and large weight percents of fluorocarbon amphiphile.

Composite cast multibilayers provided a route to the formation of multilayer, two-dimensional polymer networks.[72,74]

Myoglobins[71] and iron-oxide particles[76] have also been organized in cast multilayers. Flexibility and versatility render composite cast bilayers to be suitable for the construction of functional, ultrathin, polymeric superstructures with molecularly defined morphologies. Cast-multilayer membranes also serve as permeability controls in capsule membranes.[52-54,77]

## SELF-ASSEMBLED FILMS

The spontaneous self-assembly of monolayers on an appropriate substrate, immersed in a dilute solution of a long chain surfactant, is well documented.[12] In particular, the self-assembly of long chain thiols on gold substrates has provided a viable approach to sensor construction and permitted the electrochemical characterization of chemisorption.

The spontaneous, layer-by-layer self-assembly of polyelectrolyte/clay platelet/semiconductor composite nanostructured films, reported here, well exemplifies our proposed membrane-mimetic approach to advanced materials.[4] While the dimensions, organization, and deceptive simplicity of spontaneous self-assembly are analogous to the corresponding properties of the biological membrane, the present system offers a high degree of stability and relative ease of systematic chemical and physical manipulation. Layers of clay platelets, particularly if self-assembled under an applied potential, are likely to cap imperfections and pinholes and, thus, increase the structural integrity of the nanostructured films. Polymerization of appropriate polyelectrolyte layers in the film provides an additional means of stability and permeability control. Significantly, judicious selection of components and their layer-by-layer self-assembly will result in nanostructured films with the desired mechanical, optical, electro-optical, electrical, magnetic, and electromagnetic properties.[4]

## ACKNOWLEDGMENT

Support of this work by a grant from the United States' National Science Foundation is gratefully acknowledged.

## REFERENCES


1. E.E. Bittar, "Membrane Structure and Function," Wiley-Interscience, New York (1968).
2. J.H. Fendler, "Membrane Mimetic Chemistry," Wiley-Interscience, New York (1982).
3. D.A. Butterfield, J. Lee, S. Ganapathi, and D. Bhattacharyya, Biofunctional membranes. 4. Active-site structure and stability of an immobilized enzyme, papain, on modified polysulfone membranes studied by electron-paramagnetic-resonance and kinetics, *J. Membr. Sci.* 91:47 (1994).
4. J.H. Fendler, "Membrane-Mimetic Approach to Advanced Materials," Springer-Verlag, Berlin (1994).
5. J.N. Israelachvili, "Intermolecular & Surface Forces," Second Edition, Academic Press, San Diego, CA (1992).
6. J.L. Parker, H.K. Christenson, and B.W. Ninham, Surface force apparatus, *Rev. Sci. Instrum.* 60:3135 (1989).
7. R. Wiesendanger, "Scanning Probe Microscopy and Spectroscopy," Cambridge University Press, Cambridge (1994).
8. M. Löche and H. Möhwald, Fluorescence microscopy of monolayers, *Rev. Sci. Instr.* 55:1968 (1984).
9. S. Henon and J. Meunier, Microscope at the Brewster angle: direct observation of first-order phase transitions in monolayers, *Rev. Sci. Instr.* 62:936 (1991).
10. D. Hönig and D. Möbius, Direct visualization of monolayers at the air-water interface by Brewster angle microscopy, *J. Phys. Chem.* 95:4590 (1991).
11. J. Als-Nielsen and H. Möhwald, Synchrotron x-ray scattering studies of Langmuir films, *in*: "Handbook on Synchrotron Radiation," S. Ebashi, M. Koch, and E. Rubinstein, eds., Elsevier, The Netherlands (1991), Vol. 4.
12. A. Ulman, "An Introduction to Ultrathin Organic Films from Langmuir-Blodgett to Self-Assembly," Academic Press, Boston (1991).
13. A. Pockels, Surface tensions, *Nature* 43:437 (1891).
14. "Advances in the Applications of Membrane-Mimetic Chemistry," T.F. Yen, R.D. Gilbert, and J.H. Fendler, eds., Plenum Press, New York (1994).
15. "Molecular and Biomolecular Electronics," R.R. Birge, ed., American Chemical Society: Washington, DC (1994).
16. D.J. Robinson and J.C. Earnshaw, Initiation of aggregation in colloidal particle monolayers, *Langmuir* 9:1436 (1993).
17. G. Onoda, Polystyrene monolayers, *Phys. Rev. Lett.* 55:226 (1985).
18. Z. Hórvölgyi, S. Németh, and J.H. Fendler, Spreading of hydrophobic silica beads at water air interfaces, *Colloids Surf. A: Physicochem. Eng. Aspects* 71:327 (1993).
19. S. Németh, T.-C. Jao, and J.H. Fendler, Concentration and solvent-dependent excimer formation of 1-pyrenylmethanamine covalently attached to maleic anhydride-grafted ethylene propylene copolymers, *Macromol.* 27:5449 (1994).
20. H. Haas and H. Möhwald, Ordered protein arrays as mesophases, *Langmuir* 10:363 (1994).

21. N.A. Kotov, F.C. Meldrum, C. Wu, and J.H. Fendler, Monoparticulate layer and Langmuir-Blodgett-type multiparticulate layers of size-quantized, cadmium-sulfide clusters: a colloid-chemical approach to superlattice construction, *J. Phys. Chem.* 98:2735 (1994).
22. N.A. Kotov, F.C. Meldrum, and J.H. Fendler, Monoparticulate layers of titanium dioxide nanocrystallites with controllable interparticle distances, *J. Phys. Chem.* 98:8827 (1994).
23. N.A. Kotov, Y.Tian, F.C. Meldrum, and J.H. Fendler, Unpublished results (1995).
24. N.A. Kotov, G. Zavala, and J.H. Fendler, Langmuir-Blodgett films prepared from ferroelectric lead zirconium titanate particles, *J. Phys. Chem.* submitted for publication (1995).
25. F.C. Meldrum, N.A. Kotov, and J.H. Fendler, Preparation of particulate mono- and multilayers from surfactant-stabilized, nanosized magnetite crystallites, *J. Phys. Chem.* 98:4506 (1994).
26. X.K. Zhao, S. Xu, and J.H. Fendler, Ultrasmall magnetic particles in Langmuir-Blodgett films, *J. Phys. Chem.* 94:2573 (1990).
27. D.W. Grainger, M. Ahlers, P. Meller, R. Blankenburg, A. Reichert, H. Ringsdorf, C. Salesse, J.N. Herron, and K. Lim, Controlling binding and organization of proteins with model biomembrane systems through specific recognition, *in*: "Biomembrane Structure and Function - the State of the Art," B.P. Gaber and K.R.K. Easwaran, eds., Adenine Press, New York (1992).
28. B.R. Heywood and S. Mann, Template-directed nucleation and growth of inorganic materials, *Adv. Mater.* 6:9 (1994).
29. H.A. Lowenstam and S. Weiner, "On Biomineralization," Oxford University Press, New York (1989).
30. L. Addadi and S. Weiner, Control and design principles in biological mineralization, *Angew. Chem. Int. Ed. Engl.* 31:153 (1992).
31. H.T. Tien, "Bilayer Lipid Membranes (BLM). Theory and Practice," Marcel Dekker, New York (1974).
32. B. Hille, "Ionic Channels of Excitable Membranes," Sinaver Associates, Sunderland, MA (1984).
33. B. Sackmann and E. Neher, "Single Channel Recording," Plenum Press, New York (1983).
34. S. Kato and T. Kunitake, Molecular design of black lipid membranes (BLM) by polymerized double-chain ammonium amphiphiles, *Chem. Lett.* 261 (1991).
35. K. Hongyo, J. Joseph, R.J. Huber, and J. Janata, Experimental observation of chemically modulated admittance of supported phospholipid membranes, *Langmuir* 3:827 (1987).
36. X.K. Zhao, S. Baral, R. Rolandi, and J.H. Fendler, Semiconductor particles in bilayer lipid membranes (BLMs). Formation, characterization, and photoelectrochemistry, *J. Am. Chem. Soc.* 110:1012 (1988).
37. X.K. Zhao, P.J. Herve, and J.H. Fendler, Magnetic particulate thin films on bilayer lipid membranes (BLMs), *J. Phys. Chem.* 93:908 (1989).
38. X.K. Zhao, S. Baral, and J.H. Fendler, Electrochemical characterization of bilayer lipid membrane semiconductor junctions, *J. Phys. Chem.* 94:2043 (1990).
39. A.T. Todorov, A.G. Petrov, and J.H. Fendler, Flexoelectricity of charged and dipolar bilayer lipid membranes studied by stroboscopic interferometry, *Langmuir* 10:2344 (1994).
40. F.C. Meldrum, N.A. Kotov, J.H. Fendler, Utilization of surfactant-stabilized colloidal silver nanocrystallites in the construction of mono- and multiparticulate Langmuir-Blodgett films, *Langmuir* 10:2035 (1994).
41. G. Decher, J. Maclennan, and J. Reibel, Highly-ordered ultrathin LC multilayer films on solid substrates, *Adv. Mater.* 3:617 (1991).
42. T. Kunitake and Y. Okahata, A totally synthetic bilayer membrane, *J. Am. Chem. Soc.* 99:3860 (1977).
43. T. Kunitake, Y. Okahata, K. Tamaki, F. Kumamura, and M. Takayanagi, Formation of the bilayer membrane from a series of quaternary ammonium salts, *Chem. Lett.(Jpn)* 387 (1977).
44. J.H. Fendler, Membrane mimetic chemistry, *C&E News* 62:25 (1984)
45. J.H. Fendler, Polymerized surfactant vesicles - novel membrane mimetic systems, *Science* 223:888 (1984).
46. J.H. Fendler and E.J. Fendler, "Catalysis in Micellar and Macromolecular Systems," Academic Press, New York (1975).

47. T. Kunitake, Synthetic bilayer membranes: molecular design, self-organization, and application, *Angew. Chem. Int. Ed. Engl.* 31:709 (1992).
48. H. Ringsdorf, B. Schlarb, and J. Venzmer, Molecular architecture and function of polymeric oriented systems: models for the study of organization, surface recognition, and dynamics of biomembranes, *Angew. Chem. Int. Ed. Engl.* 27:113 (1988).
49. J.H. Fendler, Surfactant vesicles as membrane mimetic agents: characterizations and utilizations, *Acc. Chem. Res.* 13:7 (1980).
50. M. Caffrey, D. Moynihan, and J. Hogan, A database of lipid phase transition temperatures and enthalpy changes, *Chem. Phys. Lip.* 57:275 (1991).
51. S.L. Regen, J.-S. Shin, J.F. Hainfeld, and J.S. Wall, Ghost vesicles, *J. Am. Chem. Soc.* 106:5756 (1984).
52. Y. Okahata, Lipid bilayer-corked capsule membranes. Reversible, signal-receptive permeation control. *Acc. Chem. Res.* 19:57 (1986).
53. Y. Okahata, Lipid bilayer-coated capsule membranes. Reversible, signal-receptive permeation control, *Acc. Chem. Res.* 19:57 (1986).
54. Y. Okahata, K. Ariga, and T. Seki, Polymerizable lipid-corked capsule membranes. Polymerization at different positions of corking lipid bilayers on the capsule and effect of polymerization on permeation behavior, *J. Am. Chem. Soc.* 110-2495 (1988).
55. N. Nakashima, R. Ando, and T. Kunitake, Casting of synthetic bilayer membranes on glass and spectral variation of membrane-bound cyanine and merocyanine dyes, *Chem. Lett.* 1577 (1983).
56. M. Shimomura, R. Ando, and T. Kunitake, Orientation and spectral characteristics of the azobenzene chromophore in the ammonium bilayer assembly, *Ber. Bunsenges, Phys. Chem.* 87:1134 (1983).
57. N. Higashi and T. Kunitake, Immobilization of cast bilayer films by $^{60}$Co gamma-irradiation and other means, *Polymer J.* 16:583 (1984).
58. M. Shimomura and T. Kunitake, Immobilization of synthetic bilayer membranes as multilayered polymer films, *Polymer J.* 16:187 (1984).
59. T. Kunitake, A. Tsuge, and N. Nakashima, Immobilization of ammonium bilayer membranes by complexation with anionic polymers, *Chem. Lett.(Jpn)* 1783 (1984).
60. N. Nakashima, M. Kunitake, T. Kunitake, S. Tone, and T. Kajiyama, Ordered cast films of polymerized bilayer membranes, *Macromol.* 18:1515 (1985).
61. S.L. Regen, P. Kirszensztejh, and A. Singh, Polymer supported membranes, *Macromol.* 16:335 (1983).
62. O. Albrecht and A. Laschewsky, Polymer supported membranes, *Macromol.* 17:1292 (1984).
63. S.L. Regen, Z. Foltynowicz, and K. Yamaguchi, Further evidence for polymer supported membranes, *Macromol.* 17:1293 (1984).
64. N. Higashi, T. Kajiyama, T. Kunitake, W. Prass, H. Ringsdorf, and A. Takahara, Cast multibilayer films from polymerizable lipids, *Macromol.* 20:29 (1987).
65. A. Takahara, N. Higashi, T. Kunitake, and T. Kajiyama, State of aggregation and surface chemical composition of composite thin films composed of poly(vinyl alcohol) and fluorocarbon amphiphile, *Macromol.* 21(8):2443 (1988).
66. K. Fukuta, Y. Itami, R. Shimizu, and T. Kunitake, Preparation of multilayered films of poly(stearyl acrylate) using cast films of a novel fluorocarbon amphiphile as two-dimensional templates.
67. S. Asakuma and T. Kunitake, A multi-layered film of a two-dimensionally crosslinked acrylate polymer, *Chem. Lett.* 2059 (1989).
68. Y. Okahata and H. Ebato, Application of a quartz-crystal microbalance for detection of phase transitions in liquid crystals and lipid multibilayers, *Anal. Chem.* 61:2185 (1989).
69. K. Sakata and T. Kunitake, A multilayered film of an ultrathin siloxane network, *J. Chem. Soc., Chem. Commun.* 504 (1990).
70. Y. Ishikawa and T. Kunitake, Macroscopically oriented copper(II) chelates in cast multibilayer films, *J. Am. Chem. Soc.* 108:8300 (1986).
71. I. Hamachi, S. Noda, and T. Kunitake, Layered arrangement of oriented myoglobins in cast films of a phosphate bilayer membrane, *J. Am. Chem. Soc.* 112:6744 (1990).
72. T. Kunitake, Ultrathin films as biomimetic membranes, *Polymer J.* 23:613 (1991).

73. S. Asakuma, H. Okada, and T. Kunitake, Template synthesis of two-dimensional network of cross-linked acrylate polymer in a cast multibilayer film, *J. Am. Chem. Soc.* 113:1749 (1991).
74. Y. Ishikawa and T. Kunitake, Design of spatial disposition of anionic porphyrins in matrices of ammonium bilayer membranes, *J. Am. Chem. Soc.* 113:621 (1991).
75. K. Sakata and T. Kunitake, Synthesis of polysiloxane films with varied microstructures in matrices of carbazole-containing bilayer membranes, *Thin Solid Films* 210/211:26 (1992).
76. H. Okada, K. Sakata, and T. Kunitake, Formation of oriented iron oxide particles in cast multibilayer films, *Chem. Mater.* 2:89 (1990).
77. Y. Okahata, H.-J. Lim, G. Nakamura, and S. Hachiya, A large nylon capsule coated with a synthetic bilayer membrane. Permeability control of NaCl by phase transition of the dialkylammonium bilayer coating, *J. Am. Chem. Soc.* 105:4855 (1983).

# IMMOBILIZED ENZYME REACTIONS ON BEADS AND MEMBRANES

Dibakar Bhattacharyya[a,c*], Sowmya Ganapathi[a,c], Shekhar Vishwanath[a,c], Melissa Summers[a,c], and D. Allan Butterfield[b,c]

Department of [a]Chemical and Materials Engineering, and Department of [b]Chemistry, and [c]Center of Membrane Sciences
University of Kentucky, Lexington, KY 40506

## INTRODUCTION

Immobilized biocatalysts have widespread applications in areas like organic synthesis, pollution control and for diagnostic purposes. Enzymes can be immobilized on different supports like polymeric beads, gels, and membranes. Immobilization eliminates the need to separate an enzyme from the product solution and allows these expensive compounds to be reused. In addition, the thermal stability, pH stability, and storage stability of an enzyme may be increased as a result of immobilization.

Common immobilization techniques may be grouped into two classes: physical immobilization and chemical immobilization. Physical techniques include the adsorption of an enzyme onto a carrier surface and the entrapment of an enzyme within a carrier. Chemical techniques involve the formation of a covalent bond between the enzyme and the carrier. While the weaker forces of physical immobilization often help to preserve the properties of an enzyme and facilitate regeneration, covalent attachment generally contributes a greater degree of stability. The specific effects that the support has upon the enzyme, however, greatly depend upon both the support and enzyme involved in the immobilization as well as the method of immobilization used[1].

The end application of the immobilized enzyme dictates the choice of the insoluble support. Beads have been widely used in affinity chromatography and other related areas. Membranes possess the distinct advantage of very high surface area (upto 100 $m^2/g$) for enhanced enzyme loading and can be operated in the convective flow mode. Membrane bioreactors can also be easily scaled-up and have lower pressure drops compared to affinity columns. In addition, polymeric beads and membranes can be functionalized to provide strong covalent attachment of the protein. Standard methods include the cyanogen bromide, glutaraldehyde, carbodiimide and carbonyl-diimidazole reactions[2].

The effects of various immobilization methods upon the properties of an enzyme have been studied extensively. In these studies, a wide variety of both enzymes and supports have been investigated. The physical adsorption of amines onto an alumina support was examined[3]. Physical adsorption was also studied to examine the effects of support pore diameter upon the immobilization of glucose oxidase and catalase onto titania[4]. The immobilization of alcohol dehydrogenase on beads by means of both adsorption and covalent attachment was investigated[5]. The effect of spacer length upon the covalent attachment of protease onto polyacrolein microspheres was examined at length[6].

In this paper three enzymes, papain, β-galactosidase (BG) and catalase have been investigated. Catalase and papain were covalently attached on carboxylic acid modified polystyrene beads, while β-galactosidase and papain were immobilized on aldehyde modified polysulfone membranes.

Papain, a derivative of papaya latex is a sulfhydryl protease with a molecular weight of 23,900. It is used commercially in the tenderization of meat, tanning of leather, and dry cleaning[7]. The primary structure of papain includes a single polypeptide chain composed of 212 amino acids. The active site of the enzyme is located in a cleft which is situated in the center of the secondary and tertiary structures of the molecule[8].

Catalase (EC 1.11.1.6) was one of the earliest enzymes discovered. The primary animal sources of catalase are bovine liver and rat liver. Its molecular weight is around 244,000. It is comprised of a protein unit attached to Hematin (Protohematin IX). There are four hematin groups per unit of enzyme. Catalase can utilize hydrogen peroxide as both donor and acceptor. It can also utilize alkyl hydrogen peroxides as substrate and a number of organics can replace the second peroxide molecule (e.g. ethanol, nitrous acid, thiol compounds and pyrogallol). Catalase has been immobilized on a variety of supports. Do and Hossain[9] immobilized catalase in controlled pore glass and Si-Al particles and developed a model predicting intraparticle enzyme distribution. They also carried out experiments[10] immobilizing catalase in porous glass particles and concluded that immobilization was controlled by diffusion of enzyme into the pores of the support. Vasudevan and Weiland[11] conducted deactivation studies on catalase immobilized on non-porous glass beads eliminating diffusion limitations. They concluded that hydrogen peroxide deactivated the free enzyme and not the enzyme-substrate complex at high substrate concentrations. Metelitsa and others[12] immobilized catalase on polyamide film membranes and polystyrene test-tubes. The enzyme was observed to retain 11.2 % and 4.5 % of its free activity.

BG is used as a reagent to evaluate lactose content in blood and other biological fluids. It is used in milk treatment to remove lactose from milk. This is done because a large number of people are afflicted by lactose intolerance. It has a molecular weight of 540,000. It is tetrameric composed of four identical subunits of 135,000 daltons each with an active site which may be independently active. An emerging field in immobilized enzyme systems involves the use of site-directed mutagenesis to orient the active-site of the enzyme away from the immobilization support. The enzyme β-galactosidase was site-specifically modified and subsequently immobilized on the polysulfone membrane[13].

## EPR Spectroscopy

Electron paramagnetic spectroscopy is a technique that detects unpaired electrons. The absence of paramagnetism in most biomolecules necessitates the use of a spin label which is typically the nitroxide free radical. The spin label is attached to the active site of the enzyme and the spectrum of the enzyme before and after immobilization gives a quantitative depiction of the conformational changes of the active site. Previous work has demonstrated the presence of two subpopulations of immobilized papain[14,15] on flat sheet and hollow fiber polymer membranes. In this paper, an EPR investigation of papain immobilized on polystyrene beads is discussed.

This paper deals with the activity analysis of enzymes immobilized on beads and functionalized membranes. Different enzymes were utilized for this purpose. β-galactosidase, a bulky enzyme and papain, a much smaller enzyme were both immobilized on the same membrane to study effects-like crosslinking which occurs predominantly in bulkier enzymes. Catalase was immobilized on beads because the substrate, hydrogen peroxide being inorganic would not adsorb on the beads. Also, the beads were nonporous to eliminate intraporous diffusional effects. The effect of adsorption of reaction constituents onto the polymeric membrane surface could significantly influence activity measurements in specific cases. Corrections were made to amidase activity measurements of papain, taking into account the effect of product adsorption.

## MATERIALS AND METHODS

### Immobilization procedure for beads

Carboxylic acid modified non-porous polystyrene beads (Sigma prod. # CLB-4) (0.45 micron) were used as immobilization supports. They contain approximately 0.2 meq of carboxylic acid per gram of polystyrene. The activation of polystyrene was carried out by reacting the beads with 1-ethyl-3-(3-dimethylaminopropyl ) carbodiimide (EDAC) at a pH of 5.5 at 25° C for 24 hours[16]. Then catalase was added to the beads and reacted for 24 hours at a pH of 7.0 at 4° C to yield the immobilized enzyme. The excess or free enzyme was removed by repeated buffer washes and centrifugation. When the immobilized enzyme activity was to be determined, membrane filter syringes (0.22 micron) were used to prevent the polystyrene beads from entering the solution to be analyzed by the UV spectrophotometer.

The same procedure was followed for immobilization of papain on the beads.

### Control experiment

A control experiment was carried out to determine the extent of noncovalent binding of enzyme to the beads. The enzyme was added to the beads prior to activation and kept at 4° C for 24 hours. The mixture was then washed with buffer and centrifuged repeatedly. The supernatant was analyzed for protein content.

### Enzyme immobilization procedure on membranes

Commercially-available (Ultrabind US 450, Gelman Sciences) modified polysulfone membranes (MPS) were used for all enzyme immobilizations (thickness-152 μm and pore diameter-0.45 μm). These membranes contained aldehyde functional groups and the attachment of the enzyme proceeded as a Schiff's base reaction whereby the amine group of an enzyme was bound to an aldehyde group on the membrane structure. For all the immobilizations, 4.7 cm diameter membranes were used.

Papain was immobilized on the membrane by a static soak method where it was allowed to react at 4°C for 24 hours at a pH of 7.5. Subsequent treatment of the membrane with ethanol amine for 1 hour at 4°C was employed to ensure blockage of any unreacted aldehyde groups and prevent excess crosslinking. The enzyme immobilized on the membrane was then determined by means of the Bradford Protein Assay by analyzing the protein content of the supernatant

In the case of BG, the parent enzyme solution was pumped through the MPS membrane by a convective technique wherein the membrane was housed in a flow cell.

**Kinetic Measurements**

All kinetic measurements on biofunctional membranes were made in a recirculation mode. The substrate solution from a beaker was pumped through the membrane, housed in a cell and to a quartz flow-cell in the spectrophotometer, where on-line measurements were made. The solution containing product (cumulative flow through pores = 2 cc/min) was then recirculated into the beaker containing substrate.

BG catalyzes the following hydrolysis:

$$ONPG + H_2O \Rightarrow \beta\text{-Galactose} + \text{o-nitrophenol}$$

The rate of reaction was determined by measuring the change in absorbance of o-nitrophenol which absorbs strongly at 405 nm UV light.

Catalase decomposes Hydrogen peroxide to water and oxygen as shown:

$$2H_2O_2 \Rightarrow 2H_2O + O_2$$

The rate of reaction can be determined by measuring the change in absorbance of Hydrogen peroxide at 240 nm ultraviolet (UV) light. The experiments carried out on catalase was to ascertain the activity of the enzyme when it was immobilized (externally attached) on non-porous carboxylic acid modified polystyrene beads at different enzyme loadings. Homogeneous and immobilized activity measurements were made and comparisons were made between them.

Papain catalyzes the following hydrolysis of benzoyl arginine p-nitroanilide (BAPNA):

$$BAPNA \Rightarrow \text{N-benzoyl-arginine} + \text{p-nitroaniline}$$

The rate of the reaction was determined by monitoring the absorbance of p-nitroaniline at 410 nm on a UV-VIS spectrophotometer. Homogeneous and immobilized (on both the beads and membrane) kinetics of amidase activity of papain were determined. All BAPNA hydrolysis reactions were carried out at a temperature of 25°C and a pH of 7.5.

**Study of Product Adsorption onto the Biofunctional Membrane**

A study of product adsorption was conducted to determine the extent of adsorption on a membrane containing enzyme, for both papain and BG. The concentrations of p-nitroaniline and o-nitrophenol respectively, were monitored until equilibrium conditions were reached. The adsorption experiments were performed at different concentrations to obtain the adsorption isotherm, which was then utilized to correct for p-nitroaniline adsorption for all membrane reactions. The membrane used for the adsorption runs contained 3 mg papain and the experiments were performed at a temperature of 25°C. In the case of BG both substrate, ONPG and product, o-nitrophenol were measured on an HPLC column.

**Effect of Reaction Temperature**

Activities were determined for the homogeneous reaction of papain in an initial concentration of 1.0 mM BAPNA at various reaction temperatures. Similarly, activities were determined for immobilized papain by varying reaction temperature. In each case, temperatures were maintained through the use of a constant temperature water bath. Similar experiments were conducted on homogeneous and immobilized BG to measure activation energies.

### Comparison of Storage Stability

Storage stability of papain in the homogeneous and immobilized states was compared. The enzyme was stored at a temperature of 25°C, both in the homogeneous and immobilized states. The activity of the enzyme was measured periodically thereafter the same conditions as the initial reaction until the activity had decreased by a significant amount. All reactions were carried out in the temperature range of 28.5-30°C. In addition, for the membrane a Bradford Protein Assay was performed on the storage solution after a period of 21 days to determine whether the immobilized papain had leached into the storage solution.

### EPR Studies

EPR investigations were conducted on papain immobilized on the beads and the results were compared to spectra of spin-labeled papain in solution. The specific spin label used was (1-oxyl - 2,2,5,5-tetramethyl-$\Delta^3$-pyrroline-3-methyl)methanethiosulfonate (MTS from Reanal, Budapest), which specifically reacts with the sulfhydryl group at the active-site of papain. Papain was first reacted with MTS to yield spin labeled protein, which was then immobilized on the beads. The beads were activated using EDAC as described above.

## RESULTS AND DISCUSSION

The enzymes papain and catalase were immobilized on functionalized beads. It was anticipated that the immobilized activity of the enzymes on the non-porous beads would be higher than the immobilized activity on the functionalized porous membranes due to the absence of intraporous diffusion. Immobilized activities of the enzymes, papain and BG on pre-activated MPS membranes were determined. Adsorption studies of the reaction constituents on the porous MPS membrane were performed to estimate the effect of adsorption on reaction kinetics. Effect of variables like reaction temperature and storage stability were also analyzed.

### Immobilization of enzymes on beads

Table I shows the kinetics of free and immobilized catalase and the weight of enzyme immobilized on the nonporous polystyrene beads. The control experiment which was carried out to determine the extent of noncovalent binding on the polystyrene beads did not show any significant noncovalent binding occurring on the beads. The amount of enzyme immobilized increased with the increase in initial weight of the enzyme taken. This can be attributed to the increased enzyme amino group uptake by the carboxylic acid groups on the polystyrene beads. At higher initial enzyme weights, a saturation of the amount immobilized would be expected. The $K_m$ value obtained from the homogeneous experiments agrees well with literature[11]. The Michaelis-Menten constant, $K_m$ which is a measure of the affinity of the substrate for the enzyme is increased for the immobilized enzyme. This does not directly mean a decreased affinity on immobilization because the immobilized $K_m$ value is an apparent one and not the true value. As the polystyrene beads are non-porous, there can be additional resistance due to intraparticle diffusion and the reduction in activity can be attributed only to the reduction in motion of the enzyme due to immobilization.

Immobilized activity was found to increase with enzyme loading. When the enzyme is sparsely loaded on the support, the probability of multiple point attachment is increased resulting in enzyme structure deformation. Hence the activity of the enzyme is lesser. However, in the higher loading case, the possibility of multiple point attachment is reduced due to the denser packing of enzyme on the support surface leading to an enhanced activity.

Immobilization of papain on the beads also revealed a decline in activity compared to homogeneous papain results. The relative activity (RA) of the biofunctional beads was 32%, in contrast to only 8% for the biofunctional MPS membrane. Thus the beads depicted higher activity

TABLE I Kinetic parameters $K_m$ and $V_{max}$ obtained from Lineweaver-Burke plots for Catalase immobilized on nonporous polystyrene beads

| State of enzyme | Amount# immobilized, mg | $K_m$, mM | $V_{max}$, µmole/mg.min |
|---|---|---|---|
| Homogeneous | - | 31.8 ± 0.021 | 1968 ± 21.7 |
| Immobilized | 0.046 | 49.8 ± 0.019 | 751 ± 8.1 |
| Immobilized | 5.02 | 44.6 ± 0.017 | 1457 ± 16.8 |

# - The immobilizations were on approximately the same number of beads.

than the membrane at a comparable protein loading of 2.7 mg (beads) and 2.9 mg (membrane). This could be attributed to the non-porous nature of the polystyrene beads which implies no internal diffusional resistances. Enzymatic cross-linking could also be lower, leading to a greater availability of enzyme molecules for the reaction.

**EPR Results**

The MTS spin label used for EPR studies is a nitroxide spin label specific to the sulfhydryl group. In the homogeneous state it generates a three-line spectrum, typical of nitroxide labels. Progressing from MTS to MTS-papain in solution and further to MTS labeled papain on the polystyrene beads, asymmetric line-broadening of the spectra occurs. This line-broadening is characterized by the hyperfine splitting parameter, $2A_{zz}'$, which is a measure of the distance between the low-field and high-field lines of the spectrum for each subpopulation[14]. MTS-papain in the homogeneous state displays a single population with a $2A_{zz}'$ of 53 G (Figure 1-[a]). The spectrum of the beads (Figure 1-[b]) clearly shows two subpopulations, marked A and D, versus

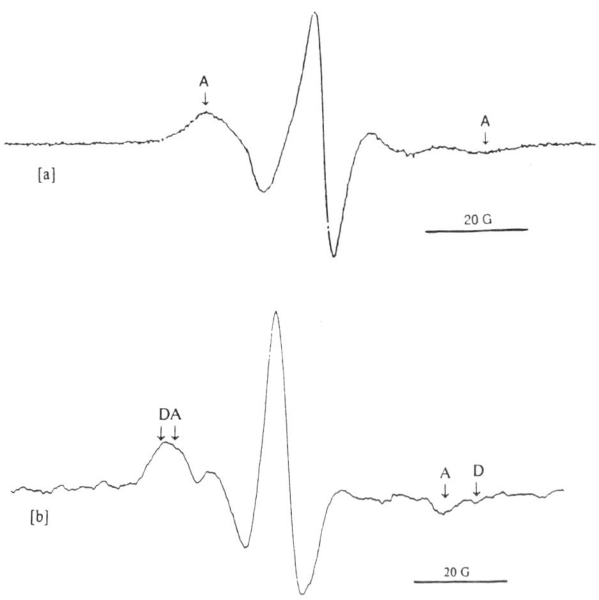


Figure 1. EPR Spectrum of MTS-papain in solution -1[a]. EPR spectrum of MTS-papain on polystyrene beads - 1[b]. A and D indicate the two subpopulations on the beads.

only one population for MTS-papain in the homogeneous state. These two are characterized by hyperfine splitting parameters of 59 G and 68 G for subpopulations A and D respectively. This spectrum is very similar to that of immobilized papain systems on polymeric membranes, where two subpopulations, one active (A) and one denatured (D) were observed[14,15].

### Immobilization of enzymes on membranes

Activities for both homogeneous and immobilized papain were determined under various conditions so that membrane loading could be examined and the kinetics, temperature effects, and storage stabilities of the two papain forms could be compared. The hydrolysis of BAPNA catalyzed by papain was employed to determine enzyme activities. In order to calculate enzyme activity, reaction data was plotted as p-nitroaniline product concentration versus time. Typical runs are shown in Figure 2 for both homogeneous and immobilized papain

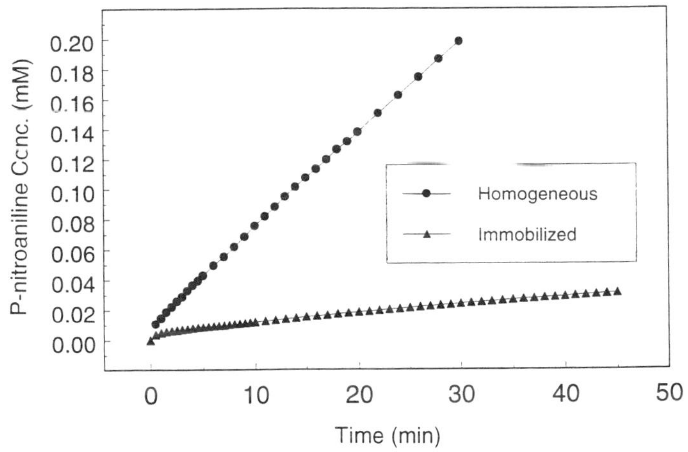


Figure 2. Progress of BAPNA hydrolysis reaction catalyzed by papain, both in the homogeneous state and on the MPS membrane.

A comparison of the activities reveals that the activity of immobilized papain is considerably lower than that of the homogeneous papain. Several factors may contribute to this decrease in activity, including denaturation of the enzyme upon immobilization, changes in active-site conformation and crowding of the enzyme molecules on the support and they are discussed in greater detail in the sections below.

### Comparison of Recirculation and Batch Reactions

Recirculation and true batch experiments were performed for the same membrane so that the results could be compared in order to ensure that the recirculation apparatus successfully models a batch reaction. An apparent activity of 1.23 μmol/g·min was obtained using the recirculation method, while experimentation using the true batch method yielded an apparent activity of 1.29 μmol/g·min. These values are in excellent agreement with each another. Hence, these results confirm that the recirculation apparatus successfully models a batch reaction.

### Effect of Loading Concentration on Apparent Activity

The normalized apparent activities were calculated for papain on MPS membranes containing three different loadings in order to determine the effect of loading concentration upon apparent activity. A comparison of the activities of immobilized papain at the three loading concentrations is presented in Figure 3. This plot reveals that low loading produces the greatest apparent activity, and apparent activities decrease with increasing loading. Similar results were obtained for immobilized papain at temperatures of 37°C[15]. Thus, for this system, larger loading concentrations are unfavorable. These results can be explained by the crowding of membrane pores which occurs when larger amounts of papain are present at high enzyme loading concentrations. Enzyme crowding blocks enzyme active sites and prevents successful binding to the BAPNA substrate thus lowering enzyme activity. In addition, the effects of diffusion cannot be ignored. Higher enzyme loading may lead to higher reaction rates; however, this situation may also result in diffusion limited reactions and thus lower apparent activities.

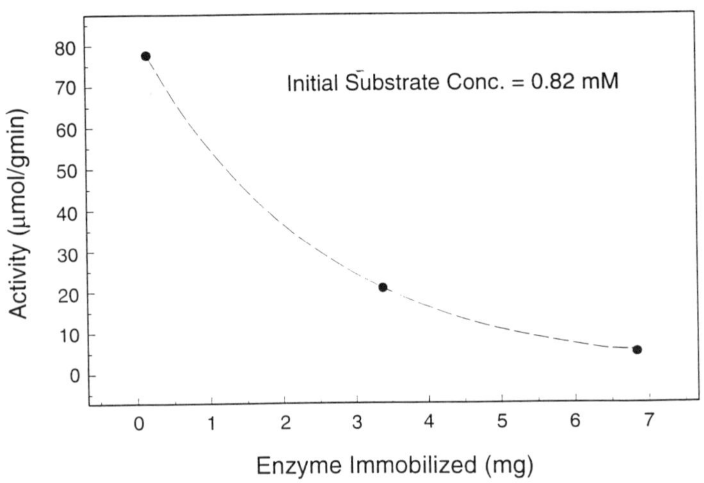


Figure 3. Effect of papain loading on activity of the immobilized enzyme. All numbers are corrected for product adsorption, assuming equilibrium adsorption.

Values of $K_m$ and $V_{max}$ were determined for papain in the homogeneous and immobilized states and are tabulated in Table II. The evaluation of these kinetic parameters allows insight into the kinetic properties of the two enzyme forms. $V_{max}$ represents the maximum activity of the enzyme i.e. the number of active sites available for catalytic action. The Michaelis-Menten constant is a measure of the enzyme's affinity for the substrate: as the value increases, the affinity of the enzyme towards the substrate decreases.

The results of these experiments suggest that the covalent attachment of the papain molecule to a macroporous membrane results in a significant reduction in the maximum attainable activity leading to an RA of 0.062. When performing a covalent immobilization, the goal is to attach the enzyme in such a manner that its active site is not deformed or blocked by the enzyme-membrane link so that the enzyme will be free to react with the substrate at the same rate as in solution. Any deviations from such an ideal situation result in a decrease in the maximum attainable activity of the enzyme. Papain attachment on the MPS membrane is a result of random immobilization. Thus, the large drop in activity suggests that a high degree of active site deformation and blockage is taking place. In addition, crosslinking of an attached enzyme with other membrane surface locations will result in active site blockage and/or denaturation. It must also be noted that the data for the immobilized papain represents apparent activities. It is interesting to note that the RAs of papain and BG were 33% and 2%, respectively at comparable loadings. This can be attributed to BG being almost thirty times bigger than papain, resulting in much higher cross-linking for BG. Thus it can be deduced that for a bulky enzyme cross-linking plays a significant role in reducing the activity of an immobilized enzyme as compared to enzymatic denaturation upon immobilization.

All the immobilizations performed on papain were random immobilizations, i.e. the active site of the enzyme was not specifically oriented away from the solid support. To illustrate this further, studies have been performed[13] on β-galactosidase, where the enzyme was site-specifically modified and then immobilized and its activity was compared to a randomly immobilized β-galactosidase's activity. The relative activity (RA) of the site-directed immobilized β-galactosidase was 87% as compared to a randomly immobilized value of 1.8%. Thus site-directed immobilization reduces crosslinking to a large extent, thereby dramatically increasing immobilized enzymatic activity.

## Product Adsorption on MPS Membrane

Preliminary investigation with papain indicated the possibility of adsorption of p-nitroaniline onto the membrane. To test this, experiments were performed in which the concentration of a p-nitroaniline solution recirculating through the MPS membrane was monitored. Figure 4 shows a plot of p-nitroaniline concentration versus time for this experiment. As depicted, the final concentration of p-nitroaniline in the beaker differs significantly from the initial concentration; thus, it is obvious that adsorption onto the biofunctional membrane is occurring and must be accounted for. Hence, further experiments were performed for several initial concentrations of p-nitroaniline and the results were recorded as the adsorption isotherm of p-nitroaniline[15].

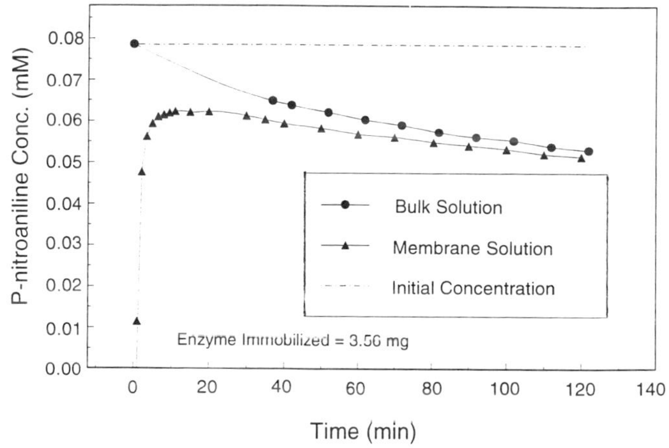


Figure 4. Kinetics of p-nitroaniline adsorption on the MPS membrane. The two curves indicate the decrease in concentration of p-nitroaniline in the beaker, with subsequent increase on the membrane. The horizontal line indicates the initial concentration used.

By assuming that all of the concentration data gathered for experiments involving the immobilized papain represented equilibrium conditions, reaction data could be corrected for the adsorption of p-nitroaniline onto the membrane, and corresponding activities and values of and could be determined. Corresponding apparent values of $K_m$ and $V_{max}$ are tabulated in Table II along with those of uncorrected immobilized papain and homogeneous papain. These results reveal that the consideration of p-nitroaniline adsorption onto the membrane decreases the value of only slightly, however, it significantly increases the value of $V_{max}$. In fact, the RA was 0.22 as opposed to the previously calculated value, 0.062. Hence, the adsorption of the p-nitroaniline product onto the membrane is in fact significant, and omitting this consideration from data reduction introduces considerable error in parameter calculation.

In the case of BG, both the product o-nitrophenol and the substrate ONPG were analyzed on an HPLC reverse phase column to ensure material balance. It was observed that the adsorption of o-nitrophenol on the MPS membrane was not significant and hence did not affect the kinetics of the reaction. Ortho-nitrophenol is completely ionized at the pH of the reaction and thus one would expect adsorption to be minimal. In contrast, for the case of papain, product p-nitroaniline is completely nonionized and hence adsorption was much higher.

Table II Michaelis-Menten parameters for homogenous papain, immobilized papain, and immobilized papain subsequent to correction for p-nitroaniline product adsorption.

| Papain Form | $V_{max}$ (μmol/g·min) | $K_m$ (mM) |
|---|---|---|
| Homogeneous | 142 | 1.53 |
| Immobilized | 8.79 | 0.40 |
| Immobilized with adsorption corrections | 33.06 | 0.37 |

### Temperature Studies

The activity of homogeneous papain and the apparent activity of immobilized papain were determined at three different reaction temperatures to assess the effect which immobilization has upon the sensitivity of the enzyme to reaction temperature. Activities and apparent activities were determined for the reaction of homogeneous and immobilized papain at varying reaction temperatures from 27°C to 53°C and the Arrhenius plot is depicted in Figure 5. The apparent activity of the immobilized papain appears to be less sensitive to reaction temperature than does the activity of the homogeneous papain. Straight line Arrhenius fits revealed that the activation energy for homogeneous papain was 36 kJ/mol ($R^2 = 0.96$) while that of the immobilized papain was 28 kJ/mol ($R^2 = 0.99$). These numbers indicate that both reaction and diffusion limitations exist in the system. If the reaction was diffusion-controlled, then it would be expected that the activation energy of the immobilized biocatalyst would be one-half of the homogeneous enzyme.

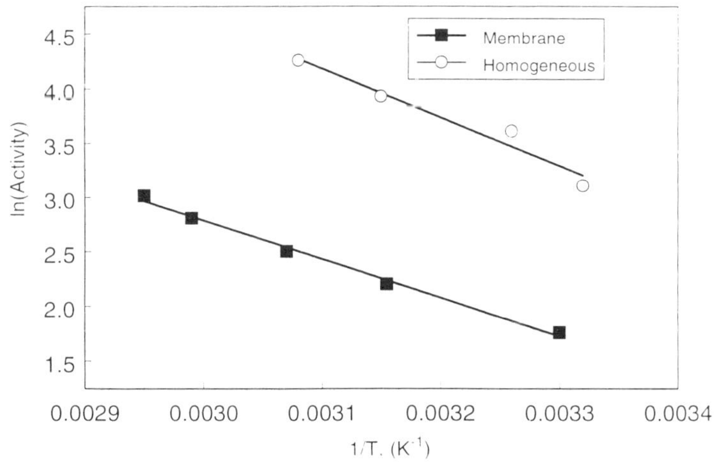


Figure 5. Arrhenius plot for homogeneous papain and papain immobilized on the MPS membrane.

Similar experiments were conducted on β-galactosidase to determine the activation energy of homogeneous and immobilized enzyme. The Arrhenius plots to determine the activation energies are shown in Figure 6. The activation energy for the homogeneous enzyme was 14.9 kJ/mol ($R^2$ = 0.97) and that of the immobilized enzyme was 41.1 kJ/mol ($R^2$ = 0.97). In this case, the activation energy of the immobilized enzyme is much higher than that of the enzyme in solution. When the enzyme is immobilized, disorientation of the active site leads to lesser performance of the enzyme and hence the enzyme's activation energy increases. Also this indicates that internal diffusion limitations were not significant for this immobilized system. This has also been verified by making use of a mathematical model[13].

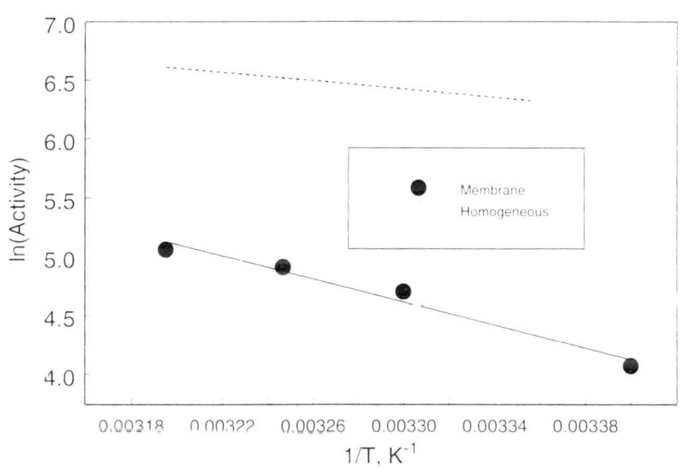


Figure 6. Arrhenius plots for homogeneous β-galactosidase and immobilized β-galactosidase on the MPS membrane.

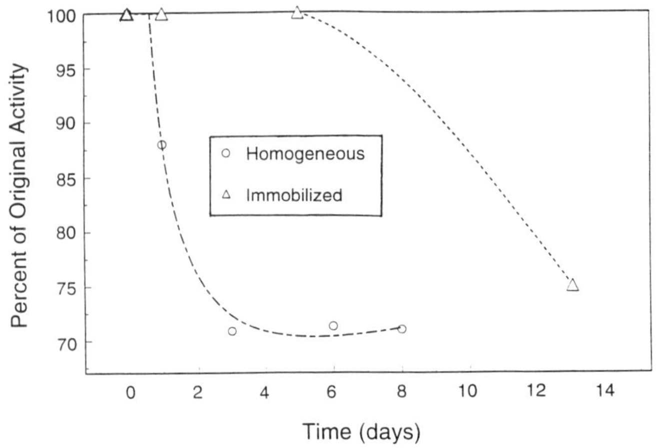


Figure 7. Storage stability of papain in the homogeneous state and in the immobilized state on the membrane. All activities are reported relative to activity on day 1 of the experiment.

**Storage Stability**

The results of the storage stability experiments are plotted in Figure 7 as percent of original activity versus time. Here, the activity or apparent activity measured immediately following preparation was taken as 100%. This plot reveals that the homogeneous form of the papain enzyme loses activity rapidly when stored at a temperature of 25°C. It was observed that the activity of homogeneous papain dropped to 70% of the original activity after only two days. In contrast, the immobilized form of the enzyme appears to be significantly more stable at a temperature of 25°C. There was no reduction in activity up to five days after enzyme immobilization, after which the activity drops gradually. Thus, the immobilization results in an increase in the storage stability of papain. Moreover, the Bradford Protein Assay revealed that the storage solution contained only 0.19 mg of papain after a period of 21 days. When compared to the 2.96 mg of papain immobilized, this quantity is insignificant and enzyme leaching can be assumed negligible.

## CONCLUSIONS

Immobilized activity measurements of catalase, β-galactosidase and papain have been reported using functionalized nonporous beads and macroporous membranes as supports. These enzymes showed a loss in activity upon immobilization which can be attributed to active site disorientation and/or partial denaturation of the enzyme upon immobilization. Catalase, which was immobilized on the beads had an RA of upto 76% in the high loading case. Thus the activity increased with loading in the case of catalase immobilized on beads.

Papain which was immobilized both on nonporous beads and the membranes showed a significantly higher activity when immobilized on beads. This can be attributed to the nonporosity of the beads which results in the enzyme being just externally attached. Membranes have the distinct advantage of high surface area, which in turn implies higher loading possibilities for the biomolecule, although the reaction might become diffusion limiting, depending on enzyme reactivity. On the other hand, for continuous operation, beads are impractical due to longer diffusion paths and higher pressure drops. Also, for membranes, convective flow through pores can be obtained.

EPR studies on papain immobilized on the functionalized beads revealed the presence of two subpopulations of the attached protein. Earlier work of papain immobilized on macroporous flat-sheet and hollow fiber membranes (Butterfield et al., 1994, Ganapathi et al., 1995) also detected two subpopulations with similar values of the hyperfine splitting parameter, $2A_{zz}'$. In this work we have established that two subpopulations of immobilized papain also exist on non-porous beads.

In the case of papain immobilized on the MPS membrane, significant adsorption of p-nitroaniline was observed. Assuming equilibrium adsorption, corrections were made to activity measurements upon which the RA increased from 6% to 22%.

The activation energy of homogeneous BG was found to be lesser than that of immobilized BG. Thus it can be inferred that the immobilized enzyme's efficiency as a catalyst is reduced upon immobilization. On the other hand, the activation energy of immobilized papain was lesser than homogeneous Papain, indicating the presence of diffusional limitations. Storage stability tests on homogeneous and immobilized papain showed the immobilized enzyme to be more stable.

## ACKNOWLEDGMENTS

This project was funded by a grant from the National Science Foundation (CTS - 9307518). One of the authors (Melissa Summers) was supported by an NSF-REU grant. We also thank Dr. L. G. Bachas of the Department of Chemistry for his help with the work on BG.

## REFERENCES


1. R. F. Taylor, "Protein Immobilization: Fundamentals and Applications", Marcel Dekker, Inc., New York (1991).
2. E. Klein, "Affinity Membranes", John Wiley & Sons, Inc., New York (1990).
3. J. Koubek, J. Volf, and J. Pasek, Adsorption of amines on alumina, *J. Catal.*, 38:385 (1975).
4. R. A. Messing, Simultaneously immobilized glucose oxidase and catalase in controlled-pore titania, *Biotechnol. Bioeng.*, 16:897 (1974).
5. P. S. Skerker and D. S. Clark, Catalytic properties and active-site structural features of immobilized horse liver alcohol dehydrogenase, *Biotechnol. Bioeng.*, 28:148 (1988).
6. T. Hayashi and Y. Ikada, Protease immobilization onto poly-acrolein microspheres, *Biotechnol. Bioeng.*, 35: 518 (1990).
7. S. Schwimmer, "Source Book of Food Enzymology", The Avi Publishing Co., Inc. (1981).
8. J. R. Whitaker, "Principles of Enzymology for the Food Sciences", Marcel Dekker, Inc., New York (1972).
9. D. D. Do and M. M. Hossain, A novel method of determination of the internal enzyme distribution within porous solid supports and the deactivation rate constant, *Biotechnol. Bioeng.*, 28: 486 (1986).
10. D. D. Do and M. M. Hossain, Determination of intrinsic parameters for immobilization reactions of catalase and amyloglucosidase in porous glass supports, *Biotechnol. Bioeng.*, 31:730 (1988).
11. P. T. Vasudevan and R. H. Weiland, Deactivation of catalase by hydrogen peroxide, *Biotechnol. Bioeng.*, 36:783 (1990).
12. D. I. Metelitsa, E. I. Plyugacheva, V. A. Artomonov and G. M. Baran, Catalytic activity of catalase adsorbed on MIFIL polyamide membranes, *Kinetics and Catalysis*, 31:1233 (1991).
13. S. Vishwanath, W. Huang, L. G. Bachas and D. Bhattacharyya, Site-directed and random enzyme immobilization on functionalized membranes: kinetic studies and models, *J. Membrane Sci.*. In press (1995).
14. D. A. Butterfield, J. Lee, S. Ganapathi and D. Bhattacharyya, Biofunctional membranes part IV. Active-site structure and stability of an immobilized enzyme, papain, on modified polysulfone membranes studied by electron paramagnetic resonance and kinetics, *J. Membrane Sci.*. 91: 47 (1994).
15. S. Ganapathi, D. A. Butterfield, D. Bhattacharyya, Flat-sheet and hollow fiber membrane bioreactors: A study of the kinetics and active-site conformational changes of immobilized papain including sorption studies of reaction constituents, *J. Chem. Tech. Biotech.*, In press (1995).
16. K. Kono, F. Tabeta and T. Takagishi, pH-responsive permeability of poly(acrylic acid) - poly(ethylenimine) complex capsule membrane, *J. Membrane Sci.*, 76:233 (1993).

# A COMPARATIVE STUDY OF THE STRUCTURAL AND KINETIC PROPERTIES OF PAPAIN IMMOBILIZED ON MODIFIED POLYSULFONE AND MODIFIED CELLULOSE ACETATE MEMBRANES

Jinbo Lee,[a] Dibakar Bhattacharyya,[b,c] and Alan D. Butterfield,[a,c]*

Departments of [a]Chemistry, and [b]Chemical and Materials Engineering, and [c]Center of Membrane Sciences
University of Kentucky, Lexington, KY 40506-0057

## ABSTRACT

In order to determine if the chemical nature of the polymeric membrane were a factor in the conformation of bound enzymes, electron paramagnetic resonance (EPR) spin labeling techniques using a short, active-site specific spin label have been employed to study the properties of a model enzyme, papain, immobilized on fully-hydrated modified polysulfone membranes, hydrophobic, or modified cellulose acetate membranes, hydrophilic. The properties of the immobilized enzymes are compared with that of the free enzyme in solution. Both polysulfone and cellulose acetate membranes provide some stability for papain, but in general hydrophoilic cellulose acetate membranes give papain properties more similar to those of papain in aqueous solution.

This study reports the development of an effective method to acquire EPR spectra of fully-hydrated, spin labeled enzymes immobilized on polymeric membranes. All these findings indicate that the EPR spin labeling technique provides a powerful method for studying immobilized enzyme systems. Furthermore, understanding the dependence of the properties of the immobilized enzyme on the characteristics of the polymeric membrane may be helpful in the selection of support matrices for particular applications of membrane-based enzyme bioreactors, affinity membranes, and biosensors.

## INTRODUCTION

The first extensive election paramogmeter resonance (EPR) spin labeling study of the active site of an enzyme immobilized on a fully-hydrated, porous membrane was recently reported from our laboratory [1]. A methanethiosulfonate spin label a short, SH-specific nitroxide spin label, was used to specifically spin label the active site of the thiol protease, papain, in solution[2] and covalently immobilized to polysulfone/aldehyde blend membranes [1]. Two populations of the spin labeled papin were observed: an active (A) and a denatured (D) population. These populations were distinguished by EPR on the basis of: population A was more mobile than population D; the former was sensitive to changes in pH and denaturants, while the latter was not; the spectrum of population A could be converted to population D upon denaturation; only population A was sensitive to paramagnetic relation effects of $K_3Fe(CN)_6$; and only population A could be spin labeled after protein immobilization onto polysulfone membranes. These results are consistent with the idea that at least two populations of enzymes were present upon direct immobilization of papain to polysulfone/aldehyde blend membranes, and that population D is extremely tight and solvent-inaccessible. The implications of these results for biofunctional membranes are profound, and thus their generalization to other membrane systems needs to be determined.

Accordingly, we wondered if these results were generalizable to different polymeric membranes. In particular, we wondered if two populations would be present on a more hydrophilic membrane, cellulose acetate. This paper presents the results of a comparison of the active site structure of papain directly immobilized on modified polysulfone and modified cellulose acetate membranes.

## EXPERIMENTAL

### Materials

The modified polysulfone, UltraBind™ US450, was purchased from Gelman Sciences. It has aldehyde functional groups on the membrane surfaces, and according to the manufacturer has a pore size of 0.45 μM and a thickness of 152.4 μM. The modified cellulose acetate membrane, MAC™ active membrane, was purchased from Amicon. It also has aldehyde functional groups on the membrane surfaces, a pore sice from 0.2 to 10 μM. Sucrose was obtained from ICN. Dried papaya latex, N-benzoyl-DL-arginine-4-nitroanilide hydrochloride (BAPNA), DL-cysteine, and guanidine hydrochloride were obtained from Sigma Chemical Co. Urea was obtained from Aldrich. The Thiol-specific spin label, (1-oxyl-2,2,5,5-tetramethyl-$\Delta^3$ -pyrroline-3-methyl) methanethiolsulfonate (MTS) was purchased from Reanal, Budapest, Hungary. The ultrafiltration membrane with a 10,0000 MW cutoff, used for concentrating MTS-labeled papain, was obtained from Millipore Co. All other chemicals and solvents are reagent grade.

### Experimental Methods

Similar EPR and kinetic studies have been performed on papain immobilized on modified polysulfone and modified cellulose acetate membranes. The experimental methods for the EPR and kinetic studies of immobilized papain on polysulfone membranes were published previously[1].

**Immobilization of MTS-labeled Papain/Papain on Polysulfone Membrane.** Papain was purified from dried papaya latex and then labeled with the MTS spin label at the active site as described previously [1-6]. In particular, the reaction of MTS with the single cysteine of papain to form a covalent disulfide is illustrated in Scheme I. MTS labeled papain was concentrated by ultrafiltration to about 3-4 mg/mL protein content, estimated by the method of Lowry et al.[7]. Modified, flat sheet, polysulfone membrane was used for all experiments. The membrane was cut into pieces with an area of 4x4 $cm^2$. The protein was immobilized by reaction of free $NH_2$ groups of lysine and arginine on the enzyme with the aldehyde groups of the modified polysulfone membranes resulting in a stable imine.

For EPR studies, one piece of membrane was incubated with 3 mL of MTS-labeled papain solution for 24-36 hours and then the membrane was washed extensively with 5P8 (5mM phosphate buffer pH 8.00) to remove the unreacted protein. The resulting membrane was cut into four 1x4 $cm^2$ pieces for enzymatic analysis.

## RESULTS

Typical EPR spectra of MTS covalently bound to the single SH group of papain in solution and directly immobilized on modified polysulfone and modified cellulose acetate membranes are shown in Figure 1. The EPR parameter $2A'_{zz}$ is a measure of spin label motion: the larger $2A'_{zz}$, the slower the molecular motion of the spin label.[1,2,8,9] This EPR parameter is larger for immobilized papain than for papain in solution indicating slower motion of the active site of immobilized papain. The EPR signals of the immobilized enzyme are weaker for the cellulose acetate membrane than for polysulfone membrane under the same enzyme loading conditions. It was also found that the same enzyme loading in the immobilization reaction and to the same size of membranes result in about 40-50 $\mu g/cm^2$ of papain on cellulose acetate membranes but about 50-60 $\mu g/cm^2$ on polysulfone membranes. Thus, the density of the aldehyde functional groups on cellulose acetate membranes may be smaller than that on polysulfone membranes or the immobilization reaction is less effective for cellulose acetate membranes than for polysulfone membranes.

In contrast to the spectrum of MTS-labeled papain in solution (Figure 1A), the EPR spectrum of the immobilized papain on polysulfone membranes is a somewhat complicated. There are two subpopulations of immobilized enzymes, which are revealed by the EPR spin labeling technique (Figure 1B); that is, the spectrum is an overlap of two three-line nitroxide spin label spectra. This is most obviously seen in the center of the spectrum where two minima exist in the $M_I = 0$ central line (Figure 1B). However, these two environments are also easily seen in the low-field $M_I = +1$ line, and upon computer amplification of the signal, in the high-field $M_I = -1$ line as well. The splitting between the high field and low field lines, $2A_{zz}'$, which has been used very successfully in studying free papain in solution[2], was also employed to analyze the spectra of the immobilized papain on polysulfone and cellulose acetate membranes. For polysulfone membrane one subpopulation (called subpopulation A), $2A_{zz}'$ is about 60 G and around 70 G for the other subpopulation (called subpopulation D). In the case of papain immobilized on cellulose acetate membranes, $2A_{zz}'$ is around 59 G for subpopulation A, and again around 70 G for subpopulation D. Thus, the spin label has less motion at the active sites of both the immobilized enzyme subpopulations than that of papain in solution, which has a $2A_{zz}'$ around 53 G. It may be that the active site cleft, at least the portion the spin label detected, of the immobilized papain has a more closed structure than that of the free papain in solution, which might due to changes in the active site structure upon immobilization, and/or steric interactions between the enzyme and the polymeric support. Within the enzyme on the same polymeric (polysulfone or cellulose acetate) membrane, the active site cleft of one subpopulation (with a $2A_{zz}'$ about 60 G or 59 G) may be more open, and therefore provide for greater motional freedom than that of the other subpopulation (with a $2A_{zz}'$ of 70 G).

The $2A_{zz}'$ values for subpopulation A are essentially the same (60 G vs 59 G) for MTS-labeled papain immobilized on both membranes (Figure 1). However, the lineshpaes of the EPR signals of immobilized papain on cellulose acetate (Figure 1C) are significantly different from that of the immobilized papain on polysulfone membrane (Figure 1B), especially in the regions of low- and mid-field lines. The negative mid-field line for the subpopulation A is stronger than that of the subpopulation D and a larger portion of the low field signals is from the subpopulation A in the case of cellulose acetate membrane (Figure 1C), while the reverse relationship is observed in the case of polysulfone membrane (Figure 1B). The conversion factor, I(D)/I(A), which can also serve as an indicator of the relative distribution of the two subpopulations, is 0.925 for cellulose acetate membranes but 1.005 for polysulfone membranes. Therefore, a higher proportion of the active subpopulation A exists on cellulose acetate membranes than on polysulfone membranes, though the cellulose membranes are less effective than polysulfone membranes in immobilizing papain.

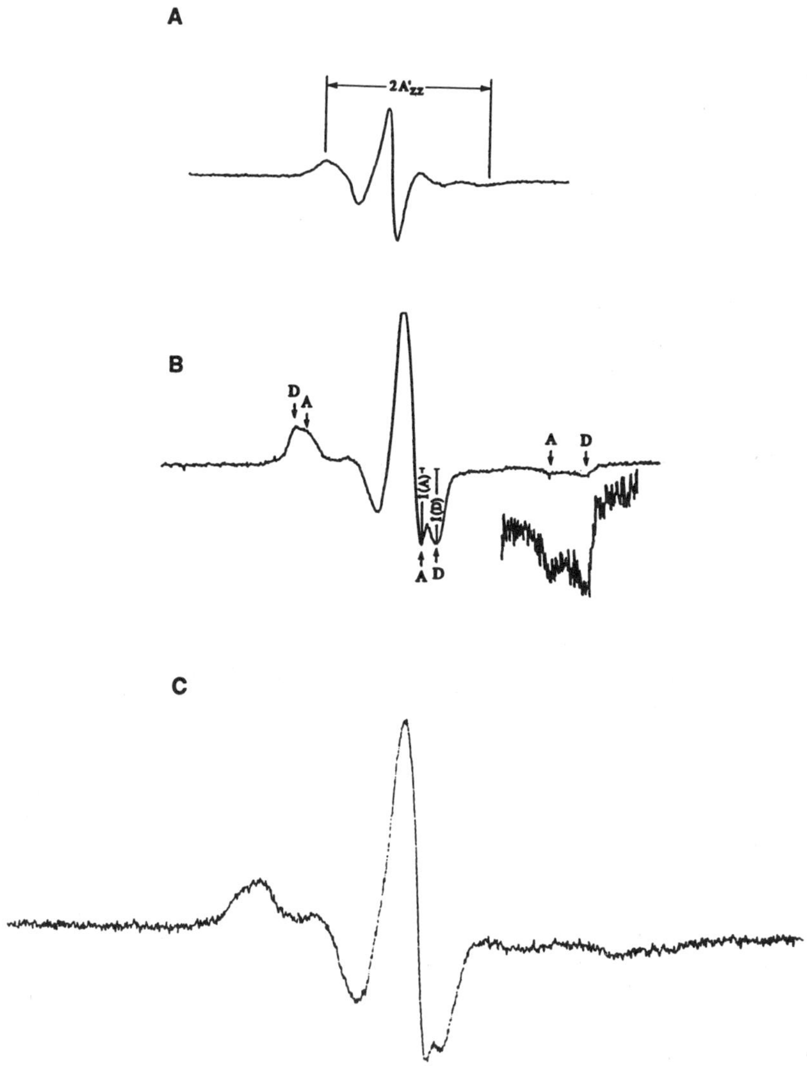


Figure 1 (A): A typical EPR spectrum of MTS-labeled papain in solution is presented. The splitting parameter, $2A_{zz}'$, is indicated. (B): A typical EPR spectrum of MTS labeled papain immobilized on modified polysulfone membranes. The signals corresponding to subpopulation A and D are indicated using arrows. The EPR signal intensities below the baseline of $M_I = 0$ center lines [I(D) and I(A) used for calculating the conversion parameter I(D)/I(A)] are also marked. Amplification of the instrumental gain demonstrates the existence of the two resonance peaks at the high field lines (inset); (C): A typical EPR spectrum of MTS labeled papain immobilized on modified cellulose acetate membranes.

Papain exhibits amidase activity[1,2]. The reaction conditions in the activity assay were first optimized. Ten minutes were found sufficient to fully activate the enzyme since the activities for longer activation time periods remained constant. Therefore, ten minute activation time was chosen for all the activity measurements of immobilized papain on polysulfone and cellulose acetate membranes.

The amidase activity follows Michaelis-Menten kinetics; however, the kinetic parameters, $V_{max}$ and $K_m$, obtained from the Lineweave-Burk plots of the experimental data may not be the true values equivalent to those obtained in homogeneous reactions, due to various diffusional resistances involved in substrate transport to and from the membrane surface and through the membrane pores or due to non-specific adsorption of substrate and product to the membrane[10]. Thus the kinetic parameters should better be treated as apparent $V_{max}$ and $K_m$, and written as $V_{max}$(app) and $K_m$(app). The apparent mean maximum velocities for papain on polysulfone (27.9 μmo l/mg.min) and cellulose acetate (49.0 μmo l/mg.min) membranes are both smaller than that of the free papain in solution (68.7 μmo l/mg.min). The higher overall $V_{max}$(app) for the immobilized papain or the cellulose acetate membrane compared to the polysulfone membrane is consistent with the higher percentage of the active subpopulation A on the former membrane compared to the later membrane, as revealed by the EPR studies. Both the $K_m$(app) values for papain immobilized on polysulfone and cellulose acetate membranes are larger than that of papain in solution, suggesting that there is indeed a conformation change in the active state of papain after immobilization. The compatible $K_m$(app) values for both membranes suggest that papain may have very similar active site conformations on both polymeric membranes, which is consistent with the EPR findings discussed above.

## pH Studies

EPR spin labeling is a sensitive method to study the active site conformation change of enzymes[8,9] including free papain in solution[2] under different pH conditions. The results of a similar study on immobilized papain on polysulfone and cellulose acetate membranes are given in Figure 2. In general, the pH dependence to $2A_{zz}$ was similar for the two respective environments, A and D, for both membranes. The pH dependence of the amidase activity of immobilized papain on cellulose acetate and polysulfone membranes are also very similar to that of free papain in solution, but the maximum reaction velocity shifted from pH 7.00 to pH 5.08 in both cases.

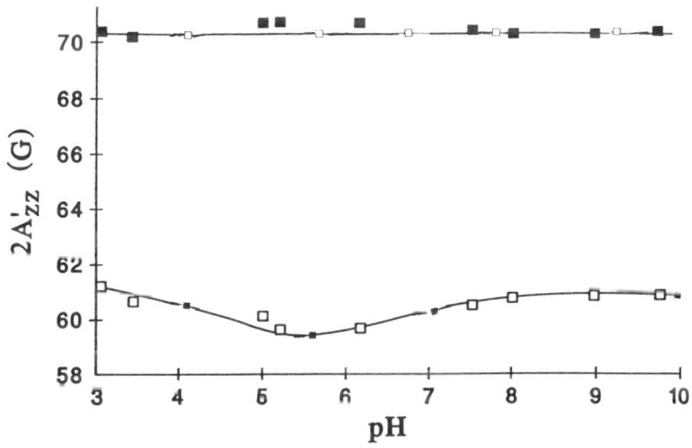


Figure 2. Effect of pH on the active site conformation of immobilized papain on polysulfone and cellulose acetate membranes. The splittings of the high and low field lines of the subpopulation A (open squares for polysulfone, closed squares for cellulose acetate) and the subpopulation D (solid squares for polysulfone, open squares for cellulose acetate) are plotted against the pH of the solutions. Each data point is the mean of three trialand the standard deviation is less than 5% of the mean.

**Denaturant Studies**

EPR studies of denaturants susch as urea and guanidine hydrochloride have shown that the active site cleft of free papain in solution might have a more closed structure as the protein became denatured.[2] The EPR spectra of immobilized papain on cellulose acetate and polysulfone membranes under different concentrations of urea solution suggest that as the urea concentration increases, subpopulation A converts into subpopulation D. If attention is focused on the negative portion (that portion of the spectrum below the base line) of the $M_I$=0 mid field lines (Figure1), it is observed that the intensity of the line corresponding to the subpopulation A decreases while the line corresponding to the subpopulation D increases as a function of urea concentration. As noted above, the relative intensity of these two lines should be a good parameter to monitor the conversion between these two subpopulations[1] and the parameter is defined in the following equation.

$$I(D)/I(A) = \frac{\text{The intensity of the negative mid-field peak of the subpopulation D}}{\text{The intensity of the negative mid-field peak of the subpopulation A}}$$

The plot of the parameter, I(D)/I(A), vs the urea concentration for papain immobilized on polysulfone membranes showed that this parameter increased from 0.925 for the control to approximately 1.6 in 6M urea. In the case of papain immobilized on cellulose acetate membranes, I(D)/I(A) changes from 0.925 for the control to 1.27 in 5M urea.

The activity measurements of papain immobilized on polysulfone membranes correlate with the EPR results. The amidase activity of papain immobilized on polysulfone membranes decreases slowly at urea concentration below 5M. Even at 5M urea, there still is 83.1% of the activity left. When the urea concentration is higher than 5M, the decrease in activity become much more dramatic and just 23.2% of the activity left at 8M urea. It is obvious that immobilization of the enzyme on polysulfone membranes enhances its stability toward the denaturation by urea.[2] However, the urea denaturation curve of papain immobilized on cellulose acetate membranes is totally different from that of papain immobilized of polysulfone membrane, but very similar to that of free papain in solution.[2] No significant improvement of the stability toward the denaturation by urea is observed in the case of papain immobilized on cellulose acetate membranes. The amidase activity drops to 25.5 % of the control even at urea concentration as low as 1M. At urea concentration higher than 2M, the remaining activities are less than 20% of the control and the immobilized enzyme is almost completely denatured in 8M urea with only 14.9% of the activity left.

Similar studies were performed in the studies of denaturation by guanidine hydrochloride. The denaturation factor I(D)/I(A) increases from about 1.0 in control to approximately 1.3 in 6M guanidine hydrochloride in the case of papain immobilized on polysulfone membranes, while for cellulose acetate membranes, I(D)/I(A) changes from 0.925 in control to 1.32 for the immobilized papain over the same guanidine hydrochloride concentration. Moreover, for the latter membrane release of spin label from the active site was observed after 6M guanidine hydrochloride addition. Even more spin label is released when the guanidine hydrochloride concentration is increased to 8M, and the measurement of I(D)/I(A) becomes impossible due to the much weaker EPR signals for the immobilized enzyme.

Significant improvement in enzyme stability toward denaturation by guanidine hydrochloride is observed for papain immobilized on both membranes, which is different from the denaturation by urea. At guanidine hydrochloride concentrations as high as 6M, the immobilized papin still has about 70% of its activity remained, compared to only 12.9% for the enzyme in solution.[2]

**Thermal Stability Studies**

Immobilization of papain on polysulfone membranes greatly increases the thermal stability of the enzyme. Immobilization on polysulfone membranes even protects the mixed disulfide bond between the MTS spin label and the active site of papain, a bond that is not stable for free papain in solution under high temperature [2], but in contrast, there is no detectable release of the spin label from the active site of the immobilized papain after incubation at 70 °C for 10 or 25 minutes. However, some of the active enzyme (subpopulation A) did denature (was converted into subpopulation D) after 10 and 25 minutes of incubation time at 70 °C. The denaturation (or conversion) parameter, I(D)/I(A), increases from 1.01 of the control to 1.18 after 10 minutes incubation at 70 °C. Further denaturation after 10 minutes at 70 °C may be very slow: there is no observable change between the EPR spectrum of the immobilized papain after 10 minute incubation at 70 °C and that after 25 minute incubation.

In contrast to papain immobilized on polysulfone membranes, no observable enhancement in thermal stability is observed when papin is immobilized on cellulose acetate membranes. Just like the spin labeled papain in solution [2], most of the spin label is released from the active site of the immobilized papain on cellulose acetate membranes under high temperature, which is due to the breakage of the disulfide bond between the cysteine residue and the spin label. The EPR signals of the immobilized papain became so weak that the denaturation factor, I(D)/I(A), was hard to evaluate.

The enhanced resistance to thermal inactivation was also observed in the amidase activity measurements. Consistent with the EPR studies, the amidase activity of the immobilized papain on polysulfone membranes stays around 60% of the control after 13 minutes of incubation at 70 °C while only 55 minutes are needed to decrease the activity of free papain to below 20% [3]. Furthermore, free papain almost loses all its activity after a 70 minute incubation at 70 °C with a bit more than 10% of the activity left. However, the immobilized papain on cellulose acetate membranes has a very similar thermal denaturation curve to that of free papain in solution; thus, there may be no improvement in thermal stability in this case.

**Reusability Studies**

One of the most significant advantages of immobilized enzymes over the native enzymes in solution is their reusability. The activity measurements were repeated 18 times for papain immobilized on polysulfone membranes and 7 times for papain on cellulose acetate membranes. The reusability curves have different characteristics in these two cases. For papain immobilized on polysulfone membranes, the amidase activity increased continuously in the first five cycles, then a relatively constant activity is observed for at least 18 cycles after an initial equilibrium period. [1] This results suggest that partition effects did play a role for substrate and/or product binding to a relatively hydrophobic polysulfone membrane. For the more hydrophilic cellulose acetate membrane, the activity actually decreases during the first three cycles and then remains fairly constant around 80% of that of the initial run.

**Storage Stability**

Papain immobilized on modified polysulfone and modified cellulose acetate blend membranes has high storage stability. The EPR spectrum of the immobilized papain has no observable change after 25 days of storage in 0.1 M phosphate buffer pH 7.00 at 4 °C on polysulfone membranes and 23 days on cellulose acetate membranes as judged by I(D)/I(A) values.

**DISCUSSION**

The EPR spin labeling technique has been shown to be very sensitive in studying the conformation of native papain in solution [2], and this study suggests that it sensitive to the conformation change of papain immobilized on polymeric membranes (modified polysulfone and modified cellulose acetate membranes). From the literature [11-17] and this study, it appears that a general trend exists that immobilized enzymes often have two major subpopulations, no matter what enzyme was immobilized and what kind of support was used. More than two subpopulations of the immobilized enzymes were also observed in a few cases [18,19]. All the immobilization matrices mentioned above, polymeric or inorganic, were in the form of beads. Although Zhuang and Butterfield [3-5] were the first to use EPR to characterize the structure of enzymes immobilized onto polymeric membranes, their systems were not ideal since the membranes were not to hydrated. Moreover, the spin label used, though specifically bound to the single cysteine in the active site was quite lengthy, causing the nitroxide group to protrude to the outside of the active site. Butterfield et al [1], using MTS, a small spin label bound to the cysteine group but confined to the active site cleft, subsequently first characterized immobilized enzymes on a fully hydrated membrane. In the current study, two major subpopulations were also found and studied for fully hydrated papain immobilized on modified polysulfone and modified cellulose acetate membranes.

The activities of the immobilized enzyme subpopulations are highly correlated with the mobility of the spin labels at the active sites of the immobilized enzymes. For all the immobilized enzymes mentioned above [11-17], the subpopulation that had faster spin label motion at the active site also had higher specific enzyme activity. Thus the rise in mobility may result in the increase of enzyme activity.

In the current study, both subpopulations of the immobilized papain on polymeric membranes (polysulfone or cellulose acetate membranes) have more restricted spin label motion at the active sites than the corresponding native form in solution. Immobilization did change the conformation of the enzyme but not homogeneously; that is, subpopulation A may have a more open active site cleft than subpopulation D since the spin label motion was faster in the former as assessed by the splitting parameter $2A_{zz}'$. The pH-dependence of the immobilized papain suggests that subpopulation A may be the active form while subpopulation D may be the denatured form of the immobilized enzyme. Denaturant (urea or guanidine hydrochloride) and thermal inactivation of the immobilized papain confirm this assumption. Subpopulation A can be converted into Subpopulation D by addition of urea or guanidine hydrochloride, the denaturants which can completely denature the immobilized papain at high concentration (total loss of its amidase activity). The conversion also occurs under high temperature. Therefore, higher spin label mobility at the active site of the immobilized enzyme does correspond to higher activity in the case of papain.

The solvent accessibility to the active site is also one of the important factors affecting the activity of hydrolytic enzymes. $K_3Fe(CN)_6$ can broaden the EPR signal of the subpopulation A but not the subpopulation D[1]. More direct evidence that the subpopulation A is the active form of the immobilized enzyme was obtained by immobilizing papain on polysulfone membrane first, then spin labeling. In this way, only the active subpopulation A can be labeled[1]. Therefore, loss of activity for the subpopulation D may partly due to the inaccessibility of solvent (and/or substrate) to its active site due to its significant conformational change.

Differences in the distribution of the active and denatured subpopulations exist when different membranes are used as the immobilization matrices. The proportion of the active subpopulation A is higher on cellulose acetate membranes than on polysulfone membranes as judged by the I(D)/I(A) factor, although the immobilization reactions are less efficient on the cellulose acetate membranes. This may partly explain the higher $V_{max}$(app) of papain immobilized on modified cellulose acetate membranes.

The properties of the immobilized papain are highly dependent on the polymeric membrane used as the immobilization matrix. Papain, immobilized on polysulfone membrane, has much higher thermal stability and stronger resistance to the inactivation by denaturants such as urea and guanidine hydrochloride. The enhancement of stability by immobilization is also observed in many other enzyme immobilization studies[15, 17, 20.] However, no significant enhancement of the thermal stability and the stability toward the denaturation by urea has been observed for the immobilized papain on cellulose acetate membranes, though the system has high storage stability and improved resistance toward the denaturation by guanidine hydrochloride, similar to those of papain immobilized on polysulfone membranes. Thus the support matrices do play important roles in affecting the properties of the immobilized enzymes. The immobilized papain on cellulose acetate membranes has higher specific enzyme activity than that on polysulfone membranes, however, this is compensated by lower stabilities under some circumstances.

Papain immobilized on polymeric (polysulfone and cellulose acetate) membranes used in this study is highly reusable. There is no significant decrease of amidase activity for many rounds of reactions, which is also an additional evidence for the stability of not only the enzyme but also the covalent bond between the enzyme and the polymer membranes.

The EPR spin labeling technique is a powerful method for the study of immobilized enzyme systems, whose investigation by optical spectroscopies is ordinarily hindered by the opacity of the supports. EPR offers the possibility to correlate active site structure of the enzyme with its kinetic performance characteristics in biofunctional membranes. In the current study, insight into the conformation of an immobilized enzyme under a variety of conditions that could be encountered in engineering processes has been gained. This insight will aid in the understanding of the rapidly growing area of biofunctional membranes in which applications of catalysis (membrane-based enzyme bioreactors) separation (affinity membranes), and analysis (biosensors) show great promise. Manipulation of the characteristics of the polymer membrane allow the experimenter a wide range of options of enzyme binding, enzyme activity, and enzyme stability.

## ACKNOWLEDGMENT

This research was supported in part by grants from the National Science Foundation (CTS-9307518).

## REFERENCES

1. D.A. Butterfield, J. Lee, S. Ganapathi, and D. Bhattacharyya, "Biofunctional membranes IV. Active site structure and stability of an immobilized enzyme, papain, on modified polysulfone membranes studied by electron paramagnetic resonance and kinetics," *J. Membr. Sci.*,91:47 (1994).

2. D.A. Butterfield, and J. Lee, "Active site structure and stability of the thiol protease, papain, studied by electron paramagnetic resonance employing a methanethiosulfonate spin label," *Arch. Biochem Biophys.*, 310:167 (1994).

3. P. Zhuang, and D.A. Butterfield, "Structural and enzymatic characterization of papain immobilized onto vinyl alcohol/vinyl butyral copolymer membrane," *J. Membr. Sci.*, 66:247 (1992).

4. P. Zhuang, and D.A. Butterfield, "Spin labeling and kinetic studies of a membrane immobilized proteolytic enzyme," *Biotech. Prog.*, 8:204 (1992).

5. P. Zhuang, and D.A. Butterfield, "Optimization of covalently coupling enzymes to polymeric membranes," *J. Appl. Polym. Sci.*, 47:1329 (1993).

6. J. Lee, M.S. Thesis (University of Kentucky), 1993.

7. O.H. Lowry, N.J. Rosebrough, A.L. Farr, and R.J. Randall, "Protein measurement with the folin phenol reagent," *J. Biol. Chem.*, 193:265 (1951).

8. L.J. Berliner, "Spin labeling in enzymology: spin labeled enzymes and proteins," Meth. Enzymol., 49:418 (1978).

9. D.A. Butterfield, "Principles of biofunctional membrane structures: alteration in the physical state of one side of the membrane by modulation of the physical state of the opposite side of the membrane," *J. Membr. Sci.*, 53:3 (1990).

10. S. Ganapathi, D.A. Butterfield and D. Bhattacharyya, "Flat sheet and hollow fiber membrane bioreactors: A study of the kinetics and active site conformational changes of immobilized papain including sorption studies of reaction constituents," *J. Chem. Technol. Biotechnol.*, in press (1995).

11. D.S. Clark, and J.E. Bailey, "Characterization of heterogeneous immobilized enzyme subpopulations using EPR spectroscopy," *Biotechnol. Bioeng.*, 26:231 (1984).

12. D.S. Clark, and J.E. Bailey, "Kinetics and EPR spectroscopy studies of immobilized chymotrypsin deactivation," *Ann. N.Y. Acad. Sci.*, 434:31 (1984).

13. G.A. Marg, G.L. Millhauser, P.S. Skerker, and D.S. Clark, "Application of EPR methods in studies of immobilized enzyme systems," *Ann. N.Y. Acad. Sci.*, 469:253 (1986).

14. J.E. Bailey, and D.S. Clark, "Electron paramagnetic resonance studies of immobilized chymotrypsin," *Meths. Enzymol.*, 135:502 (1987).

15. D.S. Clark, P.S. Skerker, E.J. Fernadez, and R.B. Jagoda, "Spectroscopic studies of structure-function relationships in free and immobilized alcohol dehydrogenase," *Ann. N.Y. Acad. Sci.*, 506:117 (1987).

16. Y. Song, G.E. Means, X. Wan, and L.J. Berliner, "A spin label study of immobilized enzyme spectral subpopulations," *Biotechnol. Bioneng.*, 40:306 (1992) .

17. R. Ulbrich, A. Schellenberger, and W. Damerau, "Studies on the thermal inactivation of immobilized enzymes," *Biotechnol. Bioeng.*, 28:511 (1986).

18. T. Asakura, H. Yoshimizu, and M. Kakizaki, "An ESR study of spin-labeld silk fibroin membranes and spin-labeled glucose oxidase immobilized in silk fibroin membranes. *Biotechnol. Bioeng.*," 35:511 (1990).

19. H. Yoshimizu and T. Asakura, "Preparation and Characterization of silk fibroin powder and its application to enzyme immobilization," *J. Applied Polymer Sci.*, 40:127 (1990).

20. S.D. Cordt, K. Vanhoof, J. Hu, G. Maesmans, M. Hendrickx, and P. Tobback, "Thermostabiltiy of soluble and immobilized α-amylase from bacillus licheniformis," *Biotechnol. and Bioeng.*, 40:396 (1992).

# CHEMICALLY MODIFIED ULTRAFILTRATION MEMBRANES FOR ENZYME FIXATION

Ulrike Artelt and Eberhard Staude

Institut für Technische Chemie
Universität Essen
D-45117 Essen, Germany

## INTRODUCTION

Ultrafiltration is a widely used and well established separation process in which asymmetric porous membranes are commonly applied. Many a textbook is listing the wide array of different applications in various branches of industry and gives extensive advice on membrane preparation and on the pros and cons of the various membrane polymers (Cheryan, 1986; Staude, 1992). Mostly, the synthetic polymers used as membrane materials are hydrophobic from their origin. Generally, this circumstance does not effect seriously the application of these types of membranes. Sometimes, however, intermolecular interactions between pol-ymer and substrate cause special problems that are referred to as fouling. Such interactions are exerted by hydrophobic forces yielding adsorption that can effectively be diminished by a chemical modification of the membrane material. This is best performed by introducing ionogenic substituents to the polymer. Due to this the hydrophobic character of the polymer becomes smaller and the adsorption is less pronounced. Such reactions sometimes cause a change to hydrophilic membranes. In any case such a chemical reaction presupposes that the polymer backbone remains intact, otherwise the polymer would lose its film-forming properties. Introducing ionogenic groups offers an additional advantage. A membrane bearing fixed charges can reject or attract solutes according to the respective sign of the matter to be ultrafiltered.

A polymer that tolerates some of the essential chemical reactions for introducing chemically reactive groups is polysulfone (PSU) (Hinke and Staude, 1991). Normally, it is applied in its original form for the preparation of ultrafiltration membranes which often

*Biofunctional Membranes*, Edited by
D.A. Butterfield, Plenum Press, New York, 1996

serve as support for manufacturing layered membranes. Reverse osmosis composite membranes are a good example. From the chemically modified types the sulfonated polysulfone (SPSU) is best known (Noshay and Robeson, 1976). Most recently other PSU modifications have been manufactured which preferably serve for scientific purposes (Rodemann and Staude, 1995). Besides polysulfone, polyamides (PA) and polyethersulfone (PES) are high potential polymers for ultrafiltration membranes. However, contrary to the wide range of chemical modification of polysulfone, PA and PES are less suited for chemical substitution regarding the requirements necessary for membrane formation. Yet, also for these polymers some reaction possibilities are available by which the performances of the membranes made from them can be influenced (Schmidt-Steffen and Staude, 1992; Düputell et al., 1994). Finally it turns out that a broad variety of substituted polymers exists from which ultrafiltration membranes can be produced. They differ in their surface properties because of the substituted groups. These different substituents can also be applied for enzyme fixation.

Enzyme fixation onto a solid carrier is executed for various purposes. This can be an enzyme electrode for analytical means (Coulet, 1992; Watanabe et al., 1991). In a larger scale fixed bed catalytic reactors in form of a column are used to prepare biochemical materials (Hartmeier, 1986). In recent years, a new reactor type using immobilized enzymes caught attention which is called enzyme membrane reactor. The term insinuates a special reactor type in which the site the catalytic reaction occurs is a membrane fitted out with a fixed enzyme. This is in contrast to the common terminology that understands an enzyme membrane reactor as a reactor device that comprises a membrane which acts in its original function in keeping the catalyst in the reactor and allowing the product to permeate. Paradigmatically named are batch or semi-batch reactors in which the enzyme is either dissolved (Butterworth et al., 1970) or attached to a soluble macromolecule (Wichmann et al., 1981). In the latter case a coenzyme was bonded to polyethylene glycol. Another type of an enzyme membrane reactor is the tubular plug flow reactor using a capillary module (Pronk et al., 1988). The soluble catalyst forms a gel-layer on the membrane surface. Very often the soluble enzyme is locked in the lumen; and the substrate enters the membrane by diffusion rather than by convection (Wu et al., 1993). The real enzyme membrane reactor combines both the functions of separation and catalytic reaction. Early in 1970 Barker and Burns proposed a dialysis device in which enzymes were fixed onto the membrane. The membranes in such an enzyme membrane reactor can either have the shape of hollow fibers (Korus and Olson, 1988) or flat sheet membranes (Basheer et al., 1993).

There are various physical and chemical means for fitting out membranes with enzymes. The most simple method is to mix the enzyme in the solution from which the membrane will be cast. This method is commonly referred to as enzyme inclusion. A membrane made this way mostly is a swollen gel-film; the transmembrane mass transfer occurs by diffusion (Mita et al., 1993). However not many enzymes sustain this procedure because of the protein-harming milieu. Next to this the enzyme can be attached to the external or internal surface of the membrane. The enzyme fixation by adsorption offers a wider range considering the properties of the enzymes and of the membranes as well (Ishimori et al., 1981). This works the better the more hydrophobic the membrane is. Two possibilities can be distinguished. Fixation can either be performed within the porous matrix of the membrane or on the membrane's surface facing the upstream side. In the latter case the enzyme gives rise to a fouling-like gel-layer (Alfani et al., 1979). The disadvantage of this type is that this layer can partly be released from the membrane surface by the tangential shear forces of the flow. Thus the enzyme can freely be dispersed in the solution.

Preventing this the more effective method is the chemical fixation by covalent bonding that is very often put up with a lost in enzyme's activity. For the indirect method a chemically inactive membrane is used; the fixation occurs by cross-linking the enzyme within the porous membrane with a respective agent (Caplan et al., 1965) that mostly is glutaraldehyde. No special demands are put on the membrane's properties. For direct immobilization the membrane polymer has to be fitted out with chemically reactive groups (Ceynowa and Adamczak, 1992). It depends on the respective sites at the outer sphere of the enzyme and on the membrane what type of reaction causes the fixation. Very common are the diazo coupling, the cyanuric chloride method or bondingf by N,N'-dicyclohexylcarbodiimide, but also many other substituents were applied successfully.

Finally, immobilization of enzymes is possible by ionic interactions provided the membrane bears fixed charged ions and the enzyme is oppositely charged. This type of fixation is comparable to adsorption. Its stability depends stronger on the accessibility of ionogenic groups at the outer sphere of the protein rather than on the charge of the entire protein that can be adjusted by the pH of the surrounding milieu.

From the above mentioned possibilities for immobilizing an enzyme only those will be used for the experiments described below that keep functioning the original properties of the ultrafiltration membrane. For that reason the formation of an enzyme gel-film atop the membrane will be excluded as well as the fixation by cross-linking the enzyme within in the pore volume of the membrane. In the former case, a priori a transmembrane transport resistance is exerted by the fouling layer built by the enzyme. It will be additionally enhanced when a solution containing macrosolutes containing macrosolutes is treated. In the other case the transport cross-section of the membrane becomes smaller, hence the transmembrane volume flow is influenced. Thus, adsorption, ionogenic interaction and covalent bonding remain for enzyme immobilization. Modification of either the membrane-forming polymer or the membrane itself for the purposes outlined above are described as following.

## MATERIALS

### Membrane Polymers

The basic membrane polymers for the investigations were PSU, PES and an aliphatic-aromatic PA. Whereas PES was only used in its original state and only a single version of the substituted PA was applied, the most membranes used for the experiments were made from modified polysulfone. For membrane preparation two ways exist. Either the membrane is cast from the respective polysulfone using an appropriate casting solution or the membrane is cast from a modified PSU whose chemically reactive substituent can be further reacted heterogeneously to yield the finally substituted membrane. This means is applied when the membrane to be expected cannot be prepared in the usual way because of insolubility of the polymer or the foreseen reaction causes polymeric cross-linking. This presumes that the membrane is not disintegrated by the reaction. The polymers used for preparing membranes are put in order according to the various enzyme fixation modes. The abbreviations used in the following are applied for the polymers and the membranes made from them as well. The analysis was carried out by IR and $^1$H-NMR spectroscopy, elementary analysis or by special methods suited for the particular substituents. In those cases when surface charges are introduced or changed at the ready membrane a nondestructive method should have the priority to a destructive one. Thus, by determining the streaming potential of the membrane the dissolution of the membrane can be avoided.

**Adsorption.** Polysulfone is a highly hydrophobic material. Ultrafiltration membranes made from it adsorb proteins from a solution to a sufficient content. The amount, however, cannot be increased by different substitution reactions as charges at the substituted hetero atoms account for the reduced hydrophobicity. Besides the membrane cast from PSU the subsequent PSU based polymer membranes were used for adsorptive interactions.
*PSU-NO$_2$*. The simplest N-containing PSU can be obtained by introducing a nitro group. The reaction is carried out using a mixture of $NH_4NO_3$ and trifluoracetyl anhydride. A degree of substitution (DS) of 0.9 can be achieved.
*NPSU.* The chloromethylated PSU (CPSU), see below, can be reacted with $NaN_3$ to yield PSU-$CH_2N_3$ that can be reduced to the aminomethylated PSU using $LiAlH_4$. The DS was determined to 0.7.
*PSU-OH(tert).* The first step for obtaining this polymer is the metalation of PSU using n-butyl lithium at 195 K according to Guiver et al. (1987). The lithiated polysulfone (PSULi) is reacted at the same temperature with acetone resulting in a PSU bearing tertiary alcohol groups, the DS was about 0.5.
*PSU-OH(sec).* The reaction was carried out in the similar way as mentioned for PSU-OH(tert), but acetaldehyde was added instead of acetone. The yield of the sec. alcoholic form of PSU was 0.48.

Using the chemically modified polyamide the membrane counts among the adsorptive type:
*PA-16.* The polyamide was hydrolyzed and consecutively reacted with hydrazine which eventually yielded a terminal hydrazide group.

**Ionogenic Interaction.** As PSU with fixed charges only cation exchange types come into consideration for enzyme immobilization. Since the conversion of saccharose by invertase (the model system in the investigations) is performed at a pH 4, the enzyme is positively charged and it remains attached when oppositely charged fixed ions are applied. From the two possible ionogenic groups (-COOH and -$SO_3$H) only the latter was chosen for the fixation purpose. As will be pointed out later for the sulfonic acid group spacer-like molecules are available that offer different distances of the charges from the polymeric backbone.
*SPSU.* The gentle sulfonation of PSU can be carried out in two different ways using the $SO_3$ /triethylphosphate complex. A solution can be applied, but also the reaction can be performed at solvent-free conditions. The latter is more appropriate for ready membranes. The DS depends on the reaction conditions, here a value of 0.65 was adjusted.
*PSU-COOH.* This polymer was prepared using the lithiated polysulfone form in solution to which solid $CO_2$ was added. The modified PSU had a DS of 0.48, it was used in the membrane form as precursor for a further reaction yielding a covalent bonding-type membrane.

**Covalent Bonding**. For covalent bonding various reactive groups are suited. The choice was restricted to those substituents that could be introduced by ordinary methods.
*CPSU.* The chloromethylation of PSU is a simple substitution reaction using a chloromethyl alkylether in the presence of a Friedel Crafts catalyst (Warshawsky et al., 1984). At optimized reaction conditions a DS of 0.98 of the polysulfone can be obtained .
*BrPSU.* According to Warshawsky et al. (1990) the CPSU can be converted via phase transfer catalysis.
*IPSU.* The exchange of Cl by I is possible by adding a suspension of NaI in acetone to a suspension of CPSU in acetone.
*EPSU.* Epoxidized PSU can be prepared using the lithiation step. To the reaction vessel containing the lithiated polysulfone glycidyl-4-oxohexylether is added which finally yields the EPSU. A DS higher than 0.33 cannot by achieved (Rodemann and Staude, 1994).

## Membranes

The polysulfone-based asymmetric ultrafiltration membranes were prepared according to the phase inversion process from a casting solution that consisted of 15% polymer, 4% LiCl as an additive and 81% N-methylpyrrolidinone. The die width was 300 μm and the precipitation bath temperature was 277 K. The ultrafiltration capabilities of the membranes were determined by measuring the pure water volume flow as well as the molecular weight cut-off. For this purpose 1% dextran solutions prepared from samples differing in their molar masses were used. Hence, solutions with a broad molar mass distribution could be applied. From the polymers listed in the preceding section membranes were cast. Some of them were used for further reactions which are graded in the same way as it was done for the membrane pol-ymers.

**Adsorption**. *PSU-NH$_2$*. The nitro group on the PSU-$NO_2$ membrane was reduced using hydrogen evolved from metallic tin by hydrochloric acid. This procedure did not harm the mechanic strength of the membrane. This membrane type can be used for a further step for enzyme immobilization by forming the diazonium salt using $NaNO_2$ and hydrochloric acid. The membrane in this state, however, cannot be isolated. The reaction with the enzyme has to be carried out immediately after the diazonium compound is formed. For sake of completeness the diazotated membrane PSU-$N_2^+$ is listed below.
*CPSU-OH.* The reaction of the chloromethylated membrane CPSU with ethanolamine is performed in aqueous solution under reflux for 24 h. This membrane was applied in this state, but it is also suited for further reactions (s. below).

**Ionogenic Interaction.** As aforementioned, for this fixation method only the sulfonic acid group was chosen. Rather than differentiating the membranes by their degree of substitution the distance of the fixed ions from the polymer chain was selected. The shortest distance was established by sulfonating the basic polymer. Longer distances can be accomplished by applying spacer-like molecules. For this reason already modified membranes came into consideration.
*SCPSU.* A CPSU membrane was reacted under reflux for 24 h with tauric acid, thus the spacer had a chain length of 4 atoms between the aromatic ring of the polysulfone backbone and the sulfonic acid group.

*SEPSU.* A chain length of 11 atoms can be achieved using the EPSU membrane for reacting with tauric acid under the same conditions as for the foregoing membrane type.

**Covalent Bonding.** Once again a wide variety of membranes can be found in this group. There are some precursor membranes that can be converted heterogeneously. By varying the substituents also the possibility is given for investigating the spacer influence on the activity of the enzyme covalently bonded to the membrane.

*PSU-*$N_2^+$. The diazotated membrane is obtained via the PSU-$NH_2$ membrane, it cannot be stored.

*CPSU-CHO.* The CPSU membrane is reacted with aminoacetaldehyde dimethyl acetal under reflux for 24 h. The final step is the conversion of the acetal group into the aldehyde group under the influence of sulfuric acid.

*EPSU-CHO.* For fixing the enzyme at a longer distance from the polymer using the aldehyde group the EPSU membrane is better suited than the CPSU. The reaction however is carried out in the same way as for the former membrane.

*PSU-Cya1.* Cyanuric chloride is a very common species for enzyme immobilization. The NPSU membrane is bearing the appropriate group for the reaction with this molecule. The reaction needs some precautions in order to prevent the hydrolysis of the remaining chloride positions which are necessary for enzyme fixation.

*PSU-Cya2.* A longer chain between the main polymer and the cyanuric chloride can be pro-vided using the CPSU-OH membrane instead of the NPSU membrane. The reaction procedure does not differ from that as in the foregoing case.

*PSU-CDI.* Next to cyanuric chloride, N,N'-dicyclohexylcarbodiimide is a well established immobilization reagent. This can be attached to the PSU-COOH membrane. Once the reaction has been finished, the enzyme immobilization has to follow immediately.

An overview of the essential parts of the chemical structure of the modified polymers for membrane preparation is given in Tab. 1 in which the types are arranged with respect to their enzyme fixation mode. The basic polymers are listed likewise.

## Enzyme and Enzyme Immobilization

The enzyme used for the investigations was invertase (E.C. 3.2.1.26) purchased from Merck, Darmstadt, FRG. It has a $K_M$ value of 22 mmol/L and a specific activity of 310 units/mg at a pH of 4.5 at 298 K. The enzyme was immobilized within the pore volume of the respective membrane in the following way. In an ultrafiltration cell the membrane was mount-ed the downstream side facing a 20 mL enzyme solution (1.0 mg/mL). This solution perme-ated the membrane at a pressure between 0.5 and 1.5 bar at 298 K. In the case of the diazonium coupling the temperature was 273 K. For adsorptive and ionic immobilization the pH of the enzyme solution was 4.0 (acetate buffer) and 8 for covalent coupling (phosphate buffer). Past the permeation of the enzyme solution, the membrane was rinsed with a pure buffer solution of 10 mL at 1 bar. Now the membrane was mounted in the normal position and some additional cleansing steps followed. The enzyme loaded membrane was conditioned at 1 bar for 24 h at 277 K and eventually stored at 277 K in a buffer solution of pH 4 that contained $NaN_3$ (0.05%) for preventing microbiological attack.

**Table 1.** Chemical structure of membrane polymers.

| covalent bonding | ionic interaction |
|---|---|
| 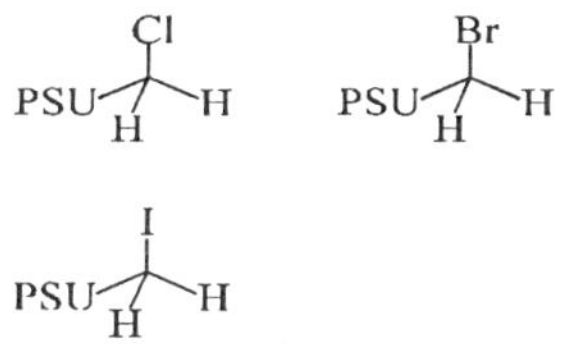 *XPSU (X = Cl, Br, I)* | $PSU\text{-}SO_3^{\ominus}$ *SPSU* |
| 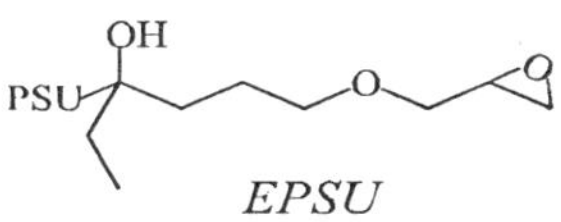 *EPSU* | 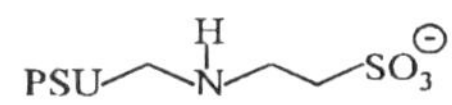 *SCPSU* |
| 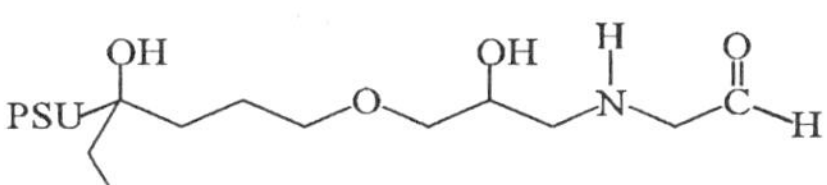 *EPSU-CHO* | 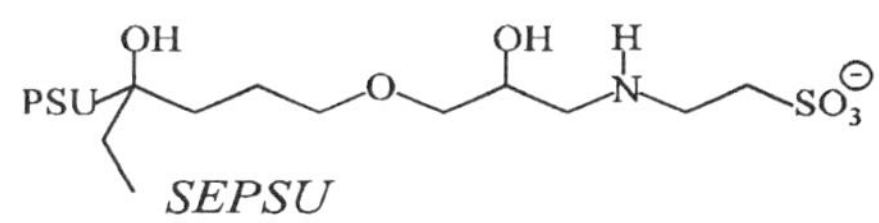 *SEPSU* |
| | **adsorptive** |
| 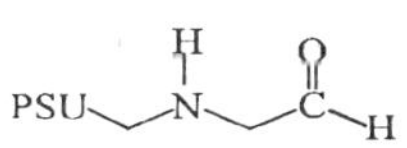 *CPSU-CHO* | polysulfone (PSU) |
| | polyethersulfone (PES) |
| | PA |
| 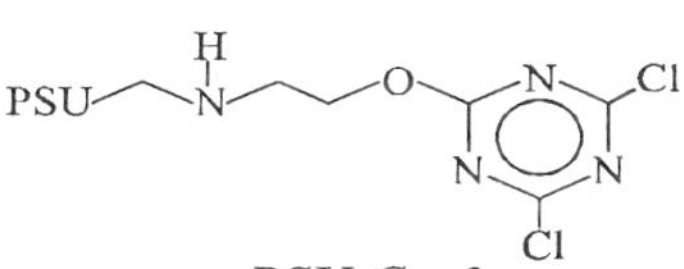 *PSU-Cya2* | *PA-16* $PA\text{-}C(=O)\text{-}NH\text{-}N_2$ |
| | *$PSU\text{-}NO_2$* |
| | *$PSU\text{-}NH_2$* |
| 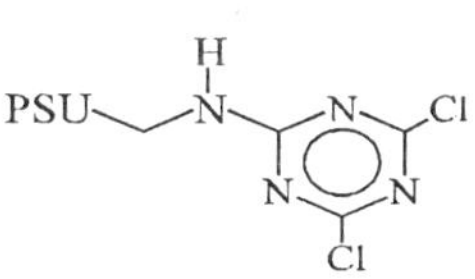 *PSU-Cya1* | 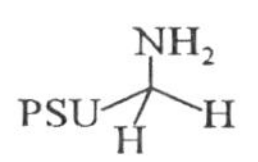 *NPSU* |
| | 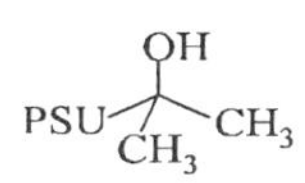 *PSU-OH(tert.)* 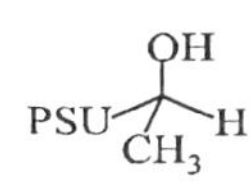 *PSU-OH(sek.)* |
| *PSU-COOH/CDI* | 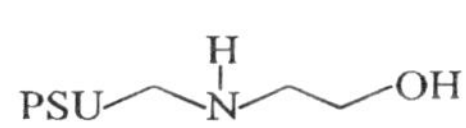 *CPSU-OH* |
| *$PSU\text{-}N_2^{\oplus}$* | |

The amount of the fixed enzyme was quantitatively determined by photometric analysis of the rinsing samples using Coomassie Brilliant Blue G 250 according to Bradford (1976), but slightly modified (Gille and Staude, 1994). The evaluated value only gives the jentire amount of the fixed enzyme per volume wet membrane. Even if it is proven that the upstream side is free of enzyme, it is not guaranteed that the enzyme is symmetrically distributed over the entire membrane neither in horizontal nor perpendicular direction. It is solely an integral value.

## RESULTS AND DISCUSSION

### Membranes

**Molecular weight cut-off.** The values are shown in Table 2. Here, the membranes are ordered according to the way the membranes are formed. Thus, if possible first the original membranes made from the modified polymer are listed followed successively by those made from them by a reaction.

As the table reveals, generally the conversion of a ready membrane does not remarkably change the transmembrane transport cross-section. This is shown for the CPSU based membranes. The precursor and those made from it (CPSU-OH, -CHO, NPSU and PSU-Cya1) have comparable MW cut-off values within the error limit. Yet any reaction influences the cut-off which mostly is shifted to lower values.

**Streaming potential.** The streaming potential of a membrane has proved to be a useful tool for characterizing porous ultrafiltration as well as microfiltration membranes; for details refer to the literature (Nyström et al., 1994; Schwarz et al., 1994). It is well known that neutral membranes like other solid materials or colloids adsorb negatively charged ions from aqueous solutions which mostly are OH ions. Due to this, a negative charge is induced on the membrane's surface. Consequently, by measuring the streaming potential a negative value is found. This effect is more pronounced when the membrane is fitted out with fixed charges. Likewise, a negative value is obtained for negatively charged membranes (cation exchange type) and a positive one for a membrane with positive charges (anion exchange type). The streaming potential increases linearly with increasing pressure. Moreover, at the same transmembrane pressure membranes with wider pores exhibit a larger streaming potential value. As a function of the pH the streaming potential depends on the dissociation capability of the fixed charges; strong acids or bases are less influenced than weak ionogenic groups. Also the isoelectric point of the membrane can be verified by this

**Table 2.** Molecular weight cut-off of the membranes.

| Membrane | Cut-off g/mol | Membrane | Cut-off g/mol | Membrane | Cut-off g/mol |
|---|---|---|---|---|---|
| PSU | 103,000 | SCPSU | 97,000 | EPSU-CHO | 95,000 |
| PES | 89,000 | NPSU | 92,000 | SEPSU | 130,000 |
| PA | 85,000 | PSU-Cya1 | 85,000 | SPSU | 117,000 |
| PA-16 | 88,000 | PSU-Cya2 | 86,000 | PSU-COOH | 120,000 |
| Cl(Br,I)PSU | 95,000 | PSU-$NO_2$ | 91,000 | PSU-OH(sec) | 113,000 |
| CPSU-OH | 96,000 | PSU-$NH_2$ | 90,000 | PSU-OH(tert) | 115,000 |
| CPSU-CHO | 95,000 | EPSU | 124,000 | | |

method. Finally, a chemical modification of a membrane that yields a change of the surface charge can ascertained very easily by this method. Fig. 1 shows these relationships. The streaming potential is plotted as a function of the pH. The membrane with the primary amine group (NPSU) is positively charged at low pH whereas the neutral membrane and both the membranes with dissociable acidic groups are positioned in the negative range. The neutral chloromethylated membrane (CPSU) was converted into the SCPSU membrane now bearing a terminal sulfonic acid group by reaction with tauric acid. This change in the surface charge is displayed by a change in the slope of the streaming potential graphs.

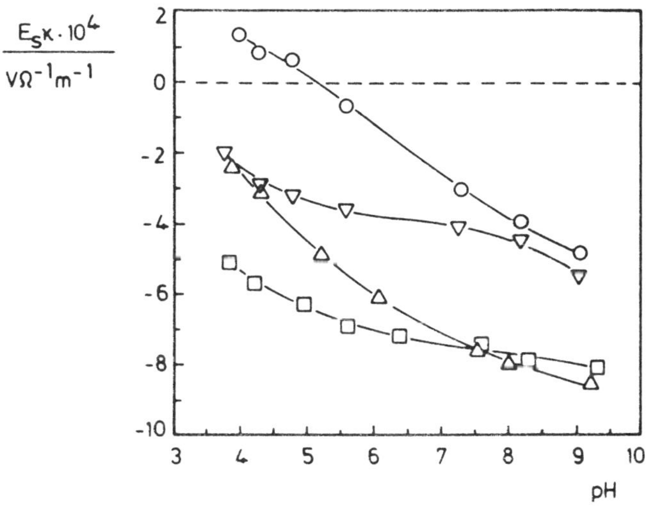


Fig. 1. Streaming potential $E_S$ κ as a function of pH for different PSU membranes. Membranes: O NPSU, ∇ PSU-COOH, Δ CPSU, □ SCPSU; solution: $10^{-3}$ *M*KCl; temp.: 293 K, pressure 1 bar.

## Enzyme Membranes

The surface properties of the different membranes clearly govern the amount of invertase attached to the membrane. There is no doubt that the group comprising the membrane types with chemically reactive substituents ranks first followed by the membranes with ionogenic groups. Adsorption gives the lowest amount. In each group differences of the invertase amount are found depending on the membranes' properties. In the adsorption group hydrophobicity plays the most influential role. From the membranes prepared from the three basic membrane polymers PSU, PES and PA, the most hydrophobic PSU membrane turns out to be best suited for adsorption. The amount was 33.1 ± 0.7 mg/cm$^3$ wet membrane. The error limit is analogical to the following values. The uptake of the more hydrophilic PA membrane was 27.3 mg/cm$^3$. Each chemical modification slightly increased the hydrophilicity of the respective membrane; consequently the enzyme uptake diminishes. The lowest value of 25.4 mg/cm$^3$ was observed for the PA-16 membrane; the -NH-NH$_2$ group accounts for the reduced adsorption capability. Better results are obtained with membranes with fixed charges. Generally, the ionogenic groups on a membrane polymer augment the membrane water content reducing the hydrophobic interaction. Nonetheless, the amount of attached invertase mounts up to the range of 39.4 to 41.0 mg/cm$^3$ regardless of the length of the spacers. Ionic interactions are more effective for enzyme immobilization; any influence of the distance between polymer matrix and enzyme of the spacers. Ionic interactions are more effective for enzyme immobilization; any influence of the distance between polymer matrix and enzyme on the immobilization can be excluded. The same was observed for the membranes with sites for covalent bonding. Thus, the enzyme amount bonded to the CPSU-CHO and EPSU-CHO membranes does not differ considerably (38.1 and 39.1 mg/cm$^3$, resp.). Also the cyanuric chloride group's reactivity is independent of the spacer length as is demonstrated for the PSU-Cya1 and PSU-Cya2 (49.1 and 49.8 mg/cm$^3$, resp.). In the first case the length of the chains are 3 and 10 atoms, and in the latter 2 and 5 atoms between the polymeric backbone and the reactive substituent. The reactivity of the substituent is the decisive factor for the mass of the immobilized invertase. In this investigation it starts with the halomethylated PSU (38.8 mg/cm$^3$), for PSU-$N_2^+$ 48.3 mg/cm$^3$ were found, but PSU-Cya2 attained the highest value with 49.8 mg/cm$^3$. As expected the coupling methods that operate best are those using diazonium salt and cyanuric chloride. The expected higher binding potential of iodine compared to that of chlorine could not be confirmed. The halomethylated membranes had the same invertase amount. In organic chemisry normally iodine is the better leaving group for substitution reactions. The values of some selected membranes are shown in Fig. 2.

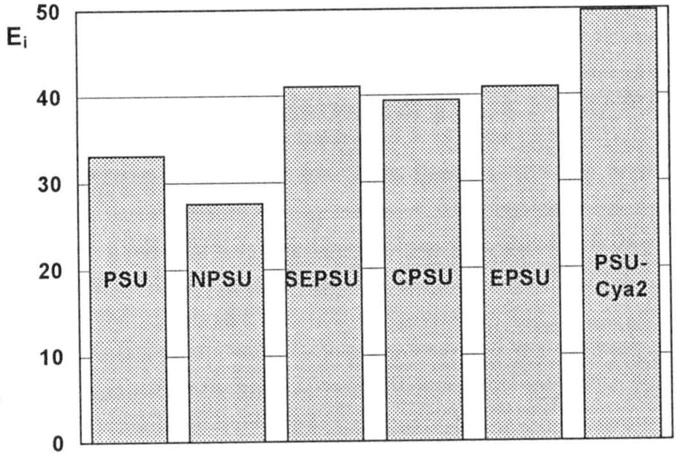


Fig. 2. Enzyme amount $E_i$ immobilized on different membranes. Adsorptive: PSU, NPSU; ionic: SEPSU; covalent: CPSU, EPSU, PSU-Cya2; $E_i$ in mg enzyme/cm$^3$ wet membrane.

Adsorption proved to be the least suitable method; the best results were achieved by covalent coupling. Despite the very different DS of the membranes bearing ionogenic groups (SPSU: 0.65; SEPSU: 0.33) the amount of invertase varied only about 2%. It is assumed that in the case of the SPSU membrane a remarkable part of the fixed charges is not approachable for the invertase. The third membrane in this group (SCPSU) contains only 4% less than the SEPSU. Contrarily, a difference of 25% was found for the membranes for covalent bonding. This is attributed to the DS and the accessibility for the entering invertase as well. The class of adsorptive membranes exhibits about the same difference.

## Saccharose Conversion

Three different sets of experiments were performed to obtain the kinetic parameters. In a batch-type arrangement the substrate entered the membrane by diffusion whereas a convective transmembrane substrate flow was established in a closed loop ultrafiltration device. In both cases pure substrate solutions were applied, inert solutes originated from the buffer solution added. In a third reaction mode, the solution contained inert macromolecules additionally. In this way the behavior of a real membrane reactor could be investigated, mostly in order to determine the influence of a fouling layer atop the membrane on the enzyme activity.

**Kinetic Parameters.** For the two different reaction conditions namely either diffusion and reaction or convection, diffusion and reaction different analytical solutions of the general mass balance equation have to be taken into consideration. Assuming 1st order reaction kinetics, in the former case the reaction can be performed either at asymmetric or symmetric conditions according as whether substrate enters the membrane from one side or from both sides. The concentration profile within the membrane is depending on the Thiele modulus for both these states. The Thiele modulus value the equivalent to a mass transfer resistance is proportional to the slope of the concentration profile. This is more pronounced for the asymmetric state. Due to this, higher effective reaction rates will be found for the symmetrically accessible membrane. The equation for the effective reaction rate is:

$$r_{eff} = -k\, c_{S,h}\, \eta_\phi = -k_{eff}\, c_{S,h} \tag{1}$$

with $\eta_{\phi,\,sym} = \tanh(\phi/2)/(\phi/2)$ and $\eta_{\phi,\,asym} = \tanh(\phi)/\phi$. Here, $c_{S,h}$ is the substrate concentration in the solution, $\eta_\phi$ is the analytical effectiveness factor describing the internal diffusion and $k_{eff}$ is the effective rate constant. The latter includes changes in the activity as well as diffusion effects since it contains the maximum reaction rate as well as the Michaelis-Menten constant. The conversion is given by

$$-\ln(1 - X) = k_{eff}\, t_V \tag{2}$$

with $t_V$ as the volume reduced time which is the product of the time interval and the membrane volume $V_M$ divided by the volume of the solution $V_L$. The latter is correspondingly corrected because of sampling.

The solution for the condition of convection, diffusion and reaction is substantially more complicated (Kotzelski and Staude, 1995). The expression for the conversion behavior is

$$1 - \frac{2\alpha e^{Bo/2}}{Bo \sinh(\alpha) + 2\alpha \cosh(\alpha)} \tag{3}$$

$\alpha = \sqrt{\frac{Bo^2}{4} + \phi_V^2}$ , $\phi_V = L\sqrt{k_{eff}/D_V}$ ≅ Thiele modulus, $D_V$ ≅ axial diffusion coefficient, and $Bo = w\ L/D_V$ ≅ Bodenstein number.

**Diffusion Experiments.** The enzyme loading procedure makes in conjunction with the asymmetric membrane structure a homogeneous enzyme distribution unlikely throughout the membrane's thickness. This assumption can be confirmed by carrying out the enzymatic reaction. For this reason, the membrane is mounted in the diffusion cell in different positions. Either the upstream side or the downstream side can face the feed. To guarantee that the saccharose diffuses the membrane only from one side, the opposite side of the membrane must be tightened. For giving the substrate restriction free access to the entire amount of enzyme in membrane, an appropriate membrane suspension becomes necessary. All the devices are equipped with a stirrer to avoid external mass transfer resistance. Thus various reaction possibilities are feasible which should enable to obtain information on the enzyme distribution. Fig. 3 presents the conversion/time curves for the three reaction modes demonstrating the anticipated non homogeneous invertase distribution. It is obvious from the figure that the catalyst is positioned to the uppermost part just beneath the active skin of the EPSU membrane indicating a very pronounced disproportion of the activities between both membrane sides. This can be expressed by the relative activities of the values for the single sides compared with that of the both-sided value. The PSU-CDI and the CPSU-CHO membranes had most extreme asymmetry. The relative activities

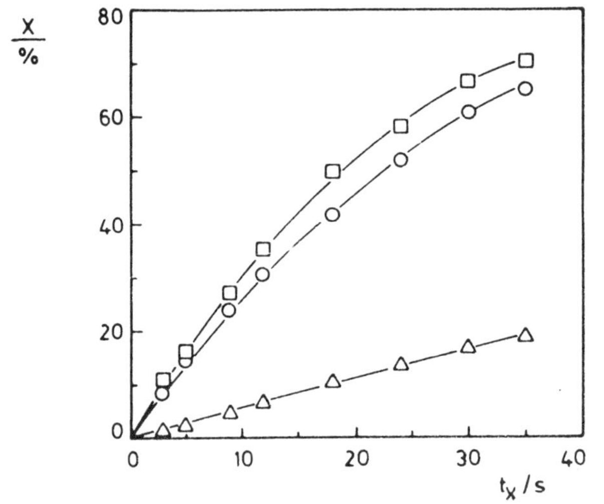


Fig. 3. Saccharose conversion as a function of time of an EPSU membrane at different positions. □ membrane facing from both sides the feed, O upstream side facing the feed, Δ downstream side facing the feed, solution: 10 m*M* saccharose, pH 4, temp.: 293 K.

were 0.89 and 0.87, resp.; for EPSU a value of 0.85 was evaluated. The most symmetric distribution exhibited the PSU-Cyal with 0.76. The values of the membranes for adsorptive immobilization were comparable to those for covalent bonding, and also rather widespread. The least difference between upstream side and downstream side yielded the membranes with ionogenic bonding with about 0.73 ± 0.02. The specific rate constants $k_{eff}'$ for the upstream side and the downstream side that reflect the true and apparent transformation of the enzyme are given in Table 3. It should be taken into consideration, however that this is an average since immobilization affects the enzyme molecules in different ways.

The experimental findings of the reaction with both-sided diffusion served for comparing the kinetics of the reaction controlled only by diffusion with those of the reaction with additional convective mass transport. The effective Michaelis-Menten constants $K_{M,eff}$ were between 123 (PSU-$NH_2$) and 180 mmol $L^{-1}$ (EPSU-CHO) for all membrane types applied for the diffusion experiments. Some values are shown in Table 4 (s. below). These values are considerably larger than the value of the native invertase of 24.5 mmol $L^{-1}$ insignificantly deviating from the value given by the supplier. It cannot be distinguished whether any structural alteration of the biocatalyst or the mass transfer by diffusion are responsible for this high constant and how strong their influences are because effective quantities comprise both parameters. A decision is only achievable by experiments under convective transmembrane substrate flow conditions.

**Table 3.** Specific rate constants.

| Membrane | upstream | downstream |
|---|---|---|
| | $k_{eff}'$ $10^2$ (L $s^{-1}$ $mmol^{-1}$) | |
| PSU | 15.37 | 2.85 |
| PSU-NH2 | 16.59 | 3.92 |
| SCPSU | 18.22 | 6.99 |
| SEPSU | 23.22 | 8.09 |
| EPSU-CHO | 24.34 | 2.62 |
| EPSU | 20.13 | 3.97 |

The catalytic constant $k_{cat}$ (definition s. Table 5) gives a statement on the catalytic quality of the enzyme membrane which is independent of the concentration of the enzyme and the diffusion of the substrate. As can be seen from Table 4, the catalytic activity can be raised by increasing the hydrophilicity of the membrane. The most hydrophobic PSU membrane exhibits the lowest activity despite the highest amount of enzyme it adsorbs. An explanation may be sought in the change of the enzyme structure. The hydrophobic groups, normally directed toward the inside of the enzyme molecule are partly oriented outwards in order to interact with the hydrophobic membrane polymer. This gradual folding up of the biocatalyst may reduce its activity. This is affected to a lower degree with more hydrophilic membrane surfaces.

Unequivocally, the activity of the membranes that immobilize the invertase by ionic interaction can be traced back to the length of the spacers. The question whether the increase in activity is a matter of the distance between polymer matrix and enzyme or the spacer allows for an increase in mobility of the enzyme cannot be answered.

For the membranes with covalent bonding of the invertase two distinct tendencies were observed. First with a longer chain of the spacer the activity of the invertase becomes enhanced as can be seen for the XPSU, EPSU and EPSU-CHO membranes. This is the same observation already found for the enzyme fixation by ionic interactions. Second, the more reactive the immobilizing substituent is the more pronounced is the loss in enzyme activity. In any case, the carbodiimide group (PSU-CDI) is more suited and widely applied for enzyme fixation than the halomethyl group (XPSU), but the catalytic constant $k_{cat}$ for the former is lower. The same is valid for the diazo coupling. The membrane PSU-$N_2^+$ with the very active enzyme coupling method exhibits surprisingly the lowest constant. With respect to the distance of the catalytic site and the membrane polymer, the CPSU-CHO and the PSU-Cyal are more or less compa able. However, they differ in their catalytic reactivity which is more pronounced taking into consideration the effective kinetic constant being related to the amount of fixed enzyme (s. last column). The preferred fixation group exhibits lower activity. The reason for this behavior can likely be found in the interaction between chemically active groups and the bonding sites of the enzyme yielding a multi point attachment. Due to this the catalytic effectiveness suffers from the probable resulting structural deformation.

**Table 4**. Kinetic data of the diffusion experiments.

| Membrane | Type of immobilization | $K_{M,\,eff}$ mmol $L^{-1}$ | $k_{cat}$ $s^{-1}$ | $k_{eff}$' $10^2$ L $s^{-1}$ $mmol^{-1}$ |
|---|---|---|---|---|
| PSU | adsorptive | 136.3 | 24.71 | 18.13 |
| PA | " | 137.4 | 32.01 | 23.27 |
| CPSU-OH | " | 150.1 | 33.56 | 22.31 |
| SPSU | ionic | 150.5 | 36.81 | 24.43 |
| SCPSU | " | 168.8 | 42.69 | 25.27 |
| SEPSU | " | 188.7 | 58.51 | 40.00 |
| XPSU | covalent | 148.6 | 30.35 | 10.42 |
| EPSU | " | 166.1 | 39.81 | 23.97 |
| EPSU-CHO | " | 180.6 | 51.31 | 28.48 |
| PSU-CDI | " | 145.3 | 26.88 | 18.49 |
| PSU-$N_2^+$ | " | 141.9 | 24.81 | 17.49 |
| CPSU-CHO | " | 141.4 | 35.59 | 25.18 |
| PSU-Cyal | " | 165.5 | 33.22 | 20.05 |

**Experiments under Convective Conditions.** Basically, the investigations of the various membranes under convective transmembrane mass transfer did not present further findings regarding their different catalytic behavior. The kinetic constants were found in nearly the same range. However, the effective Michaelis constants dropped remarkably by about one dimension and came into the range of the value measured for the native invertase in homogeneous solution which gave 24.5 mmol $L^{-1}$. From the experiments under convection with $K_{M,eff}$ constants between 26.1 (SEPSU) and 36.7 mml $L^{-1}$ (PSU) it is clearly pointed out that the large values for the diffusive experiments (see above) are related to a hindered mass transfer within the membranes. In the case of convection this means that the mass transfer resistance caused by diffusion can be disregarded, but other influences may be responsible for the somewhat larger values of the Michaelis constant namely for membranes with adsorptive and covalent fixation. For fixation by ionic interaction only slightly larger values were observed. Hence, it can be concluded that structural deformation of the enzyme originates more from adsorption or covalent bonding while the structure of the enzyme is less damaged by ionic interaction. Under these circumstances a half maximum reaction rate can be achieved by the "remaining" intact catalysts if a higher substrate concentration would be made available. This could explain the higher effective Michaelis constants experimentally found.

**Effectiveness Factors.** Contrary to catalyzed reactions in homogeneous phase transport phenomena may dominate the entire process of heterogeneously catalyzed reactions. Hence, the overall kinetic parameters include the effect of immobilization and the transport processes. Therefore they are referred to as effective parameters. Reactions of enzymes immobilized within ultrafiltration membranes exhibit the same regularities of heterogeneously catalyzed reactions in porous catalyst as far as external as well as internal diffusion are concerned. Both can contribute to the mass transfer and thus each can function as the rate determining quantity. All these effects can be summarized and related to the reaction under homogeneous conditions in form of the effectiveness factors whose definitions are shown in Table 5.

**Table 5.** Effectiveness factors in heterogeneous enzyme catalysis.

| Designation | Symbol | Definition | included influence |
|---|---|---|---|
| effectiveness factor | $\eta$ | $\frac{r_{eff} / E_i}{r_h / E_h} = \frac{k'_{eff}}{k'_h}$ | immobilization and internal diffusion |
| maximum effectiveness factor | $\eta_{max}$ | $\frac{r_{max}/ E_i}{r_{h,max}/ E_h} = \frac{k'_{cat}}{k'_{cat,h}}$ | immobilization |
| analytical effectiveness factor | $\eta_{\Phi}$ | $f(\Phi)$ | internal diffusion |

Index "eff": effective magnitude of immobilized enzyme, index "η": magnitude of enzyme in homogeneous phase, E: enzyme concentration (mmol/L).

Moreover, also the reduced reactivity of the enzyme caused by the immobilization mode plays a role is obvious from the value of the effectiveness factors. In Fig. 4 these factors are shown for some selected membranes stemming from the three different immobilization groups. The very low factors clearly display that the immobilization within the membrane drastically diminishes the efficacy of the enzyme.

The effectiveness factors also reflect the different behavior of the membranes. In the precedent sections the differences in the kinetics were already discussed. For that reason, the factors serve only for demonstrating the differences in the alternative fixation modes. According to the figure, the best value was obtained for the membrane with ionic bonding of the enzyme. This membrane is fitted out with a hydrophilic spacer. Probably, some factors effect this result. First, fixation by ionic bonding is least engaged in changing the enzymatic formation even if a multi point attachment occurs. Moreover, the well known influence of a spacer is maintaining the enzyme's activity. Finally, the hydrophilicity of the spacer decides on the catalyst's efficiency. The effect of the spacer finds its expression in the fact that in this group the value of the factor goes from 3.41 (SEPSU) to 2.46 (SCPSU) and 2.07 (SPSU). Despite the SEPSU membrane is the most efficient it shows a disadvantage in the long-term experiments. The overall stability of the activity of the membrane is less developed than that of the membranes which keep the catalyst by covalent bonding. This can be seen from Fig. 5.

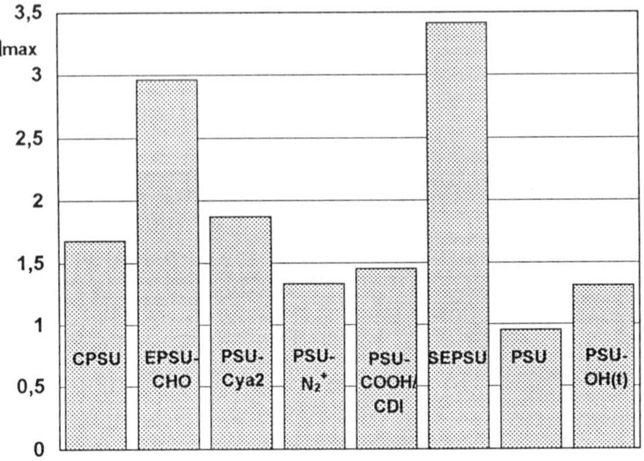


Fig. 4. Maximum effectiveness factors (%) for different membranes obtained under convective conditions.

In the sector of the membranes that covalently fix the invertase the already discussed scheme is observed for the effectiveness factors. Once again, the best result is obtained by the immobilization via the Schiff's base reaction with the reactive site distant from the membrane matrix (EPSU-CHO). Next follows the membrane fitted out with the cyanuric chloride substituent bonded to spacer. The end position holds the membrane with the very active diazonium group. On the other hand, this membrane is superior by the very large amount of attached enzyme of 48.3 mg/mL. This means the mass of invertase is less important, the better mass/reaction relationship offers the PSU-Cya2 membrane.

The lowest effectiveness factors exhibit the membranes with adsorptive attachment. This is not surprising since the enzyme uptake is lowest. Regarding the PSU membrane, its position in this field coincides with the foregoing experimental findings. This membrane adsorbs the highest amount of invertase because of the surface properties, but the hydrophobicity of the surface for its part reduces the effectiveness factor. With increasing water uptake the mass of enzyme adsorbed becomes smaller, but the value of the effectiveness factor rises. It can be said that the increase of enzyme uptake is balance by the lower value of the effectiveness factor and vice versa.

**Enzyme Stability**. Besides the activity of an immobilized enzyme within a membrane also its stability during the catalyzed reaction is a point of interest. For this reason from each category one membrane was selected. The membranes were investigated by both-sided diffusion for three hours per day for 40 days. Before each measurement the membranes were rinsed carefully. After the reaction they were stored at 278 K until use the next day. The experimental findings are shown in Fig. 5 in which the relative conversion is plotted as a function of time. The covalent bonding proves to be the favorite method with respect to stability. Within a period of about twenty days the enzyme activity remains nearly unaltered. Contrary to this, the ionic bonding seems to be less stable whereas the membranes with adsorptive bonding loose remarkably their catalytic efficiency. These observations become more evident with advancing reaction time. It is assumed that the enzyme is not kept strong enough by ionic bonding and desorption is responsible for the extremely steep drop of activity for the adsorptive membranes.

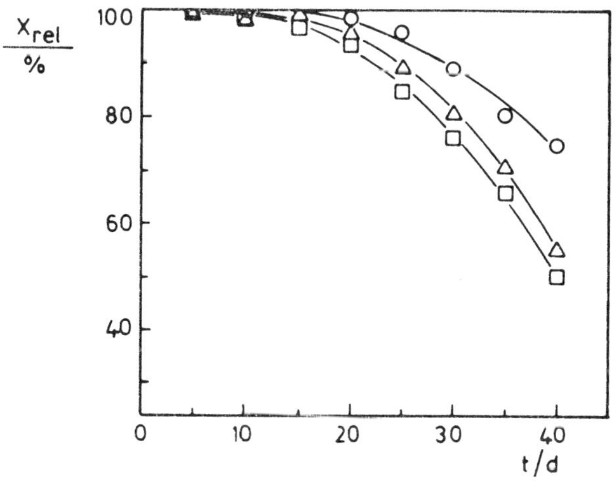


Fig. 5. Relative conversion of saccharose as a function of time for different modes of invertase immobilization. Membranes: O covalent bonding (EPSU-CHO); Δ ionic interaction (SEPSU); □ adsorptive (PSU).

## Enzyme Membrane Reactor

The membranes display the properties of real enzyme membrane reactors as was shown either for the diffusive or for the convective reaction process. The enzymatic active site is an integral part of the membrane, but the membrane acts solely as a support for the catalyst. Especially in the convective mode the membranes could not demonstrate any separating capability since the substrate was not rejected. The latter characteristic can only become manifest using a substrate being partly excluded. Dextran fractions of a molar mass of 40,000 (T 40) and 100,000 (T 100) g/mol were applied as macrosolutes. The EPSU membrane was mounted in the closed loop ultrafiltration device. The reaction was carried out at 293 K at different transmembrane pressures adjusting different residence times within the membrane. A conversion/residence time plot is shown in Fig. 6. The kinetics were not influenced by the dextran as the curve suggests. The very low membrane thickness does not allow a higher residence time, therefore the conversion of the substrate cannot exceed 40%. The membrane exhibits the behavior of a differential reactor. Adding a polymer to the feed causes concentration polarization and due to this the formation of a gel-layer atop the membrane. Even though the molecular weight cut-off of the EPSU membrane was determined to about 124,000 (s. Table 2) the fouling layer reduces the transmembrane volume flow. Consequently, the residence time increases. This increase starts even at low concentrations using the dextran with the lower molar mass. The higher molar mass dextran contributes more intensely to the fouling of the membrane. For that reason the residence time becomes larger. Now, the conversion can reach values of about 80%.

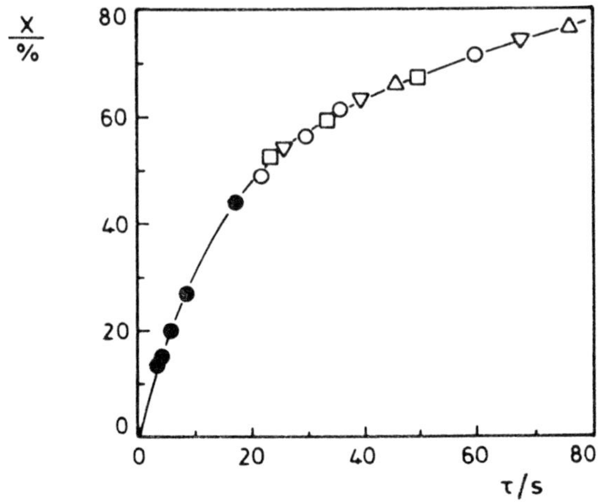


Fig. 6. Conversion/residence time relationship of an enzyme membrane reactor. Invertase immobilized on an EPSU membrane. Solutions: ● 10 m*M* pure saccharose; O s. + 0.25% dextran T 40; □ s. + 1.50% dextran T 40; ∇ s. + 0.25% dextran T 100; Δ s. + 1.50% dextran T 100; temp.: 293 K.

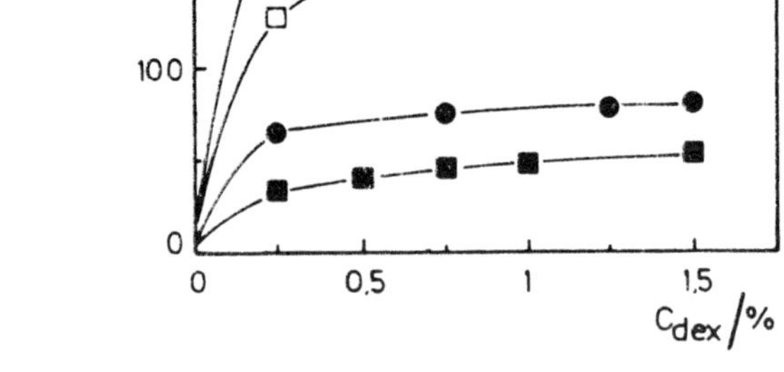


Fig. 7. Residence time as a function of dextran concentrations for different pressures. Invertase immobilized on an EPSU membrane. Solutions: saccharose (10 m*M*) +: □ dextran T 40, 1 bar; ■ dextran T 40, 3 bar; O dextran T 100, 1 bar; ● dextran T 100, 3 bar; temp.: 293 K.

In the range measured the concentration of the dextran is less influencing the residence time than the molar mass as Fig. 7 demonstrates. Here, the residence times are plotted as a function of the dextran concentration for two different pressures. As the transmembrane pressure decides the volume flow which is correlated to the residence time, the latter is larger at lower pressure.

## CONCLUSION

A real enzyme membrane reactor comprises the reaction part exerted by the biocatalyst immobilized on the membrane as well as the separation part given by the pore structure of the membrane. To realize such a device appropriate membranes are needed that are in accordance with both the goals. It depends on the separation problem what type of membranes will be chosen. This paper deals with asymmetric ultrafiltration membranes that were prepared by the commonly used phase separation process. For fixing enzymes the membrane should meet some requirements. It depends on the type of immobilization what kind of chemical modification on the membrane polymer or the membrane itself has to be carried out. Adsorption, ionic bonding or covalent bonding are the possibilities for a proper enzyme fixation. Polysulfone and its derivatives were the preferred polymers for this purpose. The modified polysulfone as well as the membranes prepared from these materials being further reacted eventually were fitted out with the appropriate substituents for attaching the invertase. A broad range of active membranes can be offered.

The adsorption of a protein on a synthetic membrane occurs by hydrophobic forces. The most hydrophobic polymer used in these investigations was PSU. All chemical reactions performed using PSU reduced the hydrophobicity. Regarding the enzyme adsorption, the leading membrane in this sector is the PSU. This, however, does not prove superior quality since the enzyme's activity becomes higher with increasing hydrophilicity. For instance, the effectiveness factor of the more hydrophilic PSU-OH(tert) is about 25% higher. The mass of invertase adsorbed reaches only 85% of that of the PSU. For all the membranes used for adsorptive bonding it pointed out that their catalytic activity is lowest compared with the membranes that attach the enzyme by ionic interaction or covalent bonding. Also the duration of fixation is not very pronounced. The release by desorption is remarkable.

Binding an enzyme by ionic interactions seems to be a really promising method. The uptake of invertase is reasonably high. Furthermore, the reactivity of the enzyme is the highest throughout the three different fixation modes. Obviously, ionic bonding does not interfere with the enzyme structure. Introducing hydrophilic spacers increases the distance of the enzyme and the polymeric backbone which also is advantageously to keep the enzyme's activity. The stability of the enzyme attached to the membrane lasts longer than in the adsorptive case, but is not as good as in the case of covalent bonding.

In this category the spectrum of differently substituted membranes was largest. Very reactive groups famed for their good binding capability like -$N_2^+$, cyanuric chloride, and carbodiimide and also less commonly used ones were introduced on the membranes. Yet, those membranes that rank first with respect to the enzyme uptake took a lower position regarding the activity. Once more the length of the spacer plays a dominant role. Fig. 8 presents the experimental findings of enzyme uptake and enzymatic reactivity related to the $k_{cat}$. Covalent bonding is the best method for stabilizing the enzyme.

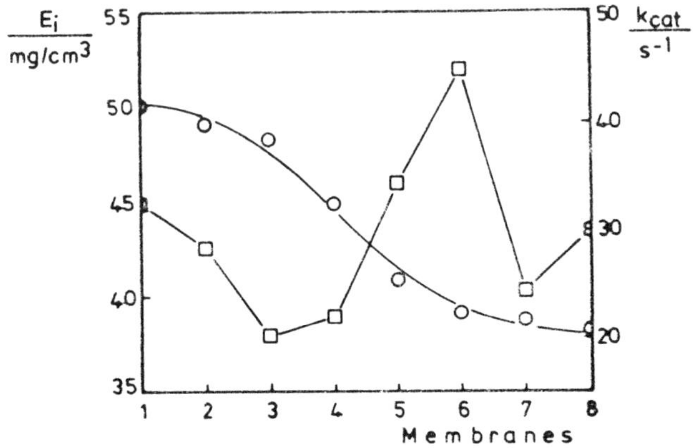


Fig. 8. Mass of invertase and reactivity ($k_{cat}$) for membranes with covalent bonding groups. Membranes: 1: PSU-Cya2; 2: PSU-Cya1; 3: PSU-$N_2^+$; 4: PSU-CDI; 5: EPSU; 6: EPSU-CHO; 7: CPSU; 8: CPSU-CHO.

The enzyme membrane reactor was realized using the EPSU membrane mounted in an ultrafiltration apparatus. The feed contained inert dextran that enabled to investigate the influence of fouling. It pointed out that the kinetics were not influenced, but the residence time increased because of the additional transmembrane mass transfer resistance. On the other hand, this yielded a higer conversion.

## REFERENCES


Alfani, F., Cantarella, M., and Greco, G., 1979, Theoretical and experimental analysis of a soluble enzyme membrane reactor, *Biotechnol. Bioeng.* 21:1421.

Barker, S.A., and Burns, R.F., 1973, Reaktor separators incorporating membrane bound enzymes, *Chem. Ind.* 18:801.

Basheer, S., Kut, O.M., Prenosil, J.E., and Bourne, J.R., 1993, Development of an enzyme membrane reactor for treatment of cyanide-containing wastewaters from the food industry, *Biotechnol. Bioeng.* 41:465.

Bradford, M.M., 1976, A rapid and sensitive method for the quantitation of microgram quantities of protein utilizing the principle of protein-dye binding, *Anal. Biochem.* 72:248.

Butterworth, T.A., Wang, D.J C., and Sinskey, A.J., 1970, Application of ultrafiltration for enzyme retention during continuous enzymatic reaction, *Biotechnol. Bioeng.* 12:615.

Caplan, S.R., Goldman, R., Silman, H.J., Kedem, O., and Katchalski, E., 1965, Papain membrane on a collodion matrix: preparation and enzymic behavior, *Science* 150:758.

Ceynowa, J., and Adamczak, P., 1992, Enzyme membrane based upon polyamide-6 for oil hydrolysis, *J. Appl. Polym. Sci.* 46:749.

Cheryan, M., 1986, "Ultrafiltration Handbook", Technomic Publishing Comp. Inc. Lancaster, PA.

Coulet, P.R., 1992, Polymeric membranes and coupled enzymes in the design of biosensors, *J. Membr. Sci.* 68:217.

Düputell, D., Staude, E., and Wyszynski, D., 1994, Sorption of human albumin and endotoxin on uncharged and charged ultrafiltration membranes made from polysulfone and polyethersulfone, *Desalination* 95:75.

Gille, M., and Staude, E., 1994, Symmetric enzyme distribution in asymmetric UF polysulfone membranes, *Biotechnol. Bioeng.* 44:557.
Guiver, M.D., and Apsimon, J.W., 1988, The modification of polysulfone by metalation, *J. Polym. Sci., Polym. Lett.* 26:123.
Hartmeier, W., 1986, "Immobilisierte Biokatalysatoren", Springer Verlag, Berlin, Heidelberg, New York, Tokio.
Hinke, E., and Staude, E., 1991, Streaming potential of microporous membranes made from homogeneously functionalized polysulfone, *J. Appl. Polym. Sci.* 42:2951.
Ishimori, Y., Karube, I., and Suzuki, S., 1981, Mechanical control of the activity of glucose oxidase immobilized on porous polyvinylchloride membrane, *Biotechnol. Bioeng.* 23:2601.
Korus, R.A., and Olson, A.C., 1977, The use of α-galactosidase and invertase in hollow fiber reactors, *Biotechnol. Bioeng.* 19:1.
Mita, D.G., Pecorella, M.A., Russo, P., Rossi, S., Canaglia, P., and Gaffete, F.S., 1993, Effect of temperature gradients on the activity of enzymes immobilized on natural polymeric matrices, *J. Membr. Sci.* 78:69.
Noshay, A., and Robeson, L.M., 1976, Sulfonated polysulfone, *J. Appl. Polym. Sci.* 20:1885.
Nyström, M., Pihlajamäki, A., Ehsani, N., 1994, Characterization of ultrafiltration membranes by simultaneous streaming potential and flux measurements, *J. Membr. Sci.* 87:245.
Pronk, W., Kerkhof, P.J.A.M., van Helden, C., and van't Riet, K., 1988, The hydrolysis of triglycerides by immobilized lipase in a hydrophilic membrane reactor, *Biotechnol. Bioeng.* 32:512.
Rodemann, K. and Staude, E., 1995, Preparation and characterization of porous polysulfone membranes with spacer-bonded N-containing groups, *J. Appl. Polym. Sci.* in press.
Rodemann, K., and Staude, E., 1994, Synthesis and characterization of affinity membranes made from polysulfone, *J. Membr. Sci.* 88:271.
Schmidt-Steffen, A., and Staude, E., 1992, Ultrafiltration membranes for chemical bonding of urease, *Biotechnol. Bioeng.* 39:725.
Schwarz, S., Jacobasch, H.-J., Wyszynski, D., and Staude, E., 1994, Electrokinetic measurements on porous polysulfone membranes, *Angew. Makromol. Chem.* 221:165.
Staude, E., 1992, "Membranen und Membranprozesse", VCH, Weinheim, New York, Basel, Cambridge.
Warshawsky, A., Deshe, A., and Gutman, N., 1984, Safe halomethylation of aromatic polymers via BCME free long chain haloalkylethers, *Br. Polym. J.* 16:234.
Warshawsky, A., Kahana, V., Deshe, A., Gottlieb, H.E., and Arad-Yellin, R., 1990, Halomethylated polysulfone: reactive intermediates to neutral and ionic film-forming polymers, *J. Polym. Sci. A* 28:2885.
Watanabe, E., Takagi, M., Takei, S., Hoshi, M., and Cao, S.-G., 1991, Development of biosensors for the simultaneous determination of sucrose and glucose, lactose and glucose, and starch and glucose, *Biotechnol. Bioeng.* 38:99.
Wichmann, R., Wandrey, Ch., Bückmann, A.F., and Kula, M.-R., 1981, Continuous enzymatic transformation in an enzyme membrane reactor with simultaneous NAD(H) regeneration, *Biotechnol. Bioeng.* 23:2789.
Wu, D.-R., Steven, M.C., and Belfort G., 1993, Kinetic resolution of racemic glycidyl butyrate using a multiphase membrane enzyme reactor: experiments and model verification, *Biotechnol.Bioeng.* 41:979.

# THERMO-INDUCED PHASE TRANSITIONS AND REGULATION OF PERMEABILITY OF BIOMIMETIC MEMBRANES

Nikolai M. Kocherginsky[1], Ke Jian Liu[2], and Harold M. Swartz[2]

[1] 1102 S. Goodwin Ave., University of Illinois, Urbana, IL 61801
[2] 308 Strasenburgh Hall, Dartmouth Medical Center, Hanover, NH 03755

For many years membrane science was essentially subdivided into two major fields: artificial membranes used mainly in separation/purification technology and biological membranes. During the last ten years however it has become evident that artificial biofunctional and biomimetic membranes, containing immobilized biological components and/or are created on the basis of principles of native biomembranes[1,2] can have very similar properties. This provides an excellent avenue to use the properties of biomembranes to develop more effective artificial membranes for separation/purification technology and to use artificial membranes to elucidate the structure and function of native membranes.

Previously we have demonstrated that nitrocellulose microfilters, impregnated with fatty acids or their esters, can be used as simple and very effective biomimetic membranes.[3,4] This conclusion is based on the following facts:

1. This membrane is a barrier with quantitatively similar selective permeability for:
a) gases, including respiratory gases oxygen and carbon dioxide,[5]
b) water and nonelectrolytes with different hydrophobicity,[6,7]
c) inorganic ions.[8]

2. Different properties (per unit of thickness) were similar to those of biomembranes, including for example:
a) electrical resistivity and specific electrical capacitance,[9]
b) permeability of fatty acids.[10,11]

3. Some other barrier properties, such as activation energy for water transport,[6] transmembrane electrical potential,[4] frequency-dependent electrical properties (impedance) also were similar to the properties of biomembranes.[9]

4. It was possible to imitate three major kinetic mechanisms of transport through biomembranes, i.e.
a) direct diffusion through the lipid phase,[5,7]
b) facilitated transport with different low-molecular carriers, such as fatty acids, analogs of phosphatidic acid, and quinones in the lipid phase,[11-14]

*Biofunctional Membranes*, Edited by
D.A. Butterfield, Plenum Press, New York, 1996

c) facilitated transport through aqueous channels with ion-exchanging groups.[6,8]

5. It was also possible to imitate two major thermodynamic mechanisms of transmembrane transport, i.e.

a) passive transport with the concentration gradient,[5-7]

b) active transport against of the concentration gradient.[10-13]

The source of energy in this last case was either another chemical substance or a physical factor such as light.[14]

6. Facilitated coupled transport of ions could proceed by either co- or counter-transport of ions with an opposite or the same charge.[10-14]

7. It was possible to get a transmembrane transport of $H^+$-ions due to the difference of transmembrane redox-potentials. Some aspects of the process were similar to the active $H^+$ - transport due to mitochondrial respiration.[13]

8. The rate of transport could be regulated by:

a) specific mechanisms such as membrane coupling,[10-14]

b) chemical factors, including organic solvents, detergents, $Ca^{2+}$, $H^+$,

c) physical factors, such as light,[14]

d) drugs, including some antiviral, antidepressants, and antibiotics.[15-17]

9. Some surface-active herbicides were able to induce self-sustained oscillations of the transmembrane electric potential in a $K^+/Na^+$ gradient. It worth it to note that these oscillations not only imitated some aspects of biological excitability, but were able to do so without any protein in the biomimetic membrane. They were determined by herbicide-induced local phase transition liquid crystal / micelles.[18]

10. Cooling-induced phase transitions of free fatty acids in these biomimetic membranes also could be used to regulate the processes of direct and facilitated transport. This transition inhibited transmembrane transport of hydrophobic nonelectrolytes and the gases $O_2$, $N_2$, and $CO_2$. The same phase transition resulted in an increase of transport of small hydrophilic ions due to the formation of microcracks, filled with aqueous solutions, and similar to pores in biomembranes.[19,20] Electrical resistivity of the system before freezing was ~ Mohm*cm$^2$ and decreased by two-three orders of magnitude after the phase transition. Temperature dependence of this electrical parameter showed hysteresis in multiple cycles of freezing-thawing, and its maximal value decreased by 10-20% after every cycle. If the temperature was kept at the point of the fatty acid phase transition, it was possible to observe fluctuations of transmembrane electric current (unpublished results).

Due to the macroscopic size and stability of biomimetic membranes the mechanism of the phase transition of these fatty acids and the influence of polymeric matrix can be easily investigated by different methods, including EPR and differential scanning calorimetry (DSC). Some illustrative results are given below.

EPR measurements were carried on using Varian-9 X-band spectrometer E-line Century Series. Temperature controller accuracy was 0.2 K. Spin probes were purchased from "Molecular Probe" company. For DSC measurements we used a Perkin-Elmer 7 Series Thermal Analysis System. The scanning rate was 2.0 K/min. Typical weight of Millipore nitrocellulose filter before and after impregnation was 0.2 mg and 0.55 mg, respectively.

7- Doxylstearic acid revealed a sharp phase transition in the bulk decanoic (capric) acid, which was much broader when the decanoic acid was immobilized in the porous polymeric support (Figures 1,2). The matrix effect was stronger with smaller pore diameters ( 0.25 μm instead of 2.5 μm).

DSC demonstrated existence of a new subclass of decanoic acid molecules in the pores (Figures 3,4). This subclass had a phase transition temperature, that was 1.5 degrees lower and possibly was able to concentrate a spin-labeled fatty acid, making the transition, determined by EPR, more gradual. Again the effect of the matrix was stronger with smaller pore diameters (Figure 5).

EPR spectra of the positively charged spin label N,N-dimethyl-N-dodecylamino-2,2,6,6-tetramethylpiperidine-1-oxyl at temperatures below phase transition could be described as a superposition of solid and liquid phase spectra (Figures 6,7), thus again supporting the idea of clustering of lipid-like substances in pores and a liquidifying effect of the nitrocellulose polymeric matrix. Temperature dependencies of both line width and intensity of EPR spectra of this probe also demonstrated a broad transition , which was different in a bilk phase and in the membrane, where it was possibly complicated also by adsorption of nitroxides on the pore surface (Figures 8, 9). Similar effects could be induced by polypeptide polymeric chains of proteins in biomembranes.

If tridecanoic or palmitic acid or a mixture of fatty acids was used for impregnation, the transition was broader than in the case of shorter decanoic acid ($C_{10}$), and the influence of the matrix in this case was less evident.

It is noteworthy that quantitative analysis of EPR spectra even for a relatively simple system like biomimetic membrane may be a serious problem. For example the radical fragment localized in hydrophobic region can be nonsensitive to the phase transition in the polar regions of membrane, and changes of EPR spectra in this case could be determined by freezing-out of nitroxides and by a local increase of its concentration at lower temperatures.

A model describing the biomimetic membrane structure is presented in Figure 10. It can be seen that the pores of the filter are filled by different liquid crystal structures, formed by the fatty acids or their esters, and the structures at the pore surface are different in comparison to the structures in the pore volume. The pore walls are covered by thin layers of water, which correspond to water channels of biomembranes. Carboxylic ion-exchanging groups, which are present as impurities on the nitrocellulose polymeric matrix, result in cation / anion selectivity of these channels. Direct measurements of water and carboxylic group content in the membrane together with knowledge of membrane porosity and pore diameter allowed us to estimate the distance from one carboxylic group to another. Though the thickness of water layers ( 25-50 ) was somewhat higher than the typical effective diameter of channels in biomembranes, the distance between ion-exchanging groups (20-25 ) was similar to that in biological channels. We believe that this is due to the coexistence of lipid-like liquid crystal media with low dielectric constant and water channels with ion-exchanging groups in them, resulting in a similarity to biomembranes. The properties of water channels can be easily modified, for example by their pretreatment in different solutions before impregnation. The method could be used for enzyme immobilization and implementing a specific catalytic function of biomimetic membranes.

Aqueous channels are absent in the usual nonmodified bilayer lipid membranes and liposomes, and special channel-forming molecules, usually proteins, need to be inserted in these membranes to initiate the channel transport in biomembranes. The nature of these channels is still obscure, and one of the possibilities is that some channels are induced by protein-lipid interactions and distortion of ideal liquid-crystal structure of lipids in contact with proteins. Modification of proteins with physiologically active substances could change the content and state of these defects in the bilayer liquid crystal structure. We can speculate that when and if the defects are big enough to be filled with water, the processes of this type could be equivalent to the opening and closing of channels in biomembranes.

We hope that biomimetic membranes described in this paper, due to their macroscopic dimensions and relative stability could be used in different devices, including membrane sensors and temperature-sensitive drug delivery vehicles, in different membrane reactors, and also as model membranes for investigations of fundamental mechanisms of biological membrane transport and its regulation.

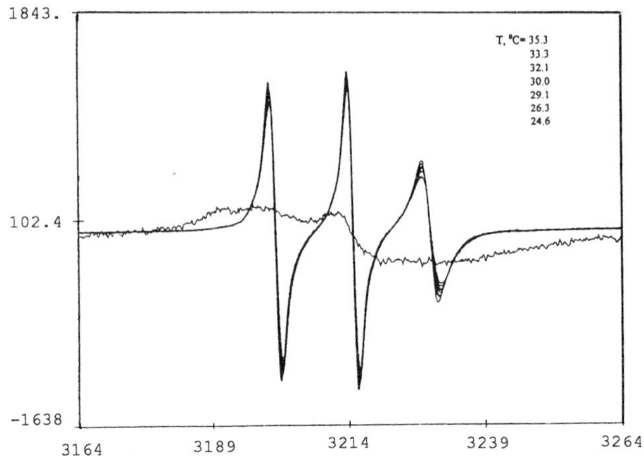


Figure 1. Typical EPR spectra of 7- doxylstearic acid in bulk phase of decanoic acid at different temperatures.

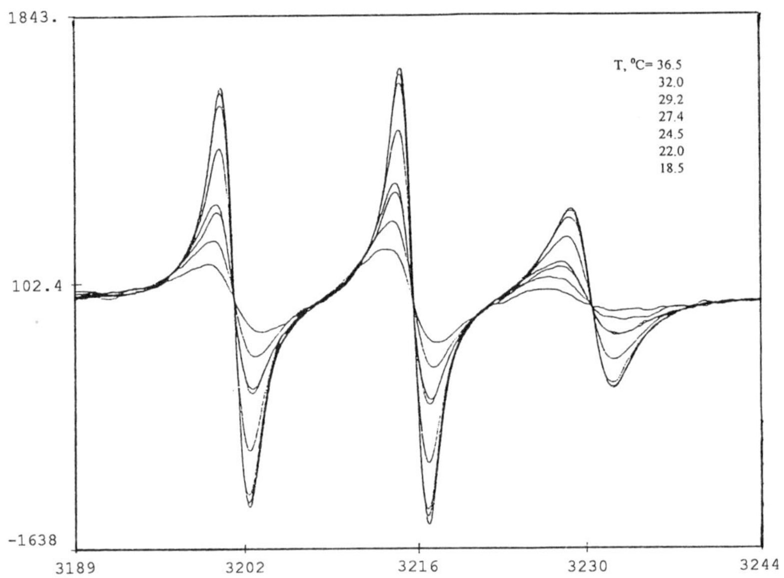


Figure 2. Typical EPR spectra of 7-doxylstearic acid in a biomimetic membrane impregnated with decanoic acid at different temperatures. Note that in the biomimetic membrane the spectrum at 18.5 $^0$C is more mobile than at 24.6 $^0$C in the pure bulk phase volume (Figure 1).

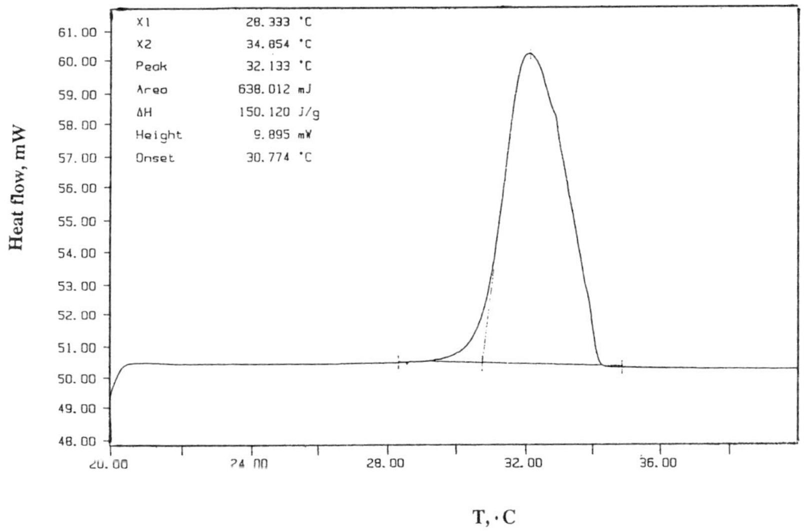


Figure 3. Differential scanning calorimetry of bulk decanoic acid.

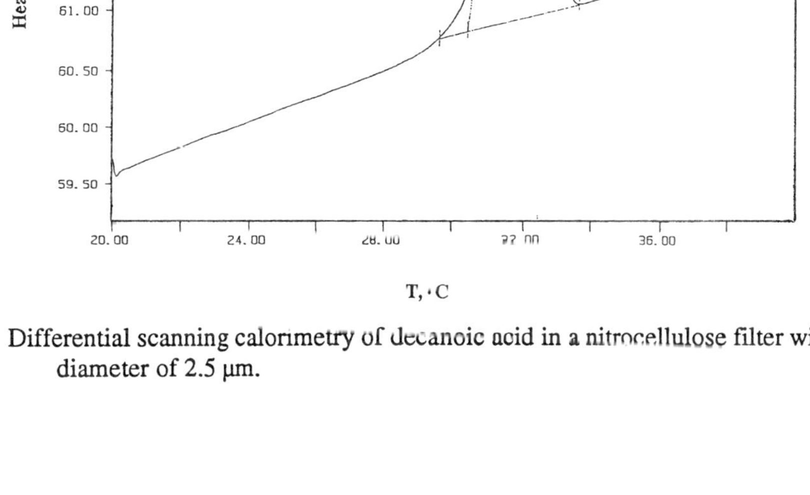


Figure 4. Differential scanning calorimetry of decanoic acid in a nitrocellulose filter with a pore diameter of 2.5 μm.

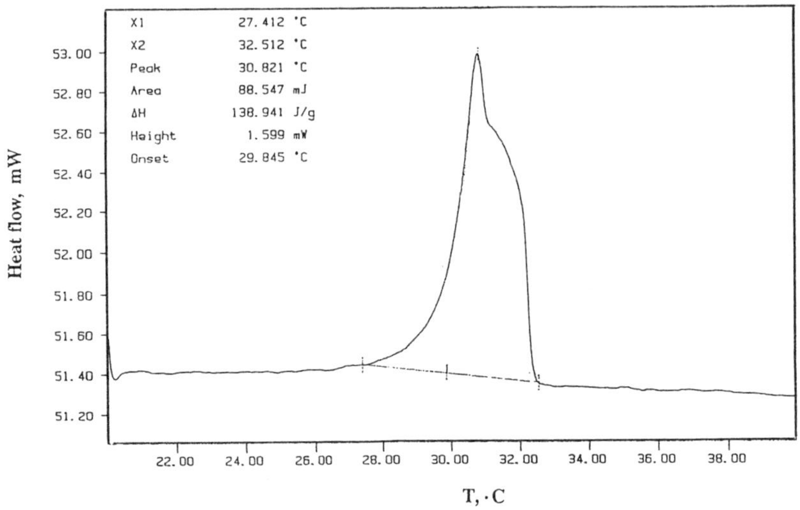


Figure 5. Differential scanning calorimetry of decanoic acid in a nitrocellulose filter with a pore diameter of 0.25 μm.

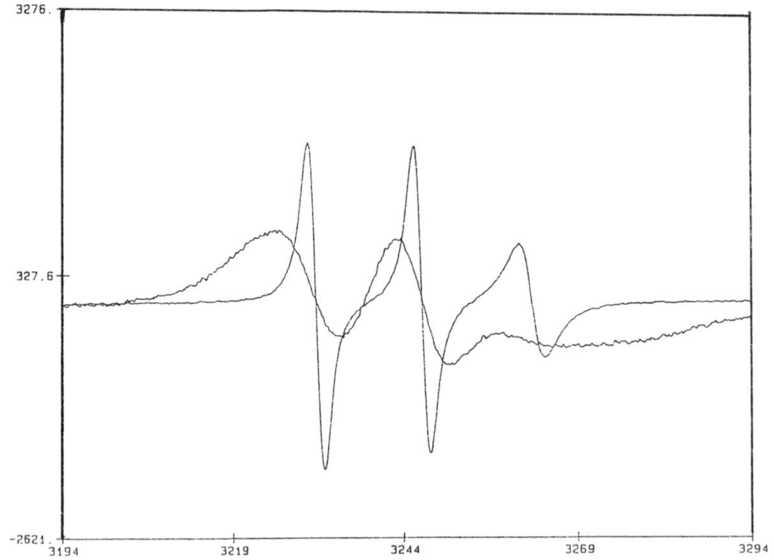


Figure 6. EPR spectra of 4-(N,N-dimethyl-N-dodecyl)ammonium-2,2,6,6-tetramethylpiperidine-1-oxyl in bulk decanoic acid above (40 $^0$C) and below (22.$^0$C ) the phase transition.

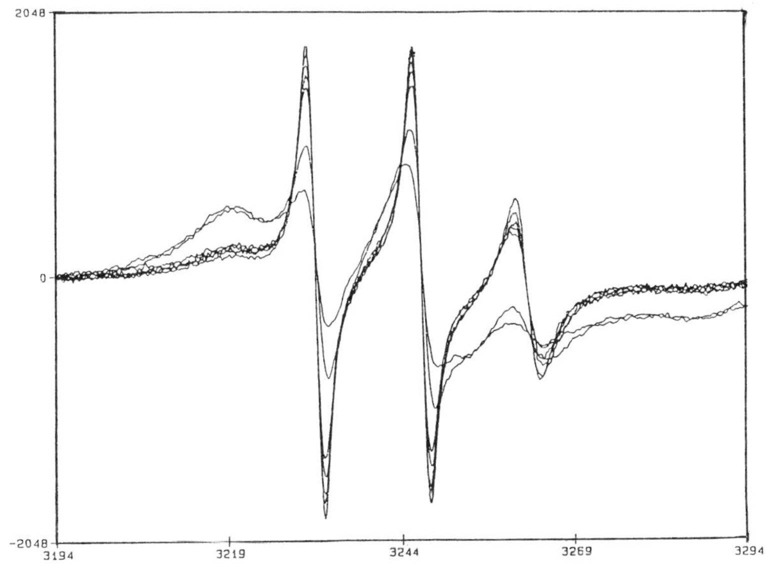


Figure 7. Typical EPR spectra of 4-(N,N-dimethyl-N-dodecyl)ammonium-2,2,6,6-tetramethylpiperidine -1-oxyl with decanoic acid in a biomimetic membrane above and below the phase transition.

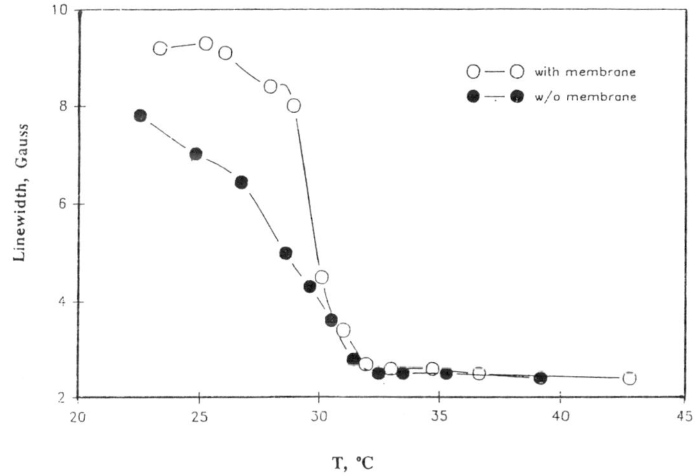


Figure 8. Temperature-dependence of line width of EPR spectra of 4 (N,N-dimethyl-N-dodecyl) ammonium-2,2,6,6-tetramethylpiperidine-1-oxyl in bulk and membrane-supported decanoic acid.

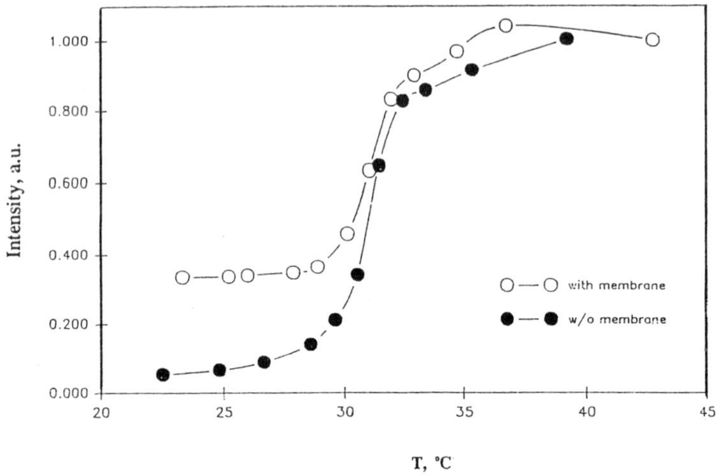


Figure 9. Temperature-dependence of intensity of the first component of EPR spectra of 4-(N,N-dimethyl-N-dodecyl) ammonium-2,2,6,6-tetramethylpiperidine-1-oxyl in bulk and membrane-supported decanoic acid.

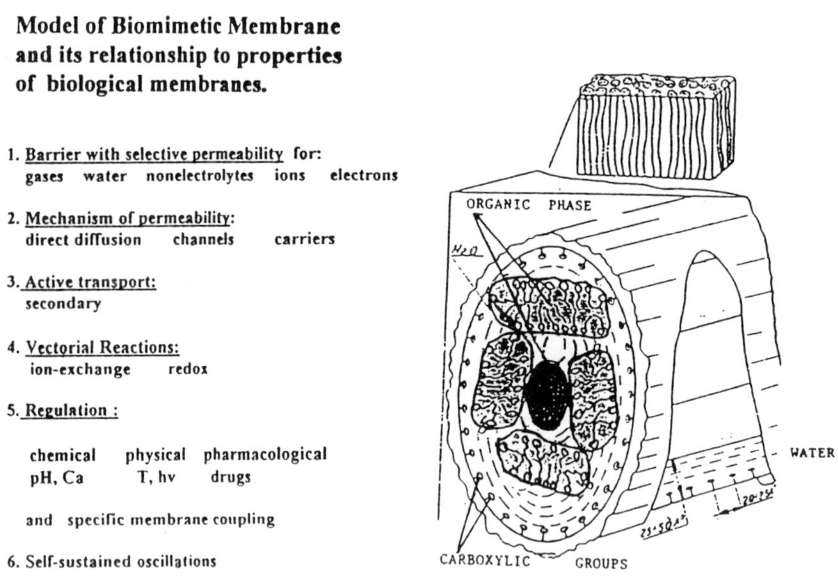


Figure 10. Short summary of properties of biological membranes which can be imitated with biomimetic membranes and scheme of the impregnated pore structure in a biomimetic membrane.

## ACKNOWLEDGEMENT

With deep acknowledgment to Miller Brewing Company for financial support of part of this work.


## REFERENCES


1. D.A. Butterfield." Biological and Synthetic Membranes," Wiley/Liss, N.Y. (1989).
2. J.H. Fendler."Membrane-Mimetic Approach to Advanced Materials," Springer-Verlag, Berlin (1994).
3. N.M. Kocherginsky, Yu. Sh. Moshkovsky, I.S. Osak, L.A. Piruzian, The model of biological membrane for the investigation of biologically active compounds. Author Certificate (Patent) USSR N# 1043564 (1983).
4. N.M. Kocherginsky, I.S. Osak, Yu. Sh. Moshkovsky, Impregnated filters as a model of biological membranes, *Biofizika* 29: 624 (1984).
5. N.M. Kocherginsky, L.E. Bromberg, G.S. Leskin, Permeability of impregnated filters to oxygen and carbon dioxide, *Russ. J. Phys. Chem.* 6: 838 (1987).
6. N.M. Kocherginsky, L.E. Bromberg, I.S. Osak, Yu. Sh. Moshkovsky, Yu.M. Popkov, Ya. B. Skuratnik, Simulation of water transport through biomembranes on impregnated filters, *Biological Membranes* 1: 1734 ( 1985).
7. N.M. Kocherginsky, L.E. Bromberg, Transport of non-electrolytes through biomimetic membranes, *Russ. J. Phys. Chem.* 62: 1112 (1988).
8. N.M. Kocherginsky , I.S. Osak, Generation of membrane electric potential on ultrafilters impregnated with fatty acids, *Russ. J. Phys. Chem.* 61: 1018 (1987).
9. N.M. Kocherginsky, I.S. Osak, L.E. Bromberg, V.A. Karyagin, G.S. Leskin, Yu. Sh. Moshkovsky, The modeling of biological membrane properties by means of filters impregnated with lipid-like substances, *J. Membrane Sci.* 30: 39 (1987).
10. N.M. Kocherginsky, I.S. Osak, Electroneutral transfer of monovalent cations with fatty acids, *Biofizika* 33: 83 (1988).
11. N.M. Kocherginsky, I.S. Osak, V.V. Deomochkin, V.L. Rubaylo, Physico-chemical mechanism of ionoforic activity of fatty acids as stimulants of transmembrane monovalent cation exchange, *Biological Membranes* 4: 838 (1987).
12. N.M. Kocherginsky, J.W. Stucki, Facilitated transport of alkaline and alkaline earth metals through liquid membranes with acidic extractants, in: 209 ACS National Meeting, Abstracts of Papers, I&IC 069, Anaheim, CA (1995).
13. N.M. Kocherginsky, M.G. Goldfeld, I.S. Osak, Electron-proton coupled transport across biomimetic polymer-supported liquid membranes, *J. Membrane Sci.* 45: 85 (1989).
14. N.M. Kocherginsky, M.G. Goldfeld, I.S. Osak, Photo-stimulated coupled transport of electrons and protons across quinone-doped liquid polymer-supported biomimetic membrane, *J. Membrane Sci.* 59: 1 (1991).
15. N.M. Kocherginsky, I.S. Osak, S.G. Tul'kes, Yu. Sh. Moshkovsky , The ion-exchange interaction of rimantadine with ultrafilters impregnated with lipid analogues, *Russ. J. Phys. Chem.* 60: 1367 (1986).

16. N.M. Kocherginsky, I.S. Osak, Yu. Sh. Moshkovsky, Physico-chemical study of the membranotoxic effects of chlorpromazine and of a group of antidepressants, *Russ. J. Phys. Chem.* 60: 1516 (1986).
17. N.M. Kocherginsky, P.N. Koroliov, V.A. Karyagin, E.V. Krasnokutskaya, A.N. Polin, V.G. Bulgakova, Investigation of physico-chemical mechanisms of gramicidin S action on model membranes, *Antibiotics and Chemotherapeutics USSR* 33: 502 (1988).
18. M. Podolak, N.M. Kocherginsky, I.S. Osak, S. Przestalski , S. Witek, Fluctuations of biomimetic membranes potential and membrane lysis, induced by surface - active glycine esters, *J. Membrane Sci.* 66: 143 (1992).
19. N.M. Kocherginsky, L.E. Bromberg, Modeling the properties of biomembranes with nitrocellulose ultrafilters impregnated with fatty acids: Effect of thermally induced phase transitions on transport of water, *Russ. J. Phys. Chem.* 62: 95 (1988).
20. N.M. Kocherginsky, L.E. Bromberg, Effect of thermally induced phase transitions on the permeability of ultrafilters impregnated with lipid-like substances, *Russ. J. Phys. Chem.* 61: 1341 (1987).

# OPTIMAL pH CONTROL IN SEQUENTIAL BIOCHEMICAL REACTION SYSTEMS VIA GENERATING A pH GRADIENT ACROSS AN IMMOBILIZED ENZYME FILM

G. Chen, J.P. Byers, R. L. Fournier, and S. Varanasi

Department of Chemical Engineering
The University of Toledo
Toledo, Ohio 43606

## ABSTRACT

Multistep synthesis reactions form the basis for the manufacture of a variety of important biochemicals. However, in many cases, the optimal operating pH is much different for each individual step of these reactions. Even so, conducting all or some of these reaction steps in the same reactor often may have technical (and/or economic) benefits, as when one of the intermediate steps has a very low product to reactant ratio at equilibrium. When it is required to perform these sequential reaction steps in the same vessel, the usual method is to use a compromise pH. However, this approach is neither satisfactory nor feasible always because the rates of most enzyme-catalyzed reactions decay exponentially as one drifts away from the enzyme's optimal pH. Thus, a technique which permits one to conduct the individual steps at their respective optimal pH and, yet, in close proximity to one another is extremely desirable. We recently proposed such a technique: a method for conducting, in a single reactor, the sequential reaction $A \Leftrightarrow B \rightarrow C$. The first step is an enzyme-catalyzed reaction that takes place in a basic environment and the second step is either an enzyme-catalyzed reaction or a fermentation that occurs in an acidic environment. The approach suggested was to coat the pellets, within which the enzyme catalyzing the $A \Leftrightarrow B$ step was immobilized, with a very thin ($< 1\mu m$) layer of urease, before suspending the pellets in an acidic bulk solution in which the step $B \rightarrow C$ takes place. Now, adding a small concentration of urea to the bulk solution allows one to sustain the necessary pH difference between the bulk solution and the interior of the pellets, thereby enabling the two steps to take place in close proximity and also at their respective optimal pH. The objective of this paper is

*Biofunctional Membranes*, Edited by
D.A. Butterfield, Plenum Press, New York, 1996

to demonstrate the feasibilty of the proposed approach. First, using batch and continuous experiments, it was shown that, when a basic environment (pH ≥ 7.5) was separated from an acidic environment (pH ≤ 4.5) by a very thin immobilized urease membrane, one can sustain the pH difference between the two environments (even in the presence of a buffer) when only a small concentration of urea (~0.01M) was maintained in the acidic environment. Next, the usefulness of this approach in the production of ethanol from xylose by the scheme:

$$\text{xylose} \underset{\text{xylose isomerase}}{\overset{\text{pH}=7.5}{\Leftrightarrow}} \text{xylulose} \underset{\text{fermentation}}{\overset{\text{pH}=4.5}{\longrightarrow}} \text{ethanol}$$

was established by demonstrating that high conversion of xylose to xylulose can be achieved when immobilized xylose isomerase pellets coated with an egg-shell layer of urease are suspended in a acidic bulk solution (pH 4.5) containing xylose and a small concentration of urea.

## 1. INTRODUCTION

Multi-step synthesis reactions form the basis for the production of a variety of chemicals. Numerous examples can easily be cited, from the production of bulk quantities of commodity chemicals such as ethylene glycol, to the much smaller quantities characteristic of pharmaceutical drug manufacture. In many of these multistep reaction sequences it may be of technical or economic benefit to combine or pair some of the reaction steps. For example, a particular reaction may be encumbered by low equilibrium conversion. There may also be process efficiency gains that lead to better economics if some of the reaction steps are paired up. Whatever the reasons, be they equilibrium constraints, technical, or economic, reaction pairing requires that two reactions in the synthesis pathway occur *simultaneously and in close proximity* to each other.

Despite its inherent advantages reaction pairing becomes difficult to implement if the conditions (temperature, pressure, pH, etc.) under which each reaction takes place are quite different from one another. Of particular interest in this paper are biosynthetic processes since they enjoy some advantages over traditional organic syntheses and because of this they are receiving increased attention. Biosynthetic steps in the reaction pathway would provide for greater specificity and greater yields of optically pure products for example. It is a well-known fact that the catalytic activity of an enzyme is limited to a narrow range of pH and that a definite pH optimum exists. The problem addressed in this paper focuses then on sequential biochemical reactions that may benefit by reaction pairing, however the optimum pH for each step in the reaction pair differs by several pH units. At first glance it appears as if the problem of maintaining separate pH environments in close proximity would present problems that would outweigh any potential benefits of reaction pairing for these particular reactions. Clearly one could envision performing the two reactions at a compromised pH and this has been a common approach described in the literature.[5, 6, 8, 9, 14, 15] However, we have developed a technology[1-3] that allows one to control and maintain the pH in the microenvironment where the reactions are occurring. This effectively allows each reaction in the pair to occur simultaneously and in close proximity to one another and at their respective optimal pH.

The basic concept of this approach is shown in Figure 1. Here we illustrate the concept for a stirred tank reactor, however the concept is applicable to a variety of reactor configurations. Consider a sequence of two biochemical reactions, for example $A \Leftrightarrow B \longrightarrow C$. The goal is to create a situation that allows each of the two reaction steps to take place simultaneously and at their respective optimal pHs. To facilitate the discussion let the first reaction, $A \Leftrightarrow B$, have an optimal pH in the basic range, and the second reaction, $B \longrightarrow C$, have an optimal pH in the acidic range (which step is basic and which step is acidic is not important). The agents that catalyze each reaction step are shown in Figure 1 as "enclosed" numbers, i.e., an "enclosed 1" refers to the first reaction step and an "enclosed 2" refers to the second reaction step. The agent that catalyzes the second reaction prefers an acidic environment and this can be easily created by simply making the bulk solution have the pH that is optimal for step 2. It should also be pointed out that the agent that promotes the second reaction step may be cells, immobilized cells, a soluble enzyme, an immobilized enzyme, or possibly even a physical step such as crystallization. In the case of an immobilized agent that catalyzes step 2, the support must not significantly alter the internal pH environment from that existing in the bulk solution.

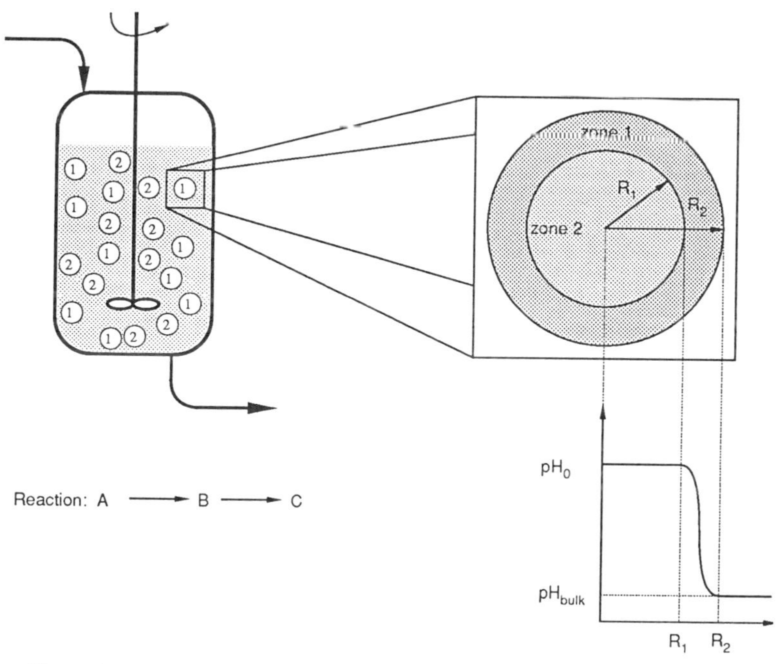


**Figure 1.** Concept of Optimal pH Control in Sequential Biochemical Reactions.

With the bulk being acidic to favor the second reaction step, we must now create a basic environment that is optimal for the agent that catalyzes the first reaction step. In this case the agent that catalyzes the first step must be immobilized and it could be an enzyme or cells. Clearly, $H^+$ ions will diffuse into these particles and attempt to diminish whatever difference in pH initially exists between these particles and the acidic bulk environment. However, by coimmobilizing in these particles a second enzyme that catalyzes a base generating reaction the $H^+$ ions that are diffusing in from the bulk can be neutralized. This is shown in the inset of Figure 1 where for convenience we define the zone I region of these particles to contain primarily the enzyme or cells that catalyze step 1. Surrounding zone I and in zone II is a thin layer containing for the most part the enzyme that catalyzes a base-generating reaction. An example of an ideal base generating reaction for zone II is the conversion of urea via the enzyme urease. When relatively small amounts of urea (< 10 mM[1]) are present in the acidic bulk solution two moles of ammonia would be produced per mole of urea consumed within the thin zone II layer containing the enzyme urease. The ammonia produced by this reaction would, in turn, neutralize the $H^+$ ions that are diffusing in from the acidic bulk solution. Thus a very sharp spatial pH gradient is generated within the particles that promote the first reaction step. By this means, it is possible to maintain optimal pH environments for the agents that catalyze each reaction step and allow these reactions to occur simultaneously and in close proximity. This approach can also be extended to include the situation where the bulk environment is basic and an acid-generating reaction occurs in zone II to maintain an acidic pH in zone I.

We have identified several biochemical reaction systems that serve to illustrate the potential application of this approach. The first example system is for the conversion of xylose to ethanol [4–13] and illustrates an example where the two steps are an enzyme reaction followed by a fermentation. Xylose, in addition to glucose, is a major constituent sugar that arises from biomass. Many yeasts that are able to efficiently convert glucose to ethanol are unable to ferment xylose or can do so only to a limited extent.[4] One particular approach that has attracted much interest is to convert the xylose to its ketose isomer, xylulose, through the action of the enzyme xylose isomerase.[5–7, 9] The same yeasts that can ferment glucose are also able to ferment xylulose to ethanol as well. Since the equilibrium conversion of xylose to xylulose is only about 20 %,[12] it is desirable to perform the xylose isomerization and xylulose fermentation simultaneously.[8] However, while the optimum pH for the isomerization is in the range of 7–8,[10] the fermentation is best carried out at a pH of 4–5. Many investigators have performed the simultaneous isomerization and fermentation at a compromised pH.[5–9] The concept illustrated in Figure 1 is readily applicable to the simultaneous isomerization and fermentation of xylose. In this case let the isomerization of xylose represent the first step ($A \Leftrightarrow B$) that prefers the basic environment. Commercially available immobilized xylose or glucose isomerase pellets (for example, Sweetzyme T ™ from Novo) can then be treated to have a thin egg shell layer of urease (see section 4 for preliminary data). These coimmobilized enzyme particles would then be suspended in the acidic fermentation broth that contains yeast. The yeast cells are now the agents that promote the second reaction step ($B \longrightarrow C$). Feed to this reactor system would then include xylose and the small amount of urea needed to operate the base-generating reaction in the immobilized xylose isomerase pellets. Xylose would diffuse from the acidic bulk solution across the thin layer of urease in zone II and be converted to xylulose in zone I. The isomerization reaction occurring in zone I would be at its optimal basic pH since any $H^+$ ions that diffused in from the acidic bulk solution would be neutralized by the base generating urea reaction that is taking place in zone II. Xylulose created in zone I then diffuses out into the bulk solution where it is converted by yeast at their optimal acidic pH into alcohol. In this manner both the isomerization and fermentation reactions can occur simultaneously and at their respective optimal pH.

This approach has also great appeal when applied to synthetic reaction pathways that employ enzymatic reactions in those reaction steps requiring high specificity and optically pure yields. An excellent example of a sequential enzyme synthesis that could benefit by this approach of optimizing the pH environments was recently described in the literature by Moo-Young et al.[14, 15] They described a 2 step enzymatic synthesis starting with pyruvate and catechol to form the amino acid 3,4-dihydroxyphenyl-L-alanine (L-dopa) followed by the conversion of L-dopa to form the catecholamine dopamine. This reaction pathway is shown in Figure 2. As pointed out by Moo-Young et al. [14] L-dopa is a basic building block for the synthetic production of morphinan alkaloids (eg., morphine, codeine), it is also a precursor for melanin (skin pigment), and it is used to treat Parkinson's disease (levodopa). Dopamine is a mammalian neurotransmitter and is a precursor to such hormones as norepinephrine and epinephrine (adrenalin).

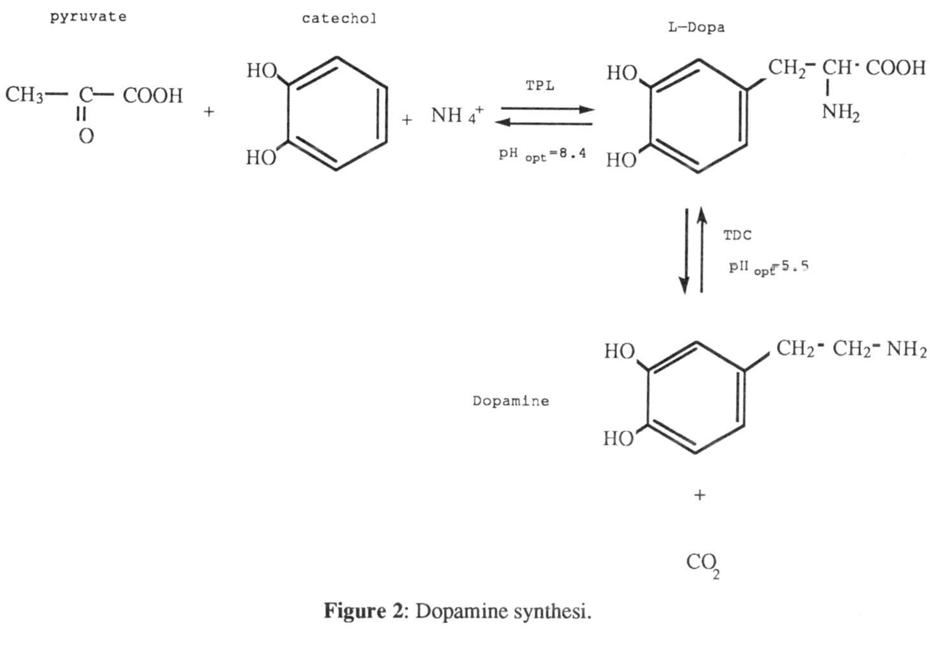


**Figure 2**: Dopamine synthesi.

As shown in Figure 2 the reaction of pyruvate and catechol is catalyzed by the enzyme tyrosine phenol lyase (TPL). The optimal pH for this reaction is 8.4.[14] This reaction is also reversible so reaction–pairing with the subsequent reaction where L-dopa is converted to dopamine would be advantageous to increase the yield. Excess substrate concentrations to drive the first step to L-dopa is not desirable since catechol at high concentrations is inhibitory to TPL and pyruvate is expensive.[14] In the second step L-dopa is irreversibly converted to dopamine through action of the enzyme tyrosine decarboxylase (TDC). The optimal pH for this step is reported to be 5.5.

Clearly, the above reaction scheme for dopamine synthesis is an ideal candidate for optimizing the pH environments of the individual steps. Here also a reactor configuration similar to that shown in Figure 1 can be employed. In this case the bulk solution would be maintained at a pH of 5.5 which is optimal for the TDC enzyme that catalyzes the second reaction step. The TDC could be immobilized or simply be in solution. TPL which catalyzes the first reaction step would be coimmobilized with a thin egg-shell layer of urease. The reactants to this system would include pyruvate, catechol, and a source for ammonium, such as ammonium chloride, as well as urea which is used to control the pH within the immobilized TPL. In this manner, both reaction steps would occur simultaneously and in close proximity and at their respective optimal pH values. It is interesting to note that for this reaction system ammonium is required in the first reaction step and 2 moles of ammonium are produced for each mole of urea that is consumed by the urease coimmobilized with the TPL. For this set of reactions some of the needed ammonium may in fact be obtained by the urease reaction.

The two examples discussed serve to illustrate the impact the proposed technolgy could have on a wide variety of biochemical syntheses.

As a first step in the evaluation of optimal pH control we have developed a mathematical model that simulates the reaction diffusion phenomena occurring in the outer layer (zone II) of the immobilized pellet shown in the inset of Figure 1. Our focus at the time the model was developed was on the simultaneous isomerization and fermentation of xylose. However, the model is readily applicable to a variety of other reaction systems. The fundamental question addressed by this model was whether or not it is possible to obtain the desired pH gradient in the proposed immobilized enzyme system. In addition to being instrumental in identifying conditions that ensure the required pH gradient, the model also provides a deeper understanding of the processes at work in the urease reaction region of the pellet (zone II). This model is fully described in our recent paper[2] and the pertinent results are summarized in section 2 below. Guided by the model predictions, we have also performed both batch and continuous studies on a flat membrane containing immobilized urease that conclusively shows that a significant pH gradient can be sustained across a thin urease layer even in the presence of pH buffers that may be present in the bulk environment. These results are summarized in section 3. After showing that it is possible to generate a significant pH gradient within an immobilized enzyme pellet, we have also shown that it is possible to conduct a second enzyme reaction within the core (zone I) of the pellet at its optimal pH of about 7.5 under conditions where the bulk pH was maintained at a value of about 4.5. These results are reported in section 4. We, then, conclude the paper with a brief overview, in section 5, of the experiments that are presently being conducted, based on the experimental results discussed in sections 3 and 4.

## 2. THE THEORETICAL MODEL

Our major interest in the modeling effort was whether or not it was possible to generate a significant pH gradient across zone II of the pellet shown in Figure 1. Accordingly we can for the most part neglect the reactions occurring in zone I and in the bulk phase and focus on just the hydrolysis of urea and the generation of the pH gradient which occurs in zone II. In addition to the reactions involving the hydrolysis of urea, and the protonation/deprotonation of ammonia and carbonic acid, another reaction was also considered in zone II. WH represents a weak acid that can act as a buffer and may result from a buffer present in the bulk solution or an acidic product generated as a result of the primary reactions occurring within the bulk or in zone I. A weak acid provides an additional pathway by which protons can be transported into zone II. A summary of the reactions occurring in the process are shown in Table 1.

**Table 1**: Chemical Reactions Occuring in Zone II of the Pellet

| Reaction | Equilibrium constant | |
|---|---|---|
| $(NH_3)_2CO + 2H_2O \overset{urease}{\rightleftharpoons} 2NH_3 + H_2CO_3$ | | (a) |
| $NH_3 + H^+ \rightleftharpoons NH_4^+$ | $K_{NH_3} = 1.79x10^{9} M^{-1}$ | (b) |
| $H_2CO_3 \rightleftharpoons HCO_3^- + H^+$ | $K_{H_2CO_3} = 4.30x10^{-7} M$ | (c) |
| $HCO_3^- \rightleftharpoons CO_3^{2-} + H^+$ | $K_{HCO_3^-} = 5.61x10^{-11} M$ | (d) |
| $WH \rightleftharpoons W^- + H^+$ | | (e) |

where $WH$ denotes a buffer.

The dissociation of $HCO_3^-$ to $CO_3^{2-}$ is ignored since it only becomes appreciable when the pH is greater than about 8.5. We have shown[2] that the differential equations that describe the reaction diffusion phenomena in zone II can be reduced to the following two coupled transport equations: one for urea, and the other for hydrogen ions (the symbols are explained in the Nomenclature):

$$\frac{D}{r^2}\frac{d}{dr}\left(r^2\frac{dC_u}{dr}\right) = v_u \tag{1}$$

$$\frac{1}{r^2}\frac{d}{dr}\left(r^2 D_{app}\frac{dC_{H^+}}{dr}\right) = v_u \tag{2}$$

In this reduced system of equations, all modes of transport of the hydrogen ions across the region where urease is immobilized are incorporated in the single transport equation of this specie via the development of an apparent diffusivity, $D_{app}$, for the hydrogen ions [2, 20]. This approach greatly simplifies the model and significantly reduces the computational effort. The rate expression, $v_u$, depends on both the local pH and the concentration of urea and couples equations 1 and 2. For the modelling effort this rate expression was based on the work of Dixon et al.[19] who carried out an extensive study of the kinetics of urease over a pH range of 3.4 to 7.8. The rate expression for urea is then given by the following equation:

$$v_u \equiv k_1 C_{SEH_2^-} + k_2 C_{SEH^{2-}} - \frac{V_{max}C_u}{K_m + C_u} \tag{3}$$

where
$$V_{max} = \frac{k_1 e_0}{\frac{C_{H^+}}{K_{SEH_3}} + 1 + \frac{K_{SEH_2}}{C_{H^+}} + \frac{K_{SEH_2}K_{SEH}}{C_{H^+}^2}} + \frac{k_2 e_0}{\frac{C_{H^+}^2}{K_{SEH_3}K_{SEH_2}} + \frac{C_{H^+}}{K_{SEH_2}} + 1 + \frac{K_{SEH}}{C_{H^+}}} \tag{4}$$

Upon substituting the rate expression and the following dimensionless parameters: $\eta = \frac{r - R_1}{R_2 - R_1}$, $\overline{C}_{H^+} = \frac{C_{H^+}}{K_{SEH_2}}$, $\overline{C}_u = \frac{C_u}{K_m}$, $\chi_1 = \frac{K_{SEH_2}}{K_{SEH_3}}$, $\chi_2 = \frac{K_{SEH}}{K_{SEH_2}}$, $V^* = \frac{k_2}{k_1}$, $\overline{D}_{app} = \frac{D_{app}}{D}$, and $\lambda = \frac{K_m}{K_{SEH_2}}$, the following equation is obtained for the hydrogen ion concentration.

$$\frac{d^2\overline{C}_{H^+}}{d\eta^2} + \left[\left(\frac{1}{\overline{D}_{app}}\right)\left(\frac{d\overline{D}_{app}}{d\eta}\right) + \left(\frac{2}{\eta + \eta_0}\right)\right]\frac{d\overline{C}_{H^+}}{d\eta} = \Phi_T^2\left(\frac{\lambda}{\overline{D}_{app}}\right)\left[\frac{\overline{C}_{H^+}^2 + V^*\,\overline{C}_{H^+}}{\chi_1\,\overline{C}_{H^+}^3 + \overline{C}_{H^+}^2 + \overline{C}_{H^+} + \chi_2}\right]\left[\frac{\overline{C}_u}{1 + \overline{C}_u}\right] \tag{5}$$

and $\Phi_T$ is the Thiele modulus defined as follows:

$$\Phi_T^2 = \frac{(R_2 - R_1)^2\, k_1\, e_0}{D\, K_m} \tag{6}$$

The urea concentration may be related to the hydrogen ion concentration by the following equation:

$$\overline{C}_u = \overline{C}_u^b - \frac{\theta}{\lambda\alpha}\left(\overline{C}_{H^+} - \overline{C}_{H^+}^b\right) \tag{7}$$

where $\theta$ and $\alpha$ are given by:

$$\theta = 1 + \frac{K_{WH}\, C_{T,WH}^b}{\left(K_{WH} + C_{H^+}\right)\left(K_{WH} + C_{H^+}^b\right)},\quad \alpha = \frac{K_{H_2CO_3}}{K_{H_2CO_3} + C_{H^+}} - \frac{2K_{NH_3}C_{H^+}}{1 + K_{NH_3}C_{H^+}} \tag{8}$$

Equation 5 was then solved using a finite difference technique.

To illustrate the solution to equation 5 it was assumed that the bulk pH was 4.5 and that the pH required for zone I of the pellet was 7.5. Therefore a 1000-fold change in the hydrogen ion concentration is required across zone II of the pellet. Figure 3 illustrates how the pH at the interface, $pH_0$, between zone II (where the urease is immobilized) and zone I (where the enzyme catalyzing $A \Leftrightarrow B$ is immobilized) changes with the Thiele modulus ($\Phi_T$), for different values of the urea concentration in the bulk phase ($C_u^b$). The increase of $pH_0$ with increasing $\Phi_T$ is to be expected since an increase in $\Phi_T$ may correspond to either a larger urease loading or an increase in the thickness of zone II of the pellet, either of which would enhance the pH difference. It is also found that the value of $\Phi_T$ required to achieve a particular value of $pH_0$, say 7.5, decreases exponentially as $C_u^b$ increases. This fact is shown in Figure 4 where the $\Phi_T$ required to generate a $pH_0$ of 7.5 is plotted against $C_u^b$. From this figure it is apparent that for $C_u^b > .01$ M, the $\Phi_T$ required to reach this target pH becomes constant. For this example, Figure 4 establishes the operating limits for the bulk urea concentrations. For $C_u^b < 10^{-4}$ M, the desired $pH_0$ cannot be attained within practical limits of $\Phi_T$. *$C_u^b > .01$ M would be wasteful since the $\Phi_T$ required to obtain the desired $pH_0$ remains constant at concentrations of urea greater than this value.*

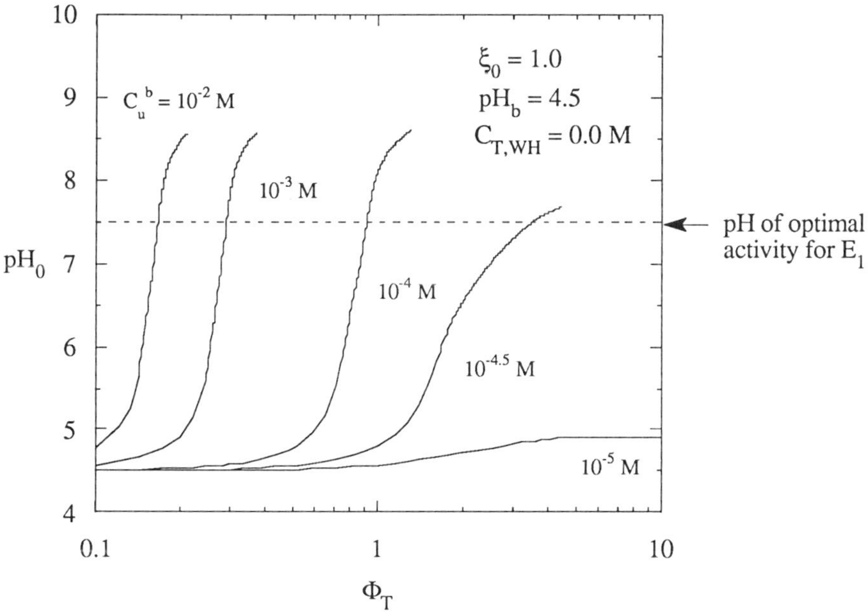


**Figure 3.** pH at the interface between zones I and II of the pellet ($pH_0$) as a function of the Thiele modulus ($\Phi_T$) for different bulk concentrations of urea ($C_u^b$). $E_1$ denotes the enzyme immobilized in zone I of the pellet.

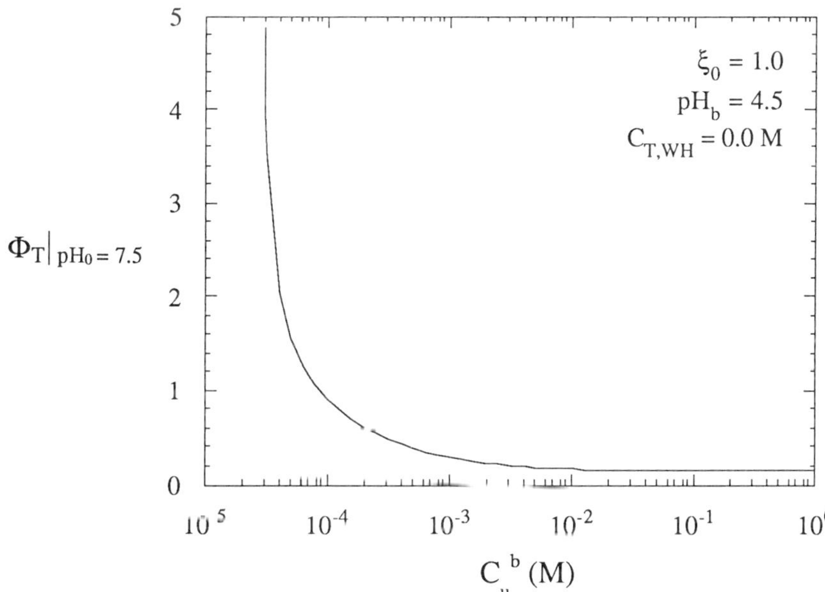


**Figure 4.** Thiele modulus ($\Phi_T$) required to attain a pH of 7.5 at the interface of Zones I and II of the pellet

Having defined from the model the urea concentration that would be required to generate the pH at the interface between zones I and II, the effect of species that are able to act as buffers ($WH$) was addressed. A buffer may be present in the bulk phase in order to control its pH, or other weak acids may simply be present in the system. Buffers affect the pH gradient generated in zone II because they provide a parallel path for protons to diffuse into the pellet.[20] In effect they operate as a shuttle bringing in protons as $WH$, releasing them at the higher pHs that exist in zone II, ($W^- + H^+$), followed by return of $W^-$ to the bulk solution where it combines with hydrogen ions to regenerate $WH$. Therefore, buffers will make it more difficult to generate the required pH gradient. Using the model, we have explored the effect of buffers on the generation of the pH gradient within zone II.

Figure 5 shows a graph of $pH_0$ vs. $\Phi_T$ with various curves corresponding to different total concentrations of buffer. In all of these cases, $C_u^b$ was .01 M and the pK of the buffer was 4.5 since the bulk pH was set at 4.5. The figure clearly shows the detrimental effect of buffers. As the buffer concentration is increased, $C_{T,WH}$, the value of $\Phi_T$ required to obtain a desired $pH_0$ becomes significantly larger. For this example buffer concentrations greater than .01 M would need to be avoided. However, the effect of the buffer could be circumvented to some extent by simply increasing the bulk urea concentration or the urease loading. *Buffer concentrations are typically in the range of 10–50 mM so the desired pH gradient is still achieveable even in the presence of buffers.*

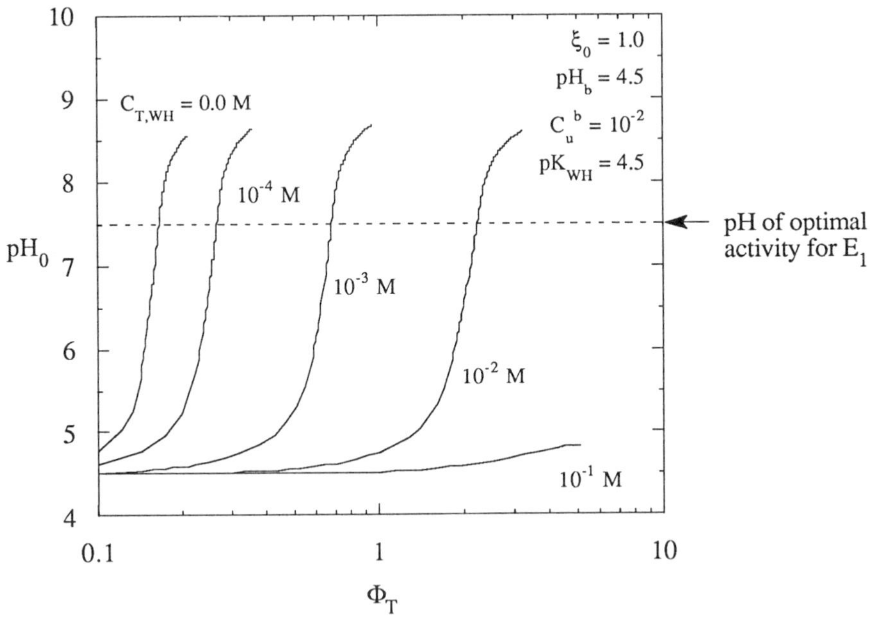


**Figure 5.** pH at the interface between zones I and II of the pellet ($pH_0$) as a function of the Thiele modulus ($\Phi_T$) for different total concentrations of buffer ($C_{T,WH}$). $E_1$ denotes the enzyme immobilized in zone I of the pellet.

## 3. DEMONSTRATION OF GENERATION OF SPATIAL PH GRADIENT IN A FLAT MEMBRANE

Having shown that it is theoretically possible to generate a spatial pH gradient of several units in the outer region of an immobilized enzyme pellet, the next step was to demonstrate this experimentally. To accomplish this the flat plate membrane reactor system shown in Figure 6 was used. This system allowed the bulk solution to operate as a continuous stirred tank reactor and provided for the measurement of the pH on either side of

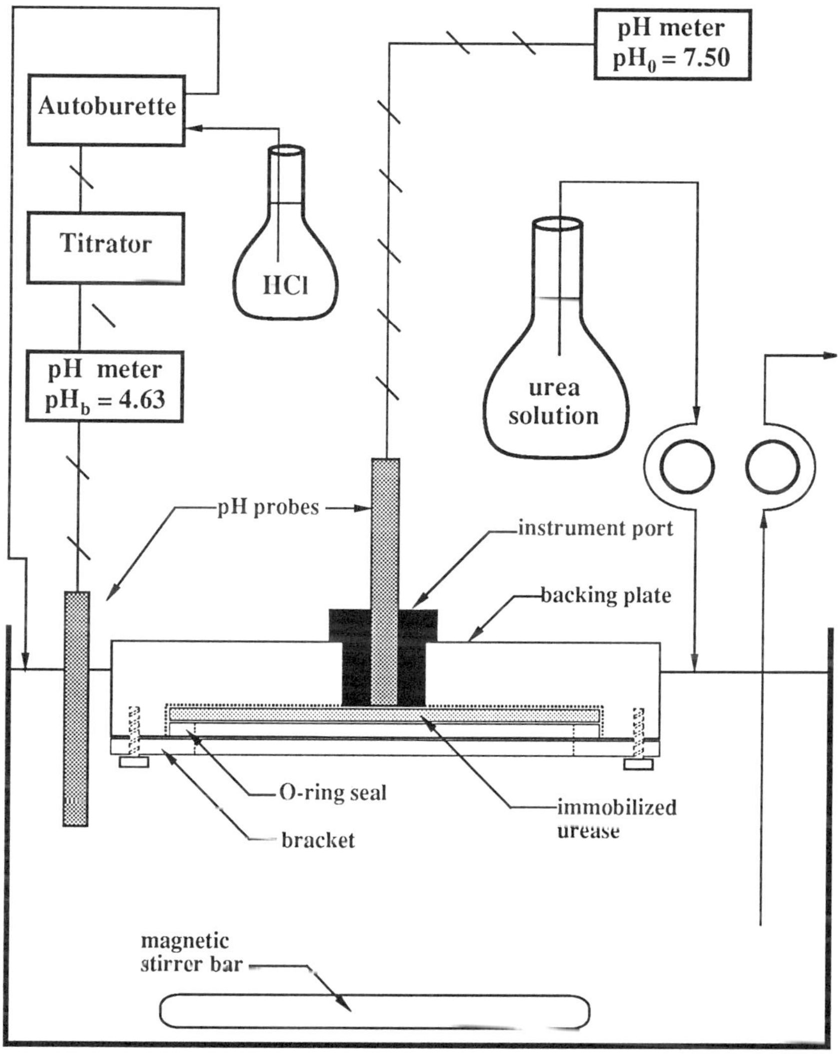


**Figure 6**: Flat plate membrane reactor system.

the membrane. A flat gluteraldehyde-activated membrane (diameter = 47 mm, thickness = .55 mm, pore size = 1 μm, provided by FMC Corp., Biosupport Materials, Pine Brook, NJ) was used to immobilize urease. One side of the membrane contacted the well-stirred bulk solution whereas the opposite face of the membrane abutted against a backing plate which was fitted with an instrument port for a surface-pH probe (supplied by Radiometer). The flat face of this electrode enabled the probe to measure the pH on the membrane surface opposite the side exposed to the bulk solution. The pH of the bulk solution was maintained at a constant value of about 4.6 using a Radiometer pH-stat system (Radiometer PHM84 Research pH meter, TTT80 Titrator, and ABU80 Autoburet).

Figure 7 shows the results of a continuous flow experiment[1] over a nine hour period during which the bulk urea concentration was increased step-wise from 0 to .033 M. The buffer that was used had a pK of 4.63 and its concentration was 10 mM. Prior to the introduction of urea into the CSTR, the pH on the opposite side of the membrane is slightly above the pH of the bulk solution. This slight difference in pH is largely due to repulsion of the hydrogen ions by the positive charge of the amine groups present on the aminated gluteraldhyde that is used to activate the membrane. Each time the bulk urea concentration was increased, the pH on the opposite face of the membrane increased to a new steady state value. These results clearly show that a pH difference of 3 units can easily be generated when the bulk urea concentration is only .01 M. This result also confirms the predictions of the theoretical model shown earlier in Figure 5, specifically that a pH difference of 3 units can be generated for a bulk urea concentration of .01 M and a buffer concentration of .01 M. As the bulk urea concentration is increased much above .01 M there is only a slight increase in the surface pH. In this case the amount of urea available for hydrolysis is in excess and the magnitude of the pH gradient is controlled solely by the rate of the hydrolysis reaction.

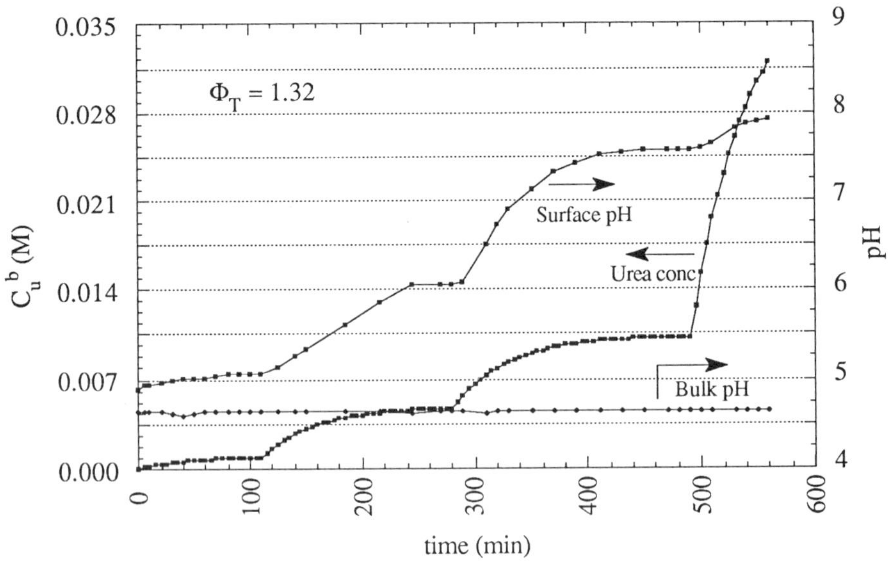


**Figure 7.**pH on the opposite face of the flat membrane ($pH_0$) as a function of the bulk concentration of urea ($C_u^b$)

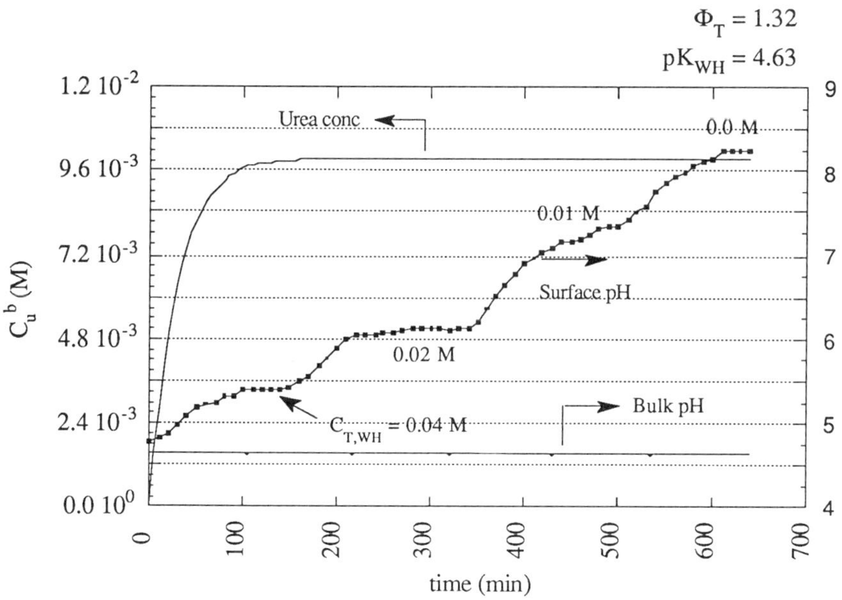


**Figure 8.**pH on the opposite face of the flat membrane ($pH_0$) as a function of the total concentration of buffer ($C_{T,WH}$)

Figure 8 summarizes the results of a continuous flow experiment[1] where the bulk buffer concentration was decreased step-wise from .04 M to 0 M. The bulk urea concentration was maintained at .01 M. As the bulk buffer concentration was decreased the pH gradient across the membrane increased as predicted by the theoretical model. In the absence of buffer the pH difference is nearly 4 units.

## 4. DEMONSTRATION OF pH CONTROL IN A COMMERICALLY PREPARED IMMOBILIZED XYLOSE ISOMERASE

The results presented in section 3 clearly show the feasibility of generating a significant pH gradient within a thin layer of immobilized urease. We have also performed a preliminary experiment that shows that the proposed pH control system allows another enzyme to have near its optimal pH activity when immersed in a bulk solution whose pH is three units less than the optimal pH for this enzyme. To accomplish this we have taken a commercially available enzyme pellet containing glucose (aka xylose) isomerase (Novo Sweetzyme T ™, optimal pH=8.0) and simply pretreated it by soaking it in a solution of urease (1 g/l). The urease becomes adsorbed within the pores of the pellet and may in fact form a thin egg-shell layer. The coimmobilized enzyme pellets were then placed in a small capillary tube forming a differential packed bed reactor that was connected to a larger reservoir containing xylose (60 g/l xylose in $10^{-2}$ M Tris buffer). This batch reactor system is shown in Figure 9. The pH of the bulk solution was maintained at either 4.6 or 7.5 using the Radiometer pH stat system. The xylulose isomerization product was detected using a Waters HPLC system.

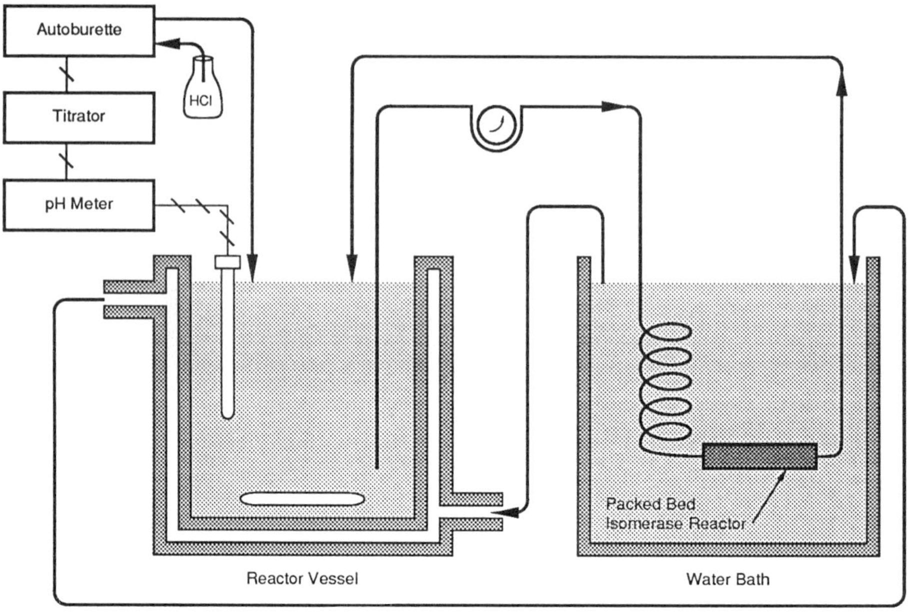


**Figure 9**: Differential batch reactor system

Three runs were made using the same urea/isomerase enzyme pellets. For the first run no urea was present in the system so we would expect that the pH throughout the enzyme pellet would also be about 4.6. Figure 10 shows that after almost 15 hours very little xylulose was formed. The xylulose concentration had only reached about 18% of its

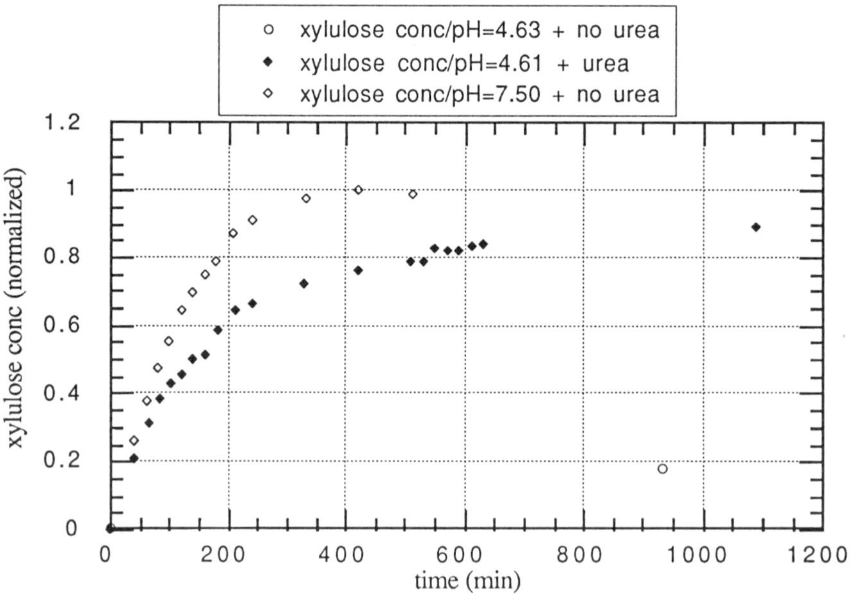


**Figure 10**: Xylulose concentrations as a function of time for differential batch studies

potential equilibrium value. This experiment was then repeated by adding urea at a concentration of .01 M with the bulk solution still at a pH of about 4.6. We would now expect a pH gradient to form within the enzyme pellet and would expect to see an increase in the glucose isomerase activity. The result in Figure 10 shows a dramatic increase in the xylulose production rate over the run without urea, clearly showing that the pH within the pellet must have been significantly higher than in the bulk solution. In this case the xylulose concentration reached its equilibrium value after only 10 hours. For the final run the bulk pH was maintained at 7.5 with no urea present in the system. This run sets the intrinsic activity of the isomerase for the case where pH effects are not a factor. As expected the initial rate for this case is higher than the second run that included urea. Some of this is due to the fact that the pH gradient that develops within the enzyme pellet results in a very thin egg shell region of reduced isomerase activity. The difference in the final equilibrium values between these last two runs may have more to do with the operation of the HPLC system.

## 5. WORK-IN-PROGRESS TO DEMONSTRATE THE INTERGATION OF PH CONTROL FOR BOTH THE STEPS OF A TWO-STEP SYNTHESIS REACTION

From the experimental results described in sections 3 and 4, it can be seen that the preliminary work on the generation of a pH gradient is very encouraging. But for the further application of this technology, a two-step synthesis reaction with each reaction step occurring at its respective optimal pH should be demonstrated. A suggested model system for this purpose is shown in Fig. 11, where salicin is first converted by β-glucosidase into saligenin

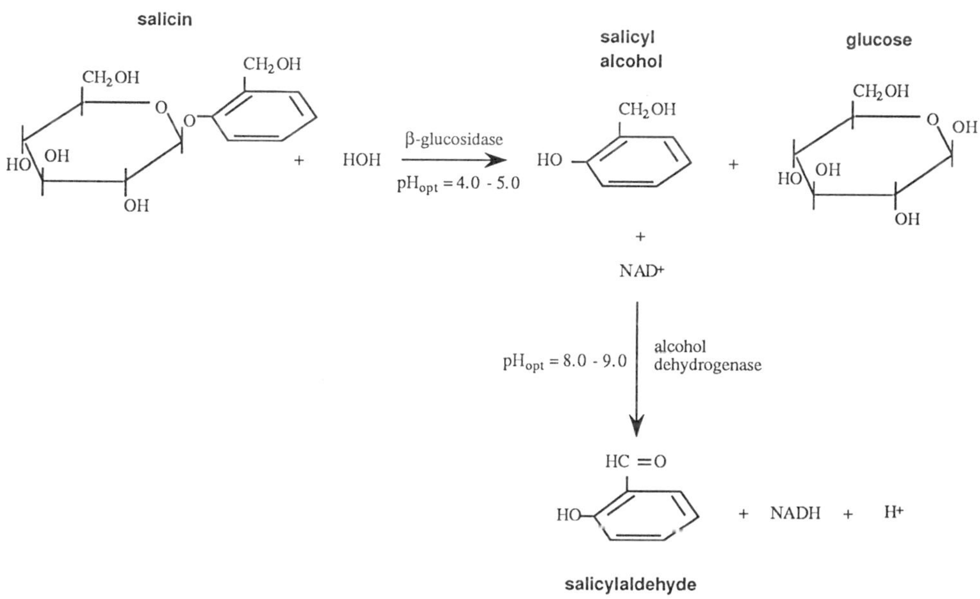


**Figure 11**: Model reaction system

and glucose, followed by the saligenin oxidation catalyzed by horse liver alcohol dehydrogenase. This model system is chosen because of its two advantages: the first is that the enzymes, substrates, intermediates, and products are all readily available from commercial sources such as Sigmal and Aldrich. Therefore there are no complicated methods required to prepare the enzymes and the enzymes can be used in their soluble form or they can be immobilized using a variety of methods. The second one is that this system represents a more general and more challenging test of the pH control since it also generates $H^+$ ions where the urea hydrolysis reaction is taking place. The suggested dual flat membrane reactor system is shown in Fig. 12 where three thin, flat membranes will be used to immobilize the enzymes. It would be a well-defined batch or continuous stirred tank membrane immobilized reactor system that would allow for precise control of the bulk pH as well as measurement of the pH drop across the membrane containing immobilized urease. The membrane holder on the left would contain the membrane with immobilized urease (generating pH gradient) and alcohol dehydrogenase (catalyzing the saligenin oxidation). The flow-through membrane holder on the right would contain the membrane with immobilized β-glucosidase (catalyzing the salicin hydrolysis). The study on this model system is in progress.

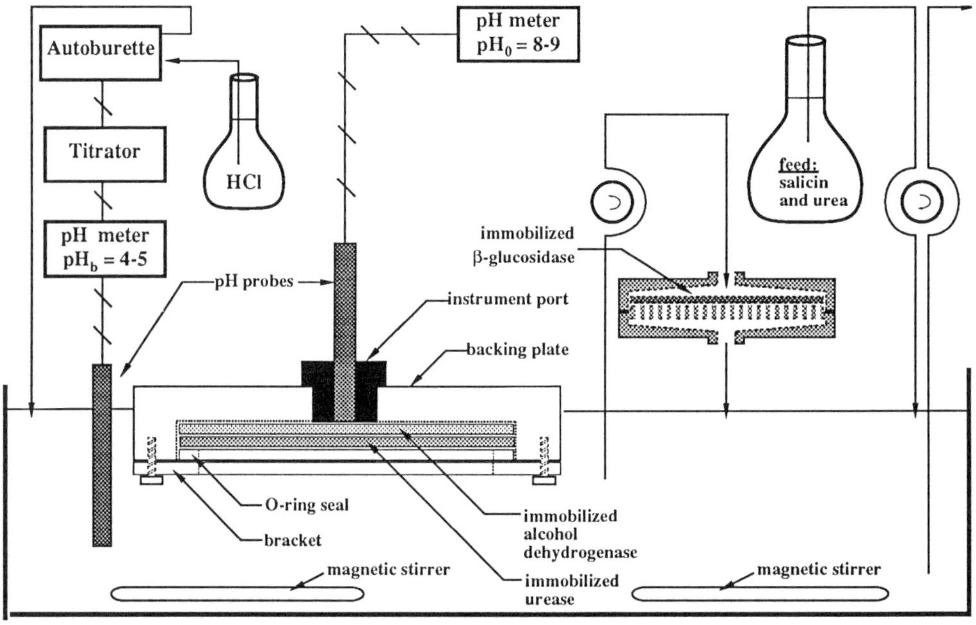


**Figure 12**: Dual flat plate membrane reactor system

## ACKNOWLEDGEMENTS


A research grant from National Science Foundation (grant # BES–9402700) in support of this work is gratefully ackowledged by the authors, R.L.F. and S.V.

## NOMENCLATURE

$C_i$ = local concentration species i

$C_i^b$ = concentration species i in the bulk liquid phase

$C_{T,i}$ = total concentration of species i

$EH_2$, $EH^-$, $E^{2-}$ = three different states of protonation of the free enzyme urease

$D$ = assumed diffusivity for all chemical species

$D_{app}$ = apparent diffusivity of hydrogen ions

$\mathcal{D}_i$ = diffusivity of species i

$e_c$ = total concentration of complexed enzyme

$e_0$ = total concentration of enzyme

$K_i$ = equilibrium constant for reaction i

$K_m$ = Michaelis constant

$k_1, k_2$ = rate constants for the dissociation of the urea-urease enzyme substrate complex into products

$\mathrm{pH}_0$ = pH at the interface between zones I and II of the immobilized enzyme pellet

$r$ = radial distance from the center of the enzyme pellet

$R_1$ = radial distance from the center of the enzyme pellet to the interface between zones I and II

$R_2$ = radial distance from the center of the pellet to the outer surface of the pellet

$s$ = substrate concentration for enzyme catalyzed reaction

$SEH_3$, $SEH_2^-$, $SEH^{2-}$, $SE^{3-}$
= four different states of protonation of the urea-urease enzyme substrate complex

$u$ = subscript denoting urea

$v_i$ = molar rate of consumption of species i per unit volume

$V_{max}$ = maximum velocity of an enzyme catalyzed reaction (eqn. 40)

$V^*$ = ratio the rate constants for the dissociation of the urea-urease enzyme substrate complex into products = $k_2/k_1$

WH = externally added buffer

$W^-$ = conjugate base of an externally added buffer

$x_i$ = mole fraction of species i

### Greek Symbols

$\delta$ = ratio of the apparent diffusivity of hydrogen ions to the diffusivity of all species = $D_{app}/D$

$\lambda$ = ratio of the Michaelis constant to $K_{SEH_2}$

$\eta$ = dimensionless radial distance across zone II of the enzyme pellet = $(r - R_1)/(R_2 - R_1)$

$\eta_0$ = ratio of the radius of zone I of the pellet to the thickness of zone II of the pellet = $R_1/(R_2 - R_1)$

$\varphi_{H^+}$ = dimensionless concentration hydrogen ions = $C_{H^+} / K_{SEH_2}$

$\varphi_u$ = dimensionless concentration of urea = $C_u / K_m$

$\chi_1$ = ratio of protonation-deprotonation equilibrium constants of the urea-urease enzyme substrate complex = $K_{SEH_2} / K_{SEH_3}$

$\chi_2$ = ratio of protonation-deprotonation equilibrium constants of the urea-urease enzyme substrate complex = $K_{SEH} / K_{SEH_2}$

$\Phi_T$ = Thiele modulus = $\left[k_1\, e_0\, (R_2 - R_1)^2 / D\, K_m\right]^{1/2}$

## REFERENCES


1. Byers, J.P., Shah, M.B., Fournier, R.L., and S. Varanasi, 1993. "Generation of a pH Gradient in an Immobilized Enzyme System.", *Biotechnol Bioeng*, **42**: 410–420

2. Byers, J.P., Fournier, R.L., and S. Varanasi, 1992. "A Feasibility Analysis of a Novel Approach for the Conversion of Xylose to Ethanol.", *Chem Eng Comm*, **112**: 165–187.

3. Fournier, R. L., S. Varanasi, and J .P. Byers, US patents: # 5,254,468 (1993) and # 5,397,700 (1995)

4. Bruinenberg, P.M., de Bot, P.H.M., van Dijken, J.P., and W.A. Scheffers, 1984. "NADH-linked aldose reductase: the key to anaerobic fermentation of xylose by yeasts.", *Appl Microbiol Biotechnol*, **19**: 256-260.

5. Chiang, L.C., Gong, C.S., Chen, L.F. and G.T. Tsao, 1981. "D-Xylulose Fermentation to Ethanol by *Saccharomyces cerevisiae*", *Appl Environ Microbiol*, **42**: 284-289.

6. Gong, C.S., Chen, L.F., Flickinger, M.C., Chiang, L.C., and G.T. Tsao, 1981. "Production of Ethanol from D-Xylose by Using D-Xylose Isomerase and Yeasts.", *Appl Environ Microbiol*, **41**: 430-436.

7. Hahn-Hägerdäl, B., Berner, S., and K. Skoog, 1986. "Improved ethanol production from xylose with glucose isomerase and *Sacchromyces cerevisiae* using the respiratory inhibitor azide.", *Appl Microbiol Biotechnol*, **24**: 287-293.

8. Hsiao, H-Y, Chiang, L-C, Chen L-F, and Tsao, G., T., 1982. *Enzyme Microb Technol*, **4**: p. 25

9. Lastick, S.M., Tucker, M.Y., Beyette, J.R., Noll, G.R., and K. Grohmann, 1989. "Simultaneous fermentation and isomerization of xylose.", *Appl Microbiol Biotechnol*, **30**: 574-579.

10. Olivier, S.P., and P.J. du Toit, 1984. "Sugar Cane Bagasse as a Possible Source of Fermentable Carbohydrates. II. Optimization of the Xylose Isomerase for the Isomerization of Xylose to Xylulose in Laboratory-Scale Units.", *Biotechnol Bioeng*, **18**: 684-699.

11. Skoog, K. and B. Hahn-Hagerdal, 1988. "Xylose Fermentation.", *Enzyme Microb Technol*, **10**: 66-80.

12. Tewari, Y.B., Steckler, D.K., and R.N. Goldberg, 1985. "Thermodynamics of the conversion of aqueous xylose to xylulose.", *Biophys Chem*, **22**: 181-185.

13. Toivola, A., Yarrow D., van den Bosch, E., van Dijken, J.P., and W.A. Scheffers, 1984. "Alcoholic Fermentation of D-Xylose by Yeasts.", *Appl Environ Microbiol*, **47**: 1221-1223.

14. Anderson, W.A., Moo–Young, M., and R.L. Legge, 1992. "Development of a Multienzyme Reactor for Dopamine Synthesis: I. Enzymology and Kinetics.", *Biotechnol Bioeng*, **39**: 781-789.

15. Anderson, W.A., Moo-Young, M., and R.L. Legge, 1992. "Development of a Multienzyme Reactor for Dopamine Sythesis: II. Reactor Enginneering and Simulation.", *Biotechnol Bioeng*, **40**: 388–395.

16. Savidge, T.A., 1984. "Enzymatic Conversion used in the Production of Penicillin and Cephlosporin.", In: Chapt. 5, E.J. Vandamme (ed.), Biotechnology of Industrial Antibiotics, Marcel Dekker, Inc., USA.

17. Ro, H.S., and H.S. Kim, 1991. "Continuous production of gluconic acid and sorbitol from sucrose using invertase and an oxidoreductase of *Zymomonas mobilis*.", *Enzyme Microb Technol*, **13**: 920–924

18. Hanson, R.L., Bembenek, K.S., Patel, R.N., and L.J. Szarka, 1992. "Transformation of Nε-CBZ-L-lysine to CBZ-L-oxylysine L-amino acid oxidase from *Providencia alcalifaciens* and L-2-hydroxy-isocaproate dehydrogenase from *Lacrobacillus confusus*.", *Appl Microbiol Biotechnol*, **37**: 599-603.

19. Dixon, N.E., Riddles, P.W., Gazzola, C., Blakeley, R.L., and B. Zerner, 1980. "Jack bean urease (EC 3.5.1.5). V. On the mechanism of action of urease on urea, formamide, acetamide, N-methylurea, and related compounds.", *Can J Biochem*, **58**: 1335-1344.

20. Ruckenstein, E. and S. Varanasi, 1984. "Acid Generating Immobilized Enzymic Reactions in Porous Media - Activity Control via Augmentation of Proton Diffusion by Weak Acids.", *Chem Eng Science*, **39**: 1185-1200.

# NON-COVALENT IMMOBILIZATION OF PAPAIN ONTO MODIFIED POLYSULFONE MEMBRANE USING AVIDIN-BIOTIN COUPLING: EPR AND KINETIC STUDIES

Anita Bhardwaj[1], Jinbo Lee[1], Kathi Glauner[1], Dibakar Bhattacharyya[2,3] and D. Allan Butterfield[1,3]*

[1]Department of Chemistry,
[2]Department of Chemical Engineering,
[3]Center of Membrane Sciences, University of Kentucky,
Lexington, KY 40506-0055

## ABSTRACT

Membrane-based bioreactors or immobilized enzymes are one example of biofunctional membranes, defined as entities composed of biological molecules attached to a polymeric matrix cast in the form of porous polymeric membranes. A detailed understanding of the structure and function of enzymes upon immobilization is essential to develop a membrane bioreactor with optimal properties involving minimal conformational changes at the active site of the enzyme. In this study, a sulfhydryl protease, papain (EC 3.4.22.2), was noncovalently immobilized onto the modified polysulfone membrane through the avidin-biotin complex. Various kinetic parameters for the amidase activity of noncovalently bound papain, using the substrate benzoyl arginine p-nitroanilide hydrochloride (BAPNA) were determined and the results were compared with those obtained from studies of papain in solution and papain directly immobilized onto the modified polysulfone membrane. As expected, there was a decrease in the enzymatic activity upon direct immobilization. However, insertion of the avidin-biotin complex as a non-covalent spacer resulted in an enzyme bioreactor which was more stable and reusable than the free enzyme.

Electron paramagnetic resonance (EPR) spin labeling techniques were used to characterize the active site conformational changes of papain immobilized on the modified polysulfone membranes through the avidin-biotin complex. This paper reports the pH dependence, reusability and storage stability studies of biofunctional membranes using avidin-biotin complex as a non-covalent spacer. A good correlation was found between the active site conformational changes and the amidase activity of papain upon noncovalent immobilization onto the modified polysulfone membrane.

*Biofunctional Membranes*, Edited by
D.A. Butterfield, Plenum Press, New York, 1996

## INTRODUCTION

Over the past few decades, electron paramagnetic resonance (EPR) spin labeling techniques have been used successfully to provide information on the active site environment of immobilized proteins[1-3]. EPR measurements do not require optically transparent samples, which makes it an excellent technique for studying active site conformational changes of enzymes immobilized on opaque support matrices. Clark and Bailey,[4-6] showed that in the case of spin- labeled α-chymotrypsin covalently immobilized on CNBr-activated sepharose, two different subpopulations of the immobilized enzymes exist in the resultant EPR spectra. The active site conformations of the immobilized enzyme corresponding to the two different subpopulations were found to be different.

Our group has also reported the presence of two EPR spectral subpopulations on studies done on papain directly immobilized on polymeric membranes.[7] We also reported the comparison of the effects of direct immobilization of papain on flat sheet modified polysulfone membrane (MPS) and hydroxyethyl cellulose (HEC) coated polyethersulfone hollow fibers.[8] As reported earlier, EPR studies showed two subpopulations of the immobilized papain on the hollow fibers. We have found that immobilization of spin-labeled enzymes, in general, have resulted in two conformationally different EPR spectral subcomponents. In addition to these findings, our past experience has shown that immobilizing an enzyme through a 6-carbon spacer enhances the enzymatic activity and minimizes the conformational changes at the active site of the enzyme.[9-11]

Most of the immobilization reactions reported earlier by our group[7] involved the direct formation of covalent bonds between the lysine residues on papain molecules and the aldehyde functional groups on modified polysulfone membranes. In all these studies, a short, sulfhydryl-specific spin label sulfhydryl specific spin label, (1-oxyl-2,2,5,5-tetramethyl-$\Delta^3$-pyrrolidine-3-methyl) methanethiosulfonate (MTS) was employed in the characterization of directly immobilized papain.

In the current study, the high binding affinity between avidin and biotin has been utilized to create a noncovalent spacer between the enzyme and the modified polymeric membrane. The bicyclic ring of biotin is necessary to bind to avidin. The biotinylation of papain was accomplished using NHS-LC-Biotin (Figure 1).

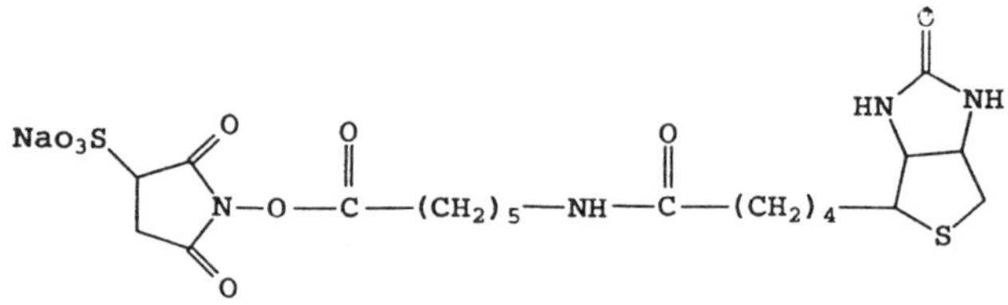


**Figure 1.** The structure of NHS-LC-biotin.

EPR spin labeling techniques were used to study the effects of biotinylation on the active site conformation and properties of papain in solution. No changes in the EPR parameters of active-site labeled biotinylated papain were found, consistent with the notion that biotin does not induce conformational changes in this enzyme.[12] The use of the avidin-biotin complex as a noncovalent spacer provided an alternative strategy for enzyme attachment. This paper reports the effect of noncovalent immobilization of papain onto the modified polysulfone membrane through the avidin-biotin complex on the structural and performance characteristics of papain.

## MATERIALS AND METHODS

### Materials

The modified polysulfone blend membrane, Ultrabind™ US450, was purchased from Gelman Sciences. According to the manufacturers, the average pore size of the membrane is 0.45 μm, and the thickness is 152.4 μm. The membrane has aldehyde functional groups on the surface available for covalent attachment. Papain, avidin (from egg white), DL-cysteine, N-benzoyl-DL-arginine-4-nitroanilide hydrochloride (BAPNA) were obtained from Sigma Chemical Company. Disodium ethylenediaminetetraacetic acid (EDTA) was obtained from Aldrich. The sulfhydryl specific spin label, (1-oxyl-2,2,5,5-tetramethyl-$\Delta^3$-pyrrolidine-3-methyl) methanethiosulfonate (MTS) was purchased from Reanal, Budapest, Hungary. The ultrafiltration membrane (10,000 MW cutoff ) used for concentrating MTS-labeled papain, was obtained from Millipore Company. NHS-LC-Biotin was purchased from Pierce. All other chemicals and solvents were reagent grade.

### Methods

A stock solution of papain was (4 mg/mL) was spin labeled, concentrated and analyzed as previously described.[13]

**Biotinylation of MTS-labeled Papain** The concentration of MTS-labeled papain was adjusted to 3 mg/mL using $NaHCO_3$ buffer, pH 8.0. Then NHS-LC-Biotin was added to 5 mL of MTS-papain solution (the mole ratio of NHS-LC-Biotin to papain was 5:1). The reaction was carried out at room temperature for 1 hour with stirring. The biotinylated MTS-labeled papain was then dialyzed extensively in 3L of 50 mM $NaHCO_3$ buffer, pH 8.0, at 0°C. The buffer was changed after every eight hours for two days. A 10,000 MW cutoff membrane was employed for dialysis.

**NonCovalent Immobilization of Papain through the Avidin-biotin Complex** A 3mL solution of avidin in 50 mM $NaHCO_3$ buffer (2mg/mL avidin) was prepared. Avidin was covalently attached to the polysulfone membrane and the system was incubated for one day at 4°C. Unreacted avidin was washed away with the 50 mM $NaHCO_3$ buffer pH 8.0. Biotinylated MTS-labeled papain solution was then added to the membrane and the system was again incubated for 24 hours at 4°C. Excess papain was washed away with 50 mM $NaHCO_3$ buffer. The resulting membrane was then cut into four 1X4 $cm^2$ pieces for EPR studies.

**EPR Spectra Acquisition** All the EPR spectra of MTS-labeled papain noncovalently immobilized on modified polysulfone membranes were recorded on a Bruker ESP-300 spectrometer with a TM rectangular cavity at room temperature using a special quartz EPR cell.[7] The membrane was taken directly from the buffer in which it was stored. Typical parameters for the EPR spectra were mw frequency, 9.78 GHz ; mw power, 24.2 mW ; modulation amplitude, 0.32 G; modulation frequency, 100 kHz ; scan width, 130G.

**Assay for the Amidase Activity of Papain Noncovalently Immobilized on Modified Polysulfone Membrane** The amidase activity assay for papain was performed as described previously.[7,8,13]

**pH dependence Studies** The buffers (0.2M) used in pH study were: acetate buffer for pH 3-5, phosphate buffer 5-8, and Tris buffer for pH 9-10. For studying the effect of pH on the active site conformation of noncovalently immobilized papain on polysulfone membrane, each 1x4 $cm^2$ piece was incubated in buffers of different pH for 30 min. at room temperature before the EPR spectrum was acquired. The distance between the high field and low field resonance peaks, 2 $A'_{ZZ}$, was used as a parameter for characterizing the motion of spin label at the papain active site.[7,13]

While measuring the papain activity under different pH conditions, the activating solutions were prepared in deionized water as described above. Each 2X2 $cm^2$ piece of polysulfone membrane with noncovalently immobilized papain was incubated in 0.5 mL buffer of different pH. The solution was incubated for about 5 minutes at 37°C. Then 9.5 mL of the activating solution was added and papain was activated for 10 minutes. Ten mL of substrate solution was then added and reacted for 20 minutes with Continuous stirring (final substrate concentration was 1mM BAPNA). The amount of 4-nitroaniline released was estimated spectrophotometrically as described above. The relative amidase activity of papain under different pH was plotted.

**Storage Stability and Reusability Studies** These studies were performed in a manner previously described.[7,13]

## RESULTS

### EPR studies

Figure 2(a) shows a typical EPR spectrum generated by free MTS spin label in solution. The three equally spaced sharp lines arise due to free and isotropic tumbling of the nitroxide moiety of the spin label. The line widths are highly sensitive to the changes in the microenvironment of the spin label. Figure 2(b) represents the effect of the microenvironment, i.e., the active site of papain, upon the spin label motion. The characteristic asymmetric broadening of the high ($M_I = -1$) and low field ($M_I = +1$) resonance lines with a consequent loss in amplitude is due to the influence of active site structure on the rotational motion of the attached spin label. Thus, the motion of the spin label may be characterized by the splitting between the low and high field resonance lines. A hyperfine splitting parameter, $2A'_{ZZ}$, can be used to measure the splitting between the high and low field lines in the EPR spectrum. This parameter is characteristic of the spin label motion; a lower value of $2A'_{ZZ}$ implies less restricted motion of the spin label and vice-versa. This parameter has been successfully used in characterizing the active site conformations of free and immobilized enzymes.[7,13]

The comparison of the active site conformation of MTS-labeled papain in aqueous solution with that of immobilized papain is shown in Fig 2(c). The spectra of MTS-labeled papain in solution showed $2A'_{ZZ}$ value of 53G, which suggests that the motion of nitroxide moiety of spin label located in the active site of papain is restricted when compared with the motion of free spin label.

The EPR spectrum (Figure 2 (c)) of papain immobilized directly onto the modified polysulfone membrane showed the presence of two subpopulations whose $2A'_{ZZ}$ values are 60G and 70G, respectively.

Various structural and kinetic properties of biotinylated papain were compared with those of the free papain previously studied extensively by our research group. It was seen that there was no significant change in the active site conformation and enzymatic activity of papain upon botinylation.[12] In Figure 2(d), a typical EPR spectrum of MTS-labeled biotinylated papain in solution is presented. The typical value of $2A'_{ZZ}$ for MTS-labeled papain in solution is about 53 G. In the case of MTS-labeled biotinylated papain in solution, the $2A'_{ZZ}$ value was found to be 54G. Thus, it may be be concluded that addition of biotin on the surface of papain molecule does not affect significantly the motion of the spin label at the active sites. Therefore, the conformation and structure of papain's active site remain essentially unaltered upon biotinylation.

When papain was immobilized onto the modified polysulfone membrane through the avidin-biotin complex, the two spectral subpopulations of the enzymes were again observed (Figure 2(e)). This spectrum resulted because of the overlap of two three-line nitroxide spin label spectra. One subpopulation had $2A'_{ZZ}$ value of 58.2 G and the other about 68.4 G. Thus, the complex formation does restrict the motion of the active site spin label for both subpopulations as expected. Also, the D subpopulation has a $2A'_{ZZ}$ value close to that of the denatured population of directly immobilized papain,[7] while the other subpopulation has a $2A'_{ZZ}$ value that is close to that of biotinylated papain (see above). Hence, subpopulation A represents the active form of immobilized papain. Comparison of $2A'_{ZZ}$ values for the active subpopulation A of immobilized enzymes reveals that the motion of the active site spin label is slightly less restricted in the case of noncovalently immobilized papain ($2A'_{ZZ}$ =58.2G) as compared to directly immobilized papain ($2A'_{ZZ}$ = 60G). This shows that a relatively more open active site structure results upon insertion of avidin-biotin complex as a noncovalent spacer between the enzyme and the polymeric membrane.

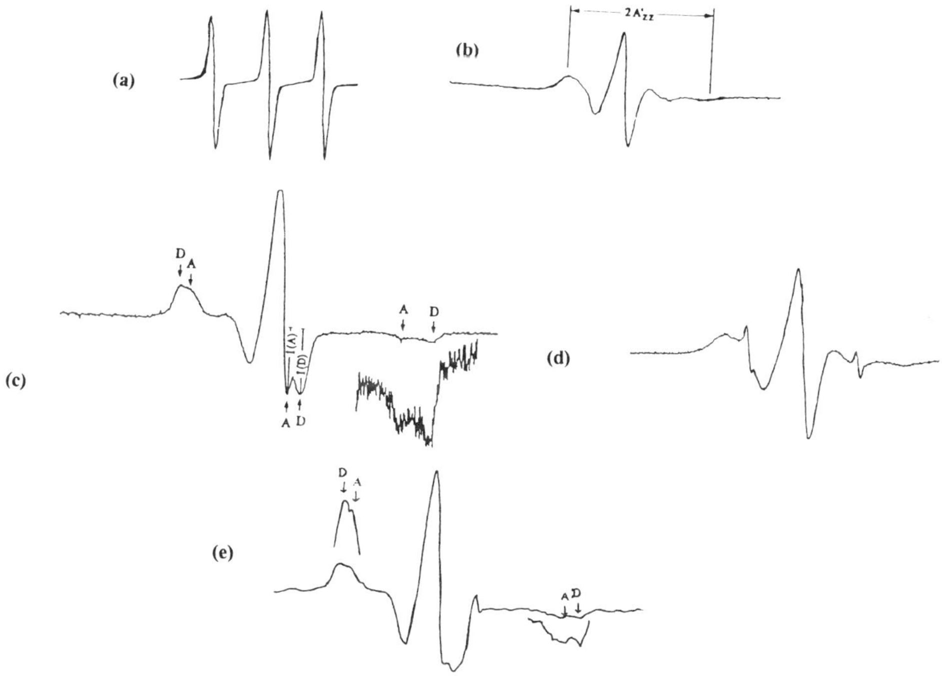


**Figure 2.** Typical EPR spectra of MTS label (a) in solution. (b) showing the effect of active site conformation on its motion. The splitting parameter, $2A'_{zz}$ is 53G. (c) bound to papain directly immobilized onto the polysulfone membrane. Two subpopulations A and D are observed. (d) bound to biotinylated papain in solution (e) bound to papain noncovalently immobilized on the polysulfone membrane via avidin-biotin complex.

## Kinetic Studies

Table I summarizes the kinetic parameters of free and immobilized papain obtained from Lineweaver-Burk plots of the data. The amidase activity of the enzyme follows Michaelis-Menten kinetics. The apparent values of the various parameters are reported. It can be seen quite clearly from the values of the various parameters that there was an increase in the value of $K_m$(app) and a decrease in the enzyme activity represented by $V_{max}$(app). It has been reported previously that the activity of an enzyme is usually reduced upon immobilization.[7] The higher values of $K_m$ (app.) observed for immobilized enzymes may be due to alterations in the active site conformation, consistent with the EPR findings.

Comparison of the kinetic parameters of papain immobilized directly and noncovalently shows that the insertion of a noncovalent spacer introduces a degree of flexibility in the enzyme which is absent in the case of direct immobilization. This result is consistent with the values $2A'_{ZZ}$ indicated above.

**Table I.** Apparent values of kinetic parameters obtained from the Lineweaver-Burk plots. Each value indicates the mean and standard deviations of three measurements

| | V max (μmol/g.min) (Apparent) | Km (mM) (Apparent) |
|---|---|---|
| Papain (in solution) | 110 ± 0. 15 | 1.2 ± 1.32 |
| Papain (in solution) after biotinylation | 100 ± 0.09 | 2.0 ± 1.24 |
| Papain immobilized directly on polysulfone membrane | 13.2± 1.41 | 12.3 ± 2.88 |
| Papain immobilized on polysulfone membrane via avidin-biotin complex | 32.3 ± 0.54 | 10.7± 2.45 |

## pH Studies

The pH of the system had almost negligible affect on the conformation of the immobilized papain (pH 5-9). The $2A'_{ZZ}$ values for the subpopulation D showed no observable changes. The $2A'_{ZZ}$ values for the active subpopulation A also remained almost unchanged in this pH range. These results suggest that the conformation of the papain active site is unaffected in this pH range.

## Reusability Studies

Figure 3 indicates that papain immobilized on the modified polysulfone membrane through avidin-biotin complex does not loose its activity for at least 20 cycles after an initial period of equilibrium. This is one of the most important advantages of a membrane-based enzyme bioreactor.

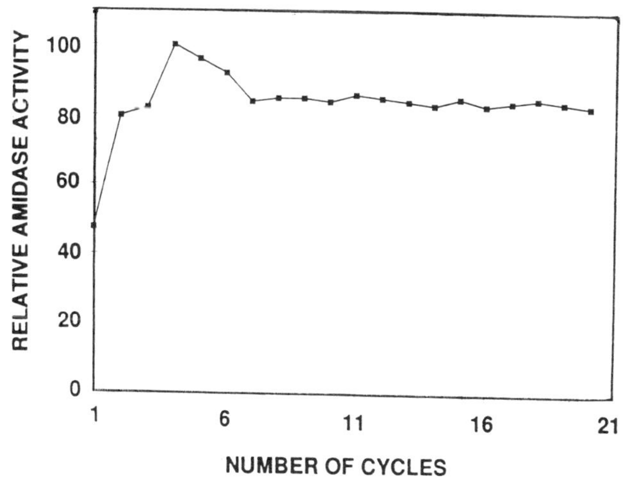


**Figure 3.** Reusability studies done on noncovalently immobilized papain

## The Effect of Storage on the Amidase Activity

The storage stability of noncovalently immobilized papain was studied by storing the system in 5mM $NH_4HCO_3$ buffer, pH 8.0 at 4°C for different time periods. There was no significant alteration in the activity and the active site conformation of the enzyme on storage.

In the EPR studies, a parameter previously employed by our group[7] to characterize immobilized enzymes, I (D) / I (A), was used to assess the change in the spectra over a period of storage. Focusing on the portion below the baseline of the spectrum, it was shown previously that the intensity of the $M_I$=0 mid field line corresponding to the active and denatured subpopulations changes with denaturing of papain. Hence, the relative intensity of these two lines, I (D)/ I(A), is a good parameter to investigate the denaturation of the enzyme upon immobilization. This parameter can be defined as

$$I(D)/I(A) = \frac{\text{the intensity of the negative mid-field peak line of the subpopulation D}}{\text{the intensity of the negative mid-field peak line of the subpopulation A}}$$

A decrease in the value of I (D)/ I(A) would correspond to a decrease in the enzymatic activity. It was observed that the enzymatic activity is retained even after 25 days of storage and that I (D)/ I(A) remains relatively constant over this time period.

## CONCLUSIONS

Although several studies of enzymes immobilized on polymeric surfaces (beads) have been reported, our group was the first to use EPR to characterize membrane-based enzyme bioreactors,[9-11] to investigate the structure and function of enzyme immobilized onto the hollow fiber membranes,[8] and, in the current study, the first to use an avidin-biotin spacer for EPR studies of biofunctional membranes.

The avidin-biotin complex, because of its high noncovalent affinity, indeed provides a practical linkage between the enzymes and the polymeric membranes as shown in this study. Since the enzyme bound with a spacer is further separated from the surface of the membrane, the enzyme, therefore, may have a higher mobility and the substrate-enzyme complex probably is more stable. However, this immobilized enzyme is not stable in the presence of denaturants because, avidin, being a protein, does not retain its conformation . Consequently, its biotin-binding capability is affected. Thus, the use of avidin-biotin complex as a noncovalent spacer is a useful method for enzyme immobilization, and one that offers an excellent alternative to direct immobilization.

## ACKNOWLEDGMENTS

This project was funded in parts by grants from the National Science Foundation (DMR-9200164 and CTS-9307518)

## REFERENCES

**1.** L.J. Berliner, S.T. Miller U.Y. Rosa and G.P. Royer, "An ESR Study of the Active Site Conformations of Free and Immobilized Trypsin". *Biochim. Biophys. Acta.* 315:195(1973)

**2.** G. A. Marg, G.L. Millhauser, P.S. Skerker and D.S.Clark, "Application of EPR Methods in Studies of Immobilized Enzyme Systems". *Ann. N. Y. Acad. Sci.*, 469:253(1986)

**3.** L.J. Berliner, "Spin Labeling in Enzymology: Spin Labeled Enzyme and Proteins". *Meth. Enzymol.*, 49:418(1978)

**4.** D.S. Clark and J.E. Bailey, "Structure-function Relationships in Immobilized Chymotrypsin Catalysis". *Biotechnol. Bioeng.*25:1027(1973)

**5.** D.S. Clark and J.E. Bailey, "Characterization of Heterogeneous Immobilized Enzyme Subpopulations using EPR Spectroscopy". *Biotechnol. Bioeng.*, 26:231(1984)

**6.** D.S. Clark and J.E. Bailey, "Kinetics and EPR Spectroscopy Studies of Immobilized Chymotrypsin Deactivation". *Ann. N. Y. Acad. Sci.*, 434:31(1984)

**7.** D. A. Butterfield, J. Lee, S. Ganapathi and D. Bhattacharyya, "Biofunctional Membranes IV. Active Site Structure and Stability of an Immobilized Enzyme, Papain, on Modified Polysulfone Membrane Studied by Electron Paramagnetic Resonance and Kinetics". *J. Membr. Sci.*, 91:47(1994)

**8.** S. Ganapathi , D. A. Butterfield and D. Bhattacharyya, "Flat Sheet and Hollow Fiber Membrane Bioreactors: A Study of Kinetics and Active Site Conformational Changes of Immobilized Papain Including Sorption Studies of Reaction Constituents". *J. Chem. Technol. Biotechnol.*(1995) (in press)

**9.** P. Zhuang and D.A. Butterfield, " Structural and Enzymatic Characteriztion of Papain Immobilized onto Vinyl Acohol/ Vinyl Butyral Copolymer Mmbrane. *J. membr. Sci.*66:247(1992)

**10.** P. Zhuang and D.A. Butterfield, "Spin Labeling and Kinetic Studies of a Membrane Immobilized Proteolytic Enzyme". *Biotech. Prog.*, 8:204(1992)

**11.** P. Zhuang and D.A. Butterfield, "Optimization of Covalently Coupling Enzymes to Polymeric Membranes". *J. Appl. Polym. Sci.*, 47:1329(1993)

**12.** J. Lee, "Electron Paramagnetic Resonance Studies of Membranes", M.S.Dissertation, University of Kentucky, 1993

**13.** D.A. Butterfield and J. Lee, "Active Site Structure and Stability of Thiol Protease, Papain, Studied by Electron Paramagnetic Resonance Employing a Methanethiosulfonate Spin Label". *Arch. Biochem. Biophys.*, Vol. 310, No.1:167 (1994)

# STRUCTURE-ACTIVITY CORRELATION OF SUBTILISIN IMMOBILIZED ON MODIFIED POLY (ETHER) SULFONE MEMBRANE BY COVALENT AND NON-COVALENT LINKAGES

Ram Subramaniam and D. Allen Butterfield

Department of Chemistry and Center of Membrane Science
University of Kentucky, Lexington, Kentucky 40506

## INTRODUCTION

Synthetic membranes are polymeric membranes and have many applications. These include reactions, separations, analysis, and medical devices. A biofunctional membrane may be defined as a synthetic membrane to which a biological molecule is bound for a specific function. Enzymes immobilized on polymeric membranes are an example of a biofunctional membrane.

Enzymes are proteins which are used as biological and chemical catalysts in a variety of reactions. The importance of using enzymes as catalysts lies in fact that the catalytic action of enzymes is specific. In favourable cases, enzymes may be easily recovered and reused. Often however, enzymes are not easily recovered from batch processes.The main disadvantage, is the instability of the enzyme in a homogeneous environment. One of the ways proposed to overcome this difficulty and be able to use the enzyme desirably is to immobilize the enzyme on a solid support, such as on a polymeric membrane and thus enhance the stability of the enzyme[1].

Previous work has been performed on papain, a sulfhydryl protease, immobilized on polymeric membranes such as poly (ether) sulfone membrane [1]. These studies demonstrated that, upon immobilizing there are two environments of the enzyme on the membrane. One of these was shown to be an active form of the enzyme, with an open active site and the other was shown to be a denatured form of the enzyme, with a closed in active site. Even though enzyme activity was lower compared to the homogenous phase, there was greater stability upon immobilization. In order, to investigate if such findings are generalizable to all other kind of proteins, the current work was conducted on subtilisin, a serine protease.

Subtilisin is a 235 amino acid residue protein[2], the active site of which is a serine located at residue 221. Subtilisin is a commercially important enzyme. It is used in the detergent industry and also to clean protein deposition on contact lenses. In the study described in the following sections, the properties of the enzyme in homogenous phase and when immobilized on polysulfone membranes are discussed. Immobilization was carried out by two methods, a) direct immobilization, leading to a covalent coupling of the ε-amino group of the lysine residues on the enzyme and the aldehyde groups of the modified PES membrane; and b) immobilization via a non-covalent linkage using avidin-biotin as the spacer between the enzyme and the membrane.

Two methods were used to characterize the enzyme – kinetics and electron paramagnetic resonance(EPR). The esterase activity of subtilisin was measured, by studying the amount of p-nitrophenol formed from p-nitrophenylacetate (PNPA). Enzyme activity was then expressed as the micromoles of product formed per milligram of protein per unit time. The active site geometry of the enzyme was studied using an EPR sensitive spin probe covalently bound at the active site of the protein molecule. The results indicate that subtilisin is a robust enzyme in solution. Direct immobilization leads to an almost total loss in the enzyme activity . On the other hand, a non-covalent spacer seems to be able to restore the enzyme activity to some extent. These studies are described in detail in the following sections.

## MATERIALS AND METHODS

### Materials

Subtilisin BPN', the spin label 4-ethoxy(fluorophosphinyl oxy) TEMPO and avidin were obtained from Sigma. Urea was obtained from BDH Chemicals Ltd. Disodium ethylenediaminetetraacetic acid (EDTA), acetonitrile, p-nitrophenylacetate, guanidine hydrochloride were all obtained from Aldrich. NHS-LC-Biotin was obtained from Pierce. The modified polyethersulfone (PES) membrane, UltraBind US450 was from Gelman Sciences. The ultrafiltration membrane with a 10,000 molecular weight cut off was obtained from Millipore. Sephadex G-25 used in the gel filtration column was purchased from Sigma. All other chemicals and solvents were reagent grade.

### Methods

**Spin labeling of Subtilisin.** Subtilisin was spin labeled with 4-ethoxy(fluorophosphynil oxy)TEMPO by a slightly modified procedure from what is previously described[3]. 10 μL of a 1M stock benzene solution of the spin label is added to 4 ml of a 4 mg/ml enzyme solution. Enzyme concentration was determined, prior to spin labeling by methods described by Lowry et al[4]. The spin label was reacted with the enzyme for 30 minutes, at which time the activity of the enzyme was found to be less than 1% of the original activity. The spin labeled enzyme was then separated from any unreacted spin label by passing through a Sephadex G-25 column and eluting with 5 mM ammonium bicarbonate buffer at pH 8. The protein peaks detected at 280 nm were pooled together and the pooled sample was then concentrated by ultrafiltration using a membrane with a 10,000 molecular weight cutoff. The concentration of the spin labeled enzyme was then determined by Lowry's method[4] and found to be between 2.7-2.8 mg/ml.

**Direct immobilization of spin labeled subtilisin on polyether sulfone membrane.** The flat sheet polyether sulfone used was modified by the manufacturer to contain aldehyde groups on its surface. The free amino groups from residues such as lysine and arginine react with the aldehyde groups to form a stable imine group[5]. To carry out the reaction, the spin labeled enzyme solution was incubated with a 4X4 $cm^2$ piece of the membrane, and stored at 4°C for 24-36 hours. After the incubation period, the membrane is washed extensively with a 5 mM phosphate buffer solution at pH 8. The difference in the protein concentrations of the solutions prior to incubation with the membrane and after incubation with the membrane was used as the amount of protein immobilized on the membrane. This was found to be typically 0.5-1.5 mg / 16 sq.cm of the flat sheet membrane.

**Acquisition of EPR spectra.** The spectra were all obtained on a Bruker ESP-300 spectrometer with a tm-8988 rectangular cavity at room temperature using a quartz cell. The parameters of acquisition were the following: microwave frequency- 9.78 GHz ; microwave power- 24.2 mW ; modulation frequency- 100 kHz ; modulation amplitude- 0.32 G ; conversion time- 10.24 ms ; scan width- 150 G . The spectra of the membrane were obtained using a special cell as previously described[1].

**Kinetic assay for Subtilisin.** The esterase activity of subtilisin was determined by a procedure modified from that described by Kezdy and Bender[6]. The substrate used was p-nitro phenyl acetate(PNPA). 36.2 mg PNPA was dissolved in 2 ml of acetonitrile and the total volume was brought up to 100 ml with deionized water. The stock substrate concentration is then 2 mM. 3 mg/ml stock enzyme concentration was prepared in 0.2 M phosphate buffer pH 8. The total volume of the assay mixture was kept at 2 ml. Michelis-Menton kinetics were verified using varying substrate concentrations from 0.1 mM - 0.4 mM and the kinetic parameters $K_m$ and $V_{max}$ were determined. The enzyme catalyses the conversion of PNPA to p-nitrophenol whose absorbance is measured at 400 nm spectrophotometrically. The molar absorptivity of the product used in all calculation was 18000 $M^{-1}cm^{-1}$[7]. The assay reaction was observed for 3 minutes and was conducted at room temperature.

**Denaturant and pH studies on Subtilisin.** EPR Denaturant Studies : 12 M stock urea and guanidine hydrochloride solutions were used as denaturants. To 0.5 ml of the spin labeled enzyme in solution varying amounts of the denaturants were added to obtain a final concentration between 0-6 M and the total volume was then brought up to 1 ml . This ensures that the enzyme concentration is the same in all the solutions. The enzyme was then incubated for 30 minutes with the denaturant prior to acquisition of the spectra. Effect of pH : 0.5 ml of the spin labeled enzyme was incubated with 0.5 ml of solutions of varying pH. For pH between 3-5, 0.1 M acetate buffer , pH between 6-8, 0.1 M phosphate buffer and between pH 9-10, 0.1 M TRIS buffer was used. The enzyme was incubated in the buffer for 5 minutes prior to acquisition of spectra. Kinetics: The procedure for studying the activity due to the effect of denaturants and pH was same as that used to obtained EPR spectra. Here the activity was assayed after the incubation period and the enzyme used was not spin labeled.

**Biotinylation of subtilisin and immobilization on PES membrane.** Subtilisin was reacted with a five times molar excess of biotin with stirring at room temperature for one hour. Reaction was carried out at pH 8 phosphate buffer. The excess unreacted biotin was separated from the enzyme by passing it down the sephadex G-25 column in a similar manner described earlier. 6 mg of avidin in 3 ml of 50 mM sodium bicarbonate buffer at pH 8 was added to a 4X4 piece of the PES membrane and incubated at 4° C for 24 hours. Unreacted avidin was rinsed away with 50 mM sodium bicarbonate buffer pH 8. The biotinylated subtilisin was then spin labeled as described before for EPR studies. The spin labeled enzyme with biotin, was then immobilized on the membrane containing avidin by incubating at 4° C for 24 hours. The unreacted enzyme was once again rinsed away by the same procedure. For kinetic studies the enzyme immobilized on the membrane was not spin labeled.

## RESULTS AND DISCUSSION

In EPR spectroscopy microwave radiation are used to induce transitions between rotational levels[9]. In a nitroxide spin label, such as the one used in this study, the unpaired electron is located largely on the nitrogen and upon coupling with the nuclear spin , three lines are observed in a EPR spectrum due to three allowed transitions. These lines are sharp lines when the spin label is in an isotropic environment. However when the motion is restricted, asymmetric line broadening occurs. The distance between the low field line and the high field line designated as $2A_{zz}$' is a measure of the mobility of the spin label. A more hindered motion implies broader lines and an increase in the observed parameter and vice versa[10-12]. In EPR spin labeling studies, the paramagnetic spin label is covalently bound to a non paramagnetic moiety. In an enzyme, the active site, located in a grove or a pocket, when attached to a spin label will restrict the motion of the spin label[8].

In doing spin labeling studies it is important to determine the number of spin labels per molecule of the protein. The spin labeling stoichiometry was found to be 0.8-0.9 spins /molecule of the subtilisin. Also since subtilisin has several serines it is important to determine that the spin label is bound only at the active site. This was accomplished by two methods. In the first, the active site was blocked by phenyl methane sulfonyl fluoride (PMSF), which is a specific inhibitor for the active site of the enzyme[13]. The enzyme was then spin labeled as described above and no spectrum was observed. In the second method the activity of the spin labeled enzyme was assayed and compared with the activity of the enzyme that was not spin labeled. The spin labeled enzyme had an activity less than 1% of the regular enzyme. Thus, both methods, along with the subtilisin : spin label stoichiometry, indicate that the active site is specifically spin labeled.

The spectrum of spin labeled subtilisin in solution is shown in Figure 1.

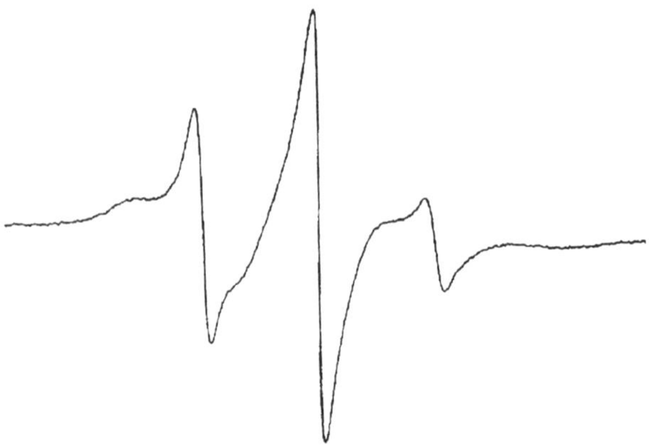

Figure 1: EPR spectrum of spin labeled subtilisin in solution

The spectrum is composed of broad lines, due to the sensitivity of the spin label to its microenvironment. The results of the kinetic studies of the enzyme are given in Table 1. The reaction with PNPA was found to be linear up to 3 minutes.

Table 1: Michelis-Menton kinetics of subtilisin in solution

| PNPA $\xrightarrow{\text{SUBTILISIN}}$ p-NITRO PHENOL | |
|---|---|
| $K_M$ | $3.834\times10^{-4}$ M |
| $V_{MAX}$ | $2.117\times10^{-4}$ M/(min-mg) |

The results of the kinetic and EPR studies of subtilisin in solution, the effect due to pH and denaturants, are respectively shown in Fig 2 and Fig 3.

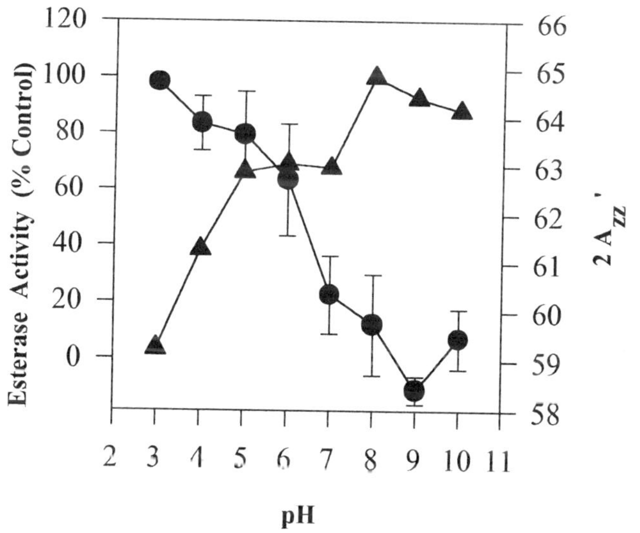


Figure 2: Effect of pH on subtilisin

The pH studies of subtilisin (Figure 2) shows that the activity of the enzyme is a maximum between pH 8 and 9. This suggests a pH optimum around a slightly basic pH. Also, the enzyme retains considerable activity at very high pH ( pH 9,10), but is greatly diminished in activity at pH below 7. The corresponding trend in the measured EPR parameter shoes a minimum at maximum activity and vice versa. This means that the spin label motion is getting more and more restricted when the activity of the enzyme is lost. This result could be interpreted as a closing in of the active site as the activity is lost, leading to a corresponding increase in 2Azz'. This is the same trend as observed for papain in solution[1]. Both of these enzymes are affected by pH by a similar mechanism. The effect of denaturants, such as urea and guanidine hydrochloride, on the activity and 2Azz',

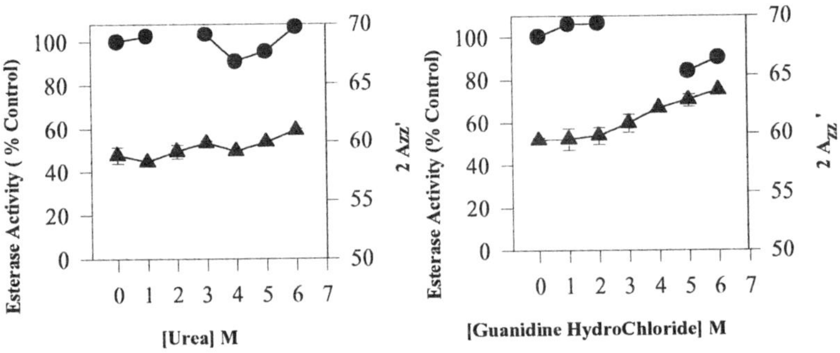


Figure 3: Effect of denaturants on subtilisin ▲ $2A_{zz}'$ ● Esterase activity

was very small (Figure 3). The fact that the enzyme remains unaffected even at high concentrations of these denaturants suggests that subtilisin is a very robust enzyme in solution. Urea and guanidine hydrochloride which are common denaturants for proteins and enzymes are incapable of denaturing subtilisin at concentrations of up to 6 M.

The EPR spectrum of spin labeled subtilisin immobilized directly on the membrane (Fig 4a) shows that there is only one population of the enzyme on the membrane. The $2A_{zz}'$ for this population is 68 G. When compared with the enzyme in solution, which has a $2A_{zz}'$ of 58 G it may be inferred that this is a denatured population. The EPR spectrum of subtilisin immobilized via avidin biotin linkage on the PES membrane shows that there are two populations of the spin labeled enzyme on the membrane ( Fig 4b)

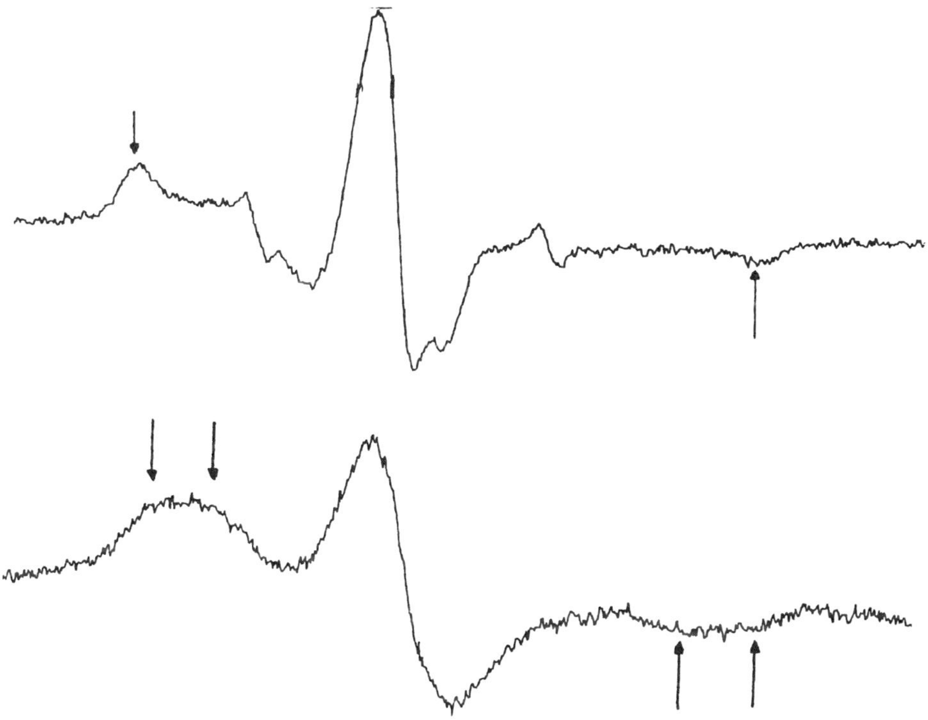

Figure 4: a)EPR spectrum of subtilisin immobilized directly on the PES membrane b) EPR spectrum of subtilisin immobilized via avidin biotin linkage

These spectra show that the two methods of immobilizing on the PES membrane lead to different kinds of attachment of the enzyme on the membrane. The results of kinetic studies of subtilisin immobilized by the two methods, along with the 2Azz values, are indicated in Table 2.

Table 2: EPR splitting parameter and activity of subtilisin under different conditions.

| | $2A_{zz}'$ (G) | Activity ( μ moles/min-gram) |
|---|---|---|
| **Subtilisin in solution** | **58** | **157** |
| **Biotinylated subtilisin** | **58** | **136** |
| **Subtilisin directly immobilized on PES membrane** | **70** | **8.7** |
| **Subtilisin immobilized via avidin-biotin linkage** | **Environment 1 60**<br>**Environment 2 70** | **13.5**<br>—— |

Both methods of immobilizing, lead to a great decrease in the activity of the enzyme as compared to the activity in solution. Biotinylation itself does not lead to a large decrease in the enzyme activity. The corresponding EPR parameter indicates that lower values are seen in solution and when immobilized via avidin-biotin. In the latter case there are two environments of the spin labeled enzyme that are seen. The splitting parameters for the two populations of the spin labeled enzyme immobilized via avidin-biotin are 60 and 70 G respectively. The activity of the enzyme is only 5% of original activity when directly immobilized on the membrane and is improved to 10% when immobilized via avidin-biotin coupling. The fact that one population has a splitting of 60 G on the membrane which is comparable to that in solution, may suggest that there is an active form of the enzyme on the membrane. The second environment is then the denatured form of the environment. These findings then indicate that subtilisin, although a robust enzyme in solution, being able to withstand high concentrations of denaturants, is considerably inactivated upon immobilization. However, it is possible to restore activity somewhat when immobilization is via a non covalent linkage such as avidin-biotin.

## ACKNOWLEDGMENTS

This project was supported in part by a grant from NSF (CTS-9307518)

## REFERENCES


1. Butterfield,D.A., Lee,J., Ganapathi,S., Battacharya,D., Journal of Membrane Sciences 91 (1994) 47-64
2. Ottesen,M., Svendsen,I., Methods in Enzymology 19 (1970) 199-215
3. Morrisett,D.J., Broomfield,A.C., J. of Bio. Chem. 247 (1972) 7224-7231
4. Lowry,O.H., Rosebrough,N.J., Farr,A.L., Randall,R.J., J. of Bio. Chem. 193 (1951) 265-275
5. Zhuang,P., Butterfield,D.A., J. Of Appl. Polym. Sci. 47 (1993) 1329-1342
6. Kezdy,J.F., Bender L.M., Biochem. 1 (1962) 1097-1106
7. Sadtler Research Laboratories 7 1684
8. Butterfield,D.A., Lee,J., Arch. Biochem. Biophys. 310 (1994) 167-171
9. Wertz and Bolton Electron Spin Resonance-Theory and Application, McGraw Hill, New York (1972)
10. Butterfield,D.A., J. Membrane Sci. 53 (1990) 3-17
11. Butterfield,D.A., Biol. Magn. Reson. 4 (1982) 1-78
12. Tse,S.H.P., Leypoldt,K.J., Gough,A.D., Biotechnol. Bioeng. 29 (1987) 69-74
13. Polgar,L., Bender,L.M., J.Amer. Chem. Soc. 88 (1966) 3153-3154

# SECTION V.

# PHARMACEUTICAL APPLICATIONS

# MOLECULAR RECOGNITION OF PHOTOIMPRINTED SURFACES

Michael Jay and Colin D. Mattingly

Division of Medicinal Chemistry and Pharmaceutics, College of Pharmacy, and
Center of Membrane Sciences
University of Kentucky, Lexington, KY 40536-0082

## INTRODUCTION

The overall goal of this work was to prepare synthetic receptor matrices that combined the properties of antibodies (specificity and selectivity of recognition) with those of synthetic polymeric membranes (stability, durability, solvent resistance, etc.). The approach taken involved imprinting molecular recognition complexes onto membrane pore surfaces using photochemistry. Fluor-containing membranes possessing photo-imprinted surfaces can be used in competitive binding scintillation proximity assays. This would overcome the two major limitations of radioimmunoassay, i.e., antibodies would be replaced by stable synthetic receptors, and the bound radioligand would not need to be separated from the unbound radioligand because only the bound radioligand would be in close enough proximity to the fluors in the membrane to induce scintillations. Furthermore, this would permit detection of beta-emitting radioligands without using a scintillation cocktail. This would have a positive economic and environmental impact by greatly reducing the generation of "mixed" (radioactive and organic) waste, the disposal of which is difficult and expensive.

## BACKGROUND

The specific recognition of molecules by bio-polymers (e.g. antibodies, enzymes, etc.) is determined by the spatial arrangement of chemical moieties around the molecule of interest (ligand). The summation of a number of noncovalent interactions between these chemical moieties (e.g. amino acid residues) and the ligand results in specific and selective binding. These noncovalent interactions include hydrogen bonding, dipole-dipole interactions, hydrophobic interactions, and a number of other attractive molecular forces. Molecular recognition chemistry is a field which has seen some remarkable advances in recent years. The work of Cram (1) is perhaps the best known; it involves the chemical synthesis of macrocyclic compounds (corands, cryptands, spherands, etc.) with cavities designed to bind specific molecular shapes. In a recent report, Johnsson et al. (2) described the synthesis of artificial, rationally designed catalytic peptides capable of specifically recognizing and decarboxylating oxaloacetate. Synthetic receptors for peptides have also been synthesized (3). While impressive results have been obtained, these approaches require tremendous efforts in synthetic organic chemistry, and a different synthetic pathway must be designed for each new molecule of interest.

Molecular recognition has also been achieved by the formation of supramolecular structures via molecular "self-assembly" (4). These aggregates rely on extensive hydrogen bonding networks similar to that observed in many biological systems. In the same vein, Rebek et al. (5) have employed "induced binding pockets" which they refer to as scorpion receptors, to form high-affinity complexes with adenine derivatives; association constants of $10^6$ $M^{-1}$ were reported.

The recognition of specific molecules by imprinted "cavities" in polymer structures has been achieved using the techniques of template synthesis and molecular imprinting. In the former method, vinyl derivatives of template molecules are polymerized and subsequently hydrolyzed, leaving a cavity in the polymer structure in the shape of the template molecule. The molecular imprinting method is similar, except that noncovalent interactions between monomers (e.g. methacrylic acid) and the "print" molecule determine the shape of the cavity formed by polymerization of the monomer and subsequent extraction of the print molecule. These approaches have a rich history dating back to 1952 when Curti and Colombo were able to partially resolve enantiomeric mixtures using "tailor-made" silica gels (6). This field began to mature with the work of Wulff (7,8) and colleagues who employed the template synthesis approach. They investigated several classes of monomers and mixtures of monomer derivatives to achieve specific molecular recognition of desired compounds. For example, they polymerized vinyl derivatives of D-glyceric acid with styrene and divinylbenzene. Hydrolysis of vinyl-containing derivatives resulted in a cavity being formed in the polymer that was selective for D-glyceric acid over the L-isomer. Later work included studies in which functional groups could be spaced at exact distances by polymerization of a template onto a rigid matrix (9). Similar studies describing selective functionalization of polymer networks were reported by Damen and Neckers (10).

Shea et al. have also been quite active in the area of template synthesis (11,12). This group has used aromatic ketones as template molecules. Ketals were copolymerized with styrene analogs; subsequent hydrolysis resulted in the liberation of the template molecule leaving a functionalized site in the polymer. The template molecules could then be re-bound to the polymer selectively over related compounds (13). Further investigations revealed that the shape of the cavity was perhaps more important than the array of the functional groups (14).

The primary limitation to the template polymerization approach is that it is only applicable to compounds that can be derivatized by polymerizable (e.g. vinyl) groups. The more general approach of molecular imprinting was introduced by Mosbach et al. (15) who have published a number of papers in which the molecular imprinting approach was used to make chromatographic supports for selective separations (16-19). These have primarily involved the use of methacrylic acid as the monomer with its carboxylate function serving as the main functional group interacting with the print molecule. Imprinted polymers made in this fashion were capable of effectively separating enantiomeric pairs (20-22). Sellergren and Shea (23) have recently described the influence of polymer morphology on the ability of molecularly imprinted polymers to resolve enantiomers. In addition, a new separation media consisting of molecularly imprinted continuous polymeric rods has recently been reported (24).

Mosbach's lab has carefully determined the conditions for producing polymers with the optimum properties; these conditions include the choice of solvent, monomer:print molecule ratios, use of cross-linkers, and method of catalysis (25). An elegant extension of this work involved the addition of catalytic groups at imprint sites in the polymer (26). In this work, imprints of N-pyridine derivatives of amino acids were made by polymerization of vinylimidazole in the presence of cobalt ions. These imprinted polymers were capable of hydrolyzing the N-[p-nitrophenol] esters of specific amino acids in the presence of others over repeated use, demonstrating the catalytic character of the system. A similar approach was used to obtain selective reductions of a steroid molecule (27).

One of the limitations of the molecular imprinting studies described above is that it is necessary to carry out the polymerization reactions in organic solvents. This could be problematic for certain print molecules, e.g. peptides. Glad et al. (28) employed mixtures of organic silanes as functional monomers that were polymerized on the surfaces of microporous silica particles in *aqueous* solution in the presence of print molecules. These were employed in the chromatographic purification of the print molecules (e.g. Safranine O and Rhodaline blue). Other inorganic matrices have the potential for serving as supports for molecular recognition sites (29). However, they would not be appropriate for the current application, i.e. scintillation proximity assays because they would attenuate the light produced by the interaction of a beta particle with a fluor. On the other hand, we have demonstrated that synthetic membrane composed of organic polymers (e.g. polysulfone) are excellent matrices for scintillation proximity assays (30).

A major consideration for using imprints in binding assays is the affinity of the substrate (print molecule) for the receptor (imprint). A recent paper by Vlatakis et al. (31) used molecularly imprinted polymers as antibody mimics for competitive binding assays. These were coined "molecular imprinted sorbent assays" (MIA) and involved the imprinting of molecules such as theophylline and diazepam onto methacrylic acid polymers. Again, the carboxyl function of methacrylic acid was the only functional group on the polymer that could participate in the recognition of the print molecules. Using this approach, they were able to obtain imprinted polymers that demonstrated strong binding affinities and cross-reactivity profiles similar to antibodies. For example, binding sites for theophylline with $K_D$ values as low as 0.35 mM were obtained, although other lower affinity sites were also observed. These imprinted polymers showed less than 1% cross reactivity with closely related compounds (e.g., xanthine, theobromine) in a profile similar to that obtained with anti-theophylline antibodies. For diazepam, $K_D$ values as low as 18 nM were obtained, and, as with theophylline, multiple binding sites of lower affinity were observed. For both compounds, MIA's were effective at concentrations low enough to be useful for therapeutic monitoring.

These studies demonstrated that imprinting approaches have the potential for creating binding sites with high enough substrate affinity and low cross-reactivity for use in specific radioassays. The affinity and selectivity of the imprinted receptors can be improved by using multiple functional groups to interact with the substrate. Our work in this area involves the creation of imprinted receptors using photochemistry. We are attempting to form imprints in the pores of synthetic membranes. This allows us to avoid carrying out a polymerization reaction in organic solvents in the presence of the print molecule. This is expected to extend the utility of imprinting studies to compounds that are insoluble or unstable in organic solvents such as peptides.

The photodecomposition of specifically designed polymers to yield reactive carbene and nitrene derivatives has been used as a method to coat hydrophobic surfaces with hydrophilic polymers (32,33). For example, Sigrist et al. (33) employed polymers containing trifluoromethyl-aryl-diazirine functions to generate carbenes which were capable of insertion into inert surfaces, e.g. polystyrene. Swanson et al. (32) used a similar approach employing polymers containing aryl azides to yield nitrenes that were used to coat polysulfone membranes. Dunkirk et al. (34) used benzophenone derivatives of polymers to yield reactive radicals which were used to coat a variety of surfaces, mainly to prevent bacterial colonization on biomaterials. Photoactivation of the benzophenone derivatives resulted in the initial formation of the singlet excited state. The lifetime of this excited state was sufficiently long to allow inter-system crossing to the triplet state which was capable of insertion into the C-H bonds of polysulfone (the polymeric membrane that we propose to use). Similar approaches have been employed to graft phospholipids onto solid supports (35).

Photoactivation of probes possessing azido groups has been used extensively by Haley et al. (36) to identify nucleotide binding proteins. More recently, bidentate photoaffinity cross-linking reagents possessing both azido and benzophenone functions have been prepared and have been demonstrated to provide greater specificity in identifying protein targets (37).

## SCINTILLATION PROXIMITY ASSAY

This term refers to the use of a fluor-containing matrix possessing a receptor (usually an antibody) for the analysis and quantification of a specific radiolabelled compound. Only those radiolabelled compounds bound to the matrix are in close enough proximity to the fluors to produce scintillations. Hart and Greenwald (38) initially described this approach for RIA's, and the approach was subsequently refined, as reported in later publications (39,40). SPA's have also been used in the detection of membrane proteins and receptors (41). The RIA work culminated with the appearance of a commercial product which employs fluor-containing micro-beads that have antibodies bound to their surface. These have been successfully employed in RIA's for a variety of ligands in which the normally required separation step has been eliminated.

Previous work in our laboratory made steps to improve scintillation proximity RIA's by using a synthetic membrane matrix to support the fluors and antibodies. Unlike the microbead system, no suspending agent needs to be added (which might lead to increased quenching in the LS counting system), and geometric detection efficiencies approaching 100% can be attained.

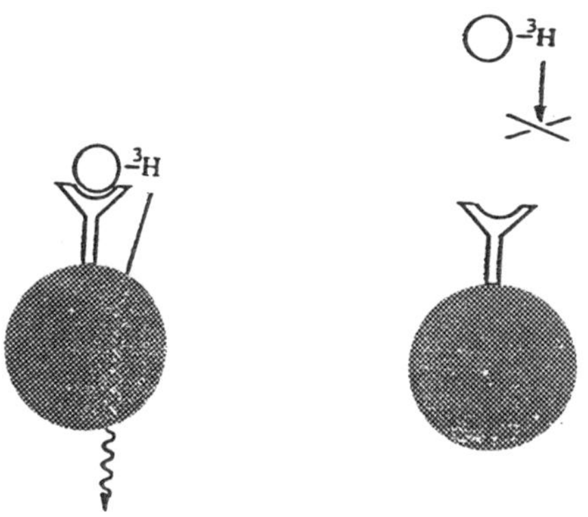


**Figure 1.** Scintillation proximity detection where only bound radioligand is in close enough proximity to the fluors in the matrix to be detected.

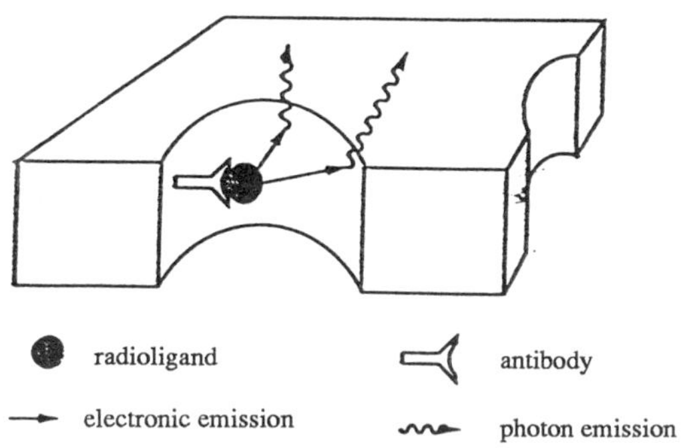


**Figure 2.** Improved geometric detection efficiency for radioligands bound within the pore structure of a microporous scintillation proximity membrane.

This is due to the fact that the antibodies are bound to the pore surfaces of the membrane and the bound radioligands are surrounded by the fluors in the membrane matrix. This is in contrast to the beads where the antibody-bound radioligands Are on the surface of the microbead resulting in a maximum geometric detection efficiency of 50%. In addition, the membrane can be manufactured in a number of different conformations. For example, we are currently interfacing SPA membranes with fiber optics for remote sensing of specific radiochemical entities.

Membranes composed of chlorinated polysulfone or PVC were prepared using phase inversion by casting the polymers, fluors and pore-forming agents in N-methyl pyrrolidinone. Water at various temperatures was used as the nonsolvent. The polysulfone membranes were then coated with a hydrophilic polymer (hydroxyethylcellulose). Secondary antibodies (directed against a primary antibody) were covalently bound to the HEC-coated membranes. When PVC membranes were used, the secondary antibodies were bound in a noncovalent manner via hydrophobic interactions. The membranes were then exposed to solutions containing a fixed amount of primary antibody and [$^{125}$I]-cyclic-AMP, and varying amounts of cyclic-AMP as depicted in the figure below:

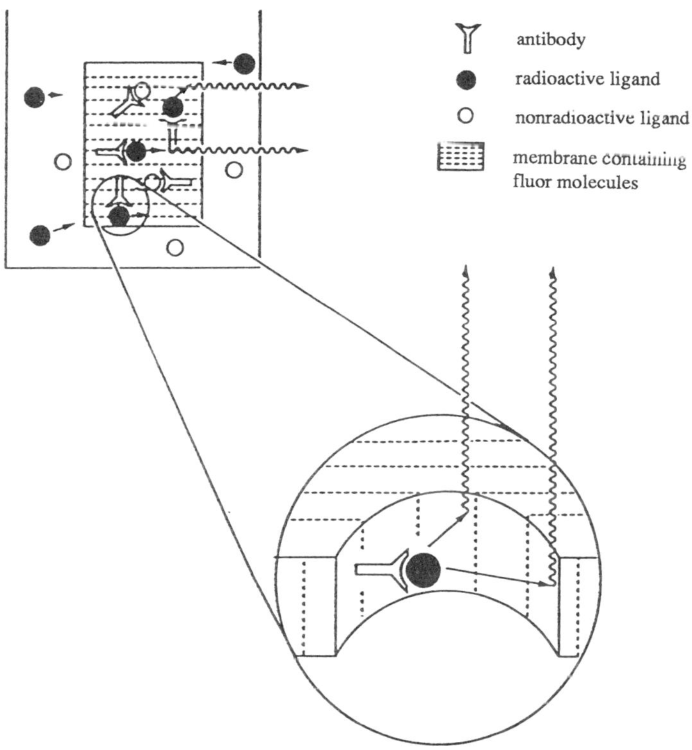


**Figure 3.**

These membranes proved to be sensitive and capable of carrying out an RIA without separating the bound from the unbound ligand. For example, by simply loading the RIA solutions with PVC membranes containing secondary antibodies into an LS counter, the following curve was generated.

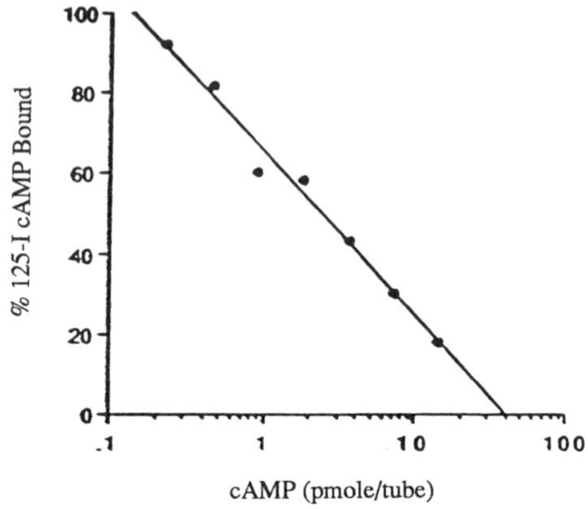


**Figure 4.** Scintillation Proximity RIA using PVC Membrane

This curve is nearly identical to the one generated from the standard RIA kit for cyclic-AMP. It should be noted that no LS cocktail was required for these measurements. We were also able to obtain information on binding kinetics and equilibrium by repeatedly counting the same sample. No sampling of the reaction mixture was required. The results are depicted in the following figure.

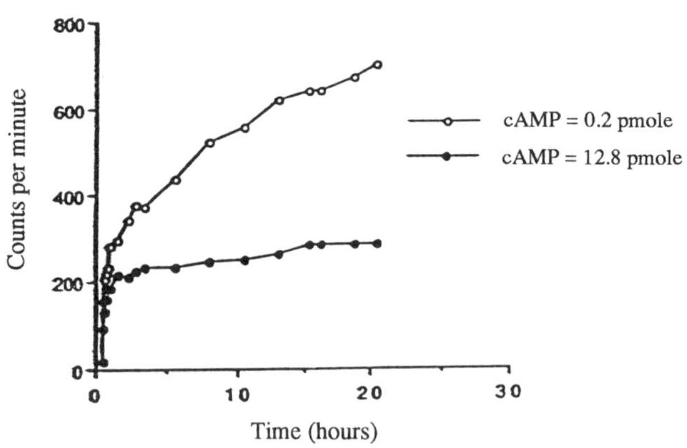


**Figure 5.** Binding of 125-I cAMP to PVC Membrane vs. Time

This experiment demonstrated another useful property of SPA membranes, i.e. they can be used to monitor reaction kinetics *in situ* (e.g. on the pore surface) in a noninvasive, nondestructive manner.

We have also applied the principle of SPA to the quantification of $^{14}CO_2$ (43). SPA membranes containing a quaternary ammonium compound were exposed to a stream of generated $^{14}CO_2$, and the scintillations emanating from the membranes were quantified in an LS counter. The membranes were also shown to be able to retain the bound activity over several days by repeatedly counting the membranes in the LS counter.

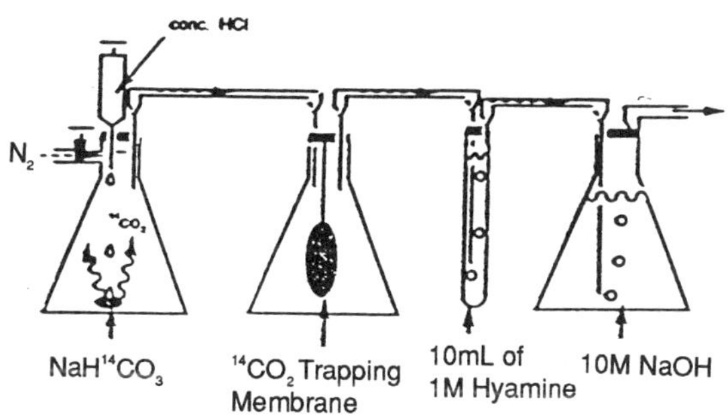


**Figure 6.** Experimental apparatus used to expose membranes to $^{14}CO_2$.

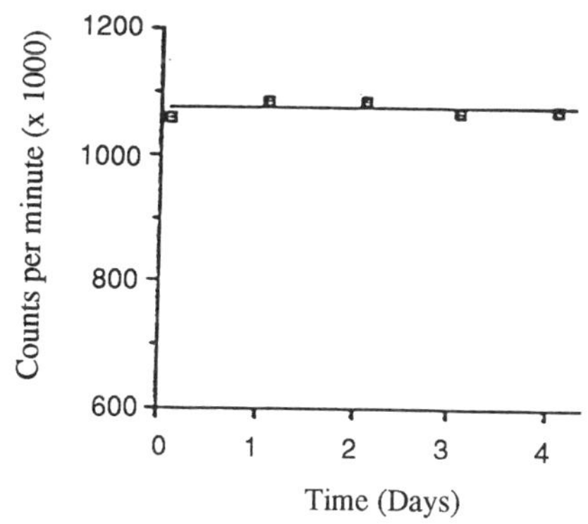


**Figure 7.** Retention of $^{14}CO$ on SPA membranes.

## PHOTOIMPRINTING OF MEMBRANES

The feasibility of the photoimprinting technique was tested by preparing solutions containing various ratios of 4-carboxybenzenesulfonazide (a PFC) to diphenylhydantoin (the print molecule) in methanol. To these solutions were added 13 mm polysulfone membranes (filter disks) possessing a pore diameter of 0.22 microns. The mixtures were allowed to equilibrate permitting the print molecules and any associated PFC's to move into the pore volumes of the membranes. The membranes were then removed from the solutions and immediately frozen in a liquid nitrogen environment. This was followed by exposing the membranes to UV radiation (254 nm) using a hand held lamp. It was expected that a few of the PFC's that were associated with the print molecules in the pores of the membrane would insert into the bonds of the pore walls following photoactivation. Most of the PFC's probably react with the solvent or insert into bonds in an unproductive manner, i.e. not resulting in the formation of photoimprinted receptors. A control membrane was prepared in an identical manner except that the original solutions contained only the PFC molecule, not the print molecule. All experiments and controls were run in triplicate. The general scheme of this photoimprinting process appears in the figure below.

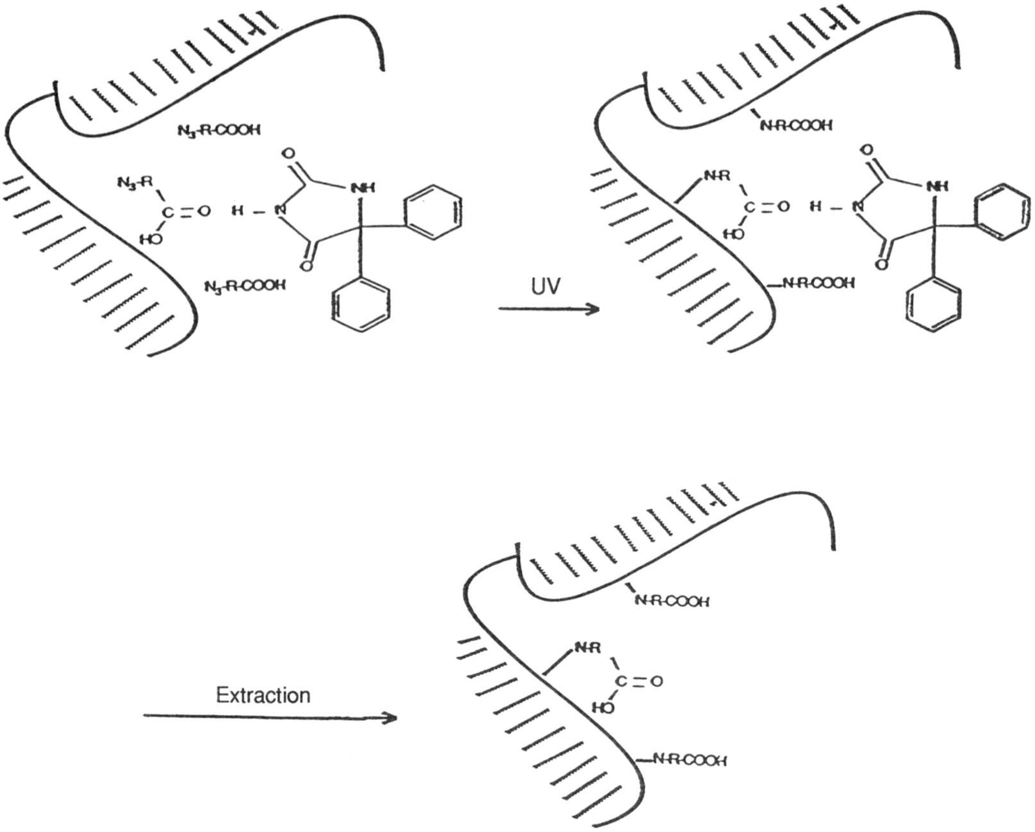


After photoactivation, the print molecules were exhaustively extracted from the membranes using acetonitrile until no print molecule could be detected in the extracts. The membranes were then added to solutions containing [$^{14}$C]-diphenylhydantoin (pH 11 buffer) and incubated for 24 hours at room temperature. The membranes were then washed with buffer and added to a scintillation vial containing a cocktail designed for counting membrane filters (Filter Count®). The disintegrations per minute (dpm) detected in the photoimprinted membranes were compared to the control membranes. The results were as follows: when a PFC/print molecule ratio of 8/1 was employed, the control membranes yielded 575 $\pm$ 19 dpm/cm$^2$ of membrane while the photoimprinted membranes yielded 1221 $\pm$ 66 dpm/cm$^2$ of membrane for an alpha value of 2.12. While this may not appear to be a highly selective, it actually compares quite favorably to the alpha values obtained for molecular imprinted polymers used in chromatographic separations (45). This was accomplished in spite of the fact that very little was done to optimize this system, and that only one type of PFC was employed to provide recognition with the print molecule.

## SUMMARY AND CONCLUSIONS

Molecules possessing a functional group on one end and a photoactivatable group on the other end (PFC's) were be mixed with print molecules allowing the functional groups of the PFC's to interact with specific chemical moieties on the print molecule. The PFC's and the print molecule will associated by non-covalent interactions similar to the way an antibody associates with an antigen. These associations occurred in the pore volume of a synthetic polymeric membrane. This system was then be exposed to UV light which converted the photoactivatable groups on the PFC's to highly reactive species (nitrenes) that were capable of inserting into the bonds of the pore surface of the membrane while the other end of the molecule was still associated with the print molecule. Extraction of the print molecule yielded a cavity in the shape of the print molecule, i.e. it would yield an "imprint" on the pore surface. The membrane was then added to a solution containing radioactive print. The beta particles emitted by the radioactive print molecules bound to the membrane could be detected by the fluors in the membrane matrix. Thus, these membranes could be used in competitive binding radioassays where *only the radioactive print molecules bound to the receptor would be detected*, since these would be in close enough ***proximity*** to the fluors to induce ***scintillations***. The scintillations can be quantified in a liquid scintillation counter without using a cocktail. Radioactive print molecules that were not bound to the receptors would be too far away from the fluors to induce scintillations due to the limited range of most beta particles, especially those emitted by tritium, or the internal conversion electrons emitted by $^{125}$I.

## REFERENCES


1. M Newcomb, JL Toner, RC Helgeson and DL Cram. Host-Guest Complexation. Chiral recognition in transport as a molecular basis for a catalytic resolving machine. *J Am Chem Soc* 101:4941(1979).

2. K Johnson, RK Allemann, H Widmer and SA Benner. Synthesis, structure and activity of artificial rationally designed catalytic polypeptides. *Nature* 365:530 (1993).

3. SD Erickson, JA Simon and WC Still. Practical synthesis of a highly enantioselctive receptor for peptides. *J Org Chem* 58:1305 (1993).

4. CT Seto, JP Mathias and GM Whitesides. Molecular self-assembly through hydrogen bonding: aggregation of five molecules to form a discrete supramolecular structure. *J Am Chem Soc* 115:1321 (1993).

5. MM Conn, G Deslongchamps, J de Mendoza and J Rebek. Convergent functional groups. 13. High-affinity complexation of adenosine derivatives within induced binding pockets. *J Am Chem Soc* 115:3548 (1993).

6. R Curti and U Colombo. Chromatography of stereoisomers with "tailor made" compounds. *J Am Chem Soc* 74:3961 (1952).

7. G Wulff, A Sarhan and K Zabrocki. Enzyme-analogue built polymers and their use for the resolution of racemates. *Tetrahedron Lett* No. 44, 4329 (1973)

8. G Wulff, W Vesper, R Grobe-Einsler and A Sarhan. Enzyme-analogue built polymers: On the synthesis of polymers containing chiral activities and their use for the resolution of racemates. *Makromol Chem* 178:2799 (1977).

9. G Wulff, B Heide and G Helfmeier. Molecular recognition through the exact placement of functional groups on rigid matrices via a template approach. *J Am Chem Soc* 108:1089 (1986).

10. J Damen and DC Neckers. Memory of synthesized vinyl polymers for their origins. *J Org Chem* 45:1328 (1980).

11. KJ Shea and EA Thompson. Template synthesis of macromolecules. Selective functionalization of an organic polymer. *J Org Chem* 43:4253 (1978).

12. KJ Shea, EA Thompson, SD Pandey and PS Beauchamp. Template synthesis of macromolecules. Synthesis and chemistry of functionalized macroporous polydivinylbenzene. *J Am Chem Soc* 102:3149 (1980).

13. KJ Shea and TK Dougherty. Molecular recognition on synthetic amorphous surfaces. The influence of functional group positioning on the effectiveness of molecular recognition. *J Am Chem Soc* 108:1091 (1986).

14. KJ Shea and DY Sasaki. On the control of microenvironment shape of functionalized network polymers prepared by template polymerization. *J Am Chem Soc* 111:3442 (1989).

15. R Arshady and K Mosbach. Synthesis of substrate-selective polymers by host-guest polymerization. *Makromol Chem* 182:687 (1981).

16. O Norrlow, MO Mansson and K Mosbach. Improved chromatogrphy: prearranged distances between boronate groups by the molecular imprinting approach. *J Chromatog* 396:374 (1987).

17. DJ O'Shannessy, LI Andersson and K Mosbach. Molecular recognition in synthetic polymers. *J Mol Recognition* 2:1 (1989)

18. B Ekberg and K Mosbach. Molecular imrpinting: a technique for producing specific separation materials. *Tibtech* 7:92 (1989).

19. LI Andersson, DJ O'Shannessy and K Mosbach. Molecular recognition in synthetic polymers: preparation of chiral stationary phases by molecular imprinting of amino acids. *J Chromatog* 513:167 (1990).

20. LI Andersson and K Mosbach. Enantiomeric resolution on molecularly imprinted polymers prepared with only non-covalent and non-ionic interactions. *J Chromatog* 516:313 (1990).

21. LI Andersson, A Miyabayashi, DJ O'Shannessy and K Mosbach. Enantiomeric resolution of amino acids on molecularly imprinted polymers as monitored by potentiometric measurements. *J Chromatog* 516:323 (1990).

22. L Fischer, R Muller, B Ekberg and K Mosbach. Direct enantioseparation of b-adrebergic blockers using a chiral stationary phase prepared by molecular imprinting. *J Am Chem Soc* 113:9358 (1991).

23. B Sellergren and K Shea. Influence of polymer morphology on the ability of imprinted network polymers to resolve enantiomers. *J Chromatog* 635:31 (1993).

24. J Matsui, T Kato, T Takeuchi, M Suzuki, K Yokoyama, E Tamiya and I Karube. Molecular recognition in continuous polymer rods prepared by a molecular imprinting technique. *Analyt Chem* 65:2223 (1993).

25. DJ O'Shannessy, B Ekberg, LI Andersson and K Mosbach. Recent advances in the preparation and use of molecularly imprinted polymers for enatiomeric resolution of amino acid derivatives. *J Chromatog* 470:391 (1989).

26. A Leonhardt and K Mosbach. Enzyme mimicking polymers exhibiting specific substrate binding and catalytic functions. *React Polymer* 6:285 (1987).

27. SE Bystrom, A Borje and B Akerman. Selective reduction of steroid 3- and 17-ketones using $LiAlH_4$ activated template polymers. *J Am Chem Soc* 115:2081 (1993).

28. M Glad, O Norrlow, B Sellergren, N Siegbahn and K Mosbach. Use of silane monomers for molecular imprinting and enzymes entrapment in polysiloxane-coated porous silica. *J Chromatog* 347:11 (1985).

29. S Mann. Molecular tectonics in biomineralization and biomimetic materials chemistry. *Nature* 365:499 (1993).

30. RK Mansfield. Ph.D. Dissertation, University of Kentucky, 1992.

31. G Vlatakis, LI Anderson, R Muller and K Mosbach. Drug assay using antibody mimics made by molecular imprinting. *Nature* 361:645 (1993).

32. MJ Swanson, SG Dunkirk, JA Pietig and PE Guire. Use photochemistry to modify membranes. *Chemtech* 624 (Oct 1992).

33. H Sigrist, H Guo and B Wegmuller. Light-dependent covalent immobilization of biomolecules on inert surfaces. *Bio/Technology* 10:1026 (1992).

34. SG Dunkirk, SL Gregg, LW Duran, JD Monthls, JE Haapala, JA Marcy, DL Clapper, RA Amos and PE Guire. Photochemical coatings for the prevention of bacterial colonization. *J Biomaterials Applications* 6:131 (1991).

35. F Borle, M Sanger and H Sigrist. Planar bilayer membranes from photoactivable phospholipids. *Biochim Biophys Acta* 106:144 (1991).

36. RL Potter and BE Haley. Photoaffinity labeling of nucleotide binding sites with 8-azidopurine analogs: Techniques and applications. *Meth Enzymol* 91:613 (1983)

37. K Rajagopalan, AJ Chavan, BE Haley and DS Watt. Synthesis and application of bidentate photoaffinity cross-lining reagents. *J Biol Chem* 268:14230 (1993).

38. HE Hart and EB Greenwald. A unique radioassay using scintillation proximity beads. *Mol Immunol* 16:265 (1979).

39. S Udenfriend, LG Gerber, L Brink and S Spector. Scintillation proximity radioimmunoassay utilizing I-125 labelled ligands. *Proc Natl Acad Sci USA* 82:8672 (1985).

40. S Udenfriend, L Gerber and N Nelson. Scintillation proximity assay: A sensitive and continuous isotopic method for monitoring ligand/receptor and antigen/antibody interactions. *Analyt Bichem* 161:494 (1987).

41. N Nelson. A novel method for detection of receptors and membrane proteins by scintillation proximity assay. *Analyt Biochem* 165:287 (1987).

# DESIGN OF MEMBRANE-BASED BIOARTIFICIAL ORGANS

Frank T. Gentile, Edward J. Doherty, Rebecca H. Li, David H. Rein, and Dwaine F. Emerich

CytoTherapeutics, Inc.
2 Richmond Square
Providence, Rhode Island 02906

## INTRODUCTION

The goal of encapsulated cell therapy research is to develop implants containing living xenogeneic cells to treat serious and disabling human conditions. The concept is straightforward: cells or small clusters of tissue are surrounded by a selective membrane barrier which admits oxygen and required metabolites, releases bioactive cell secretions but restricts the transport of the larger cytotoxic agents of the body's immune defense system. Use of a selective membrane both eliminates the need for chronic immuno-suppression in the host and allows cells to be obtained from non-human sources, thus avoiding the cell-sourcing constraints which have limited the clinical application of generally successful investigative trials of *unencapsulated* cell transplantation for chronic pain[1], Parkinson's disease[2], and type I diabetes.[3-5] Cross-species immunoisolated cell therapy has been validated in small and large animal models of chronic pain[6,7], Parkinson's disease[8,9], and Type 1 diabetes[10-12], and is under active investigation by several groups in animal models of Huntington's[13], ALS[14] and Alzheimer's[15-19]. In addition, the first encapsulated therapy using xenografts in humans has been performed in chronic pain[20] and ALS[21].

Recent reports have demonstrated the effectiveness of small diameter, ultra-filtration (UF) grade, hollow fibers as immunoisolation devices for the protection of allogeneic[22] and xenogeneic[8-25] transplantation of cells. Although of varying composition, these devices typically consist of cells imbedded in an extracellular matrix (usually a hydrogel) and enclosed in an immunoisolation membrane (See Figure 1). In addition to providing biocompatibility and permitting retrieval of appropriately designed implants, the role of the membrane portion of the device is to restrict the passage of large immune-system molecules and shed antigens, while permitting relatively unhindered passage of small molecules required for cell metabolism and release

*Biofunctional Membranes*, Edited by
D.A. Butterfield, Plenum Press, New York, 1996

of bioactive products to the host. Here we examine some of the factors which affect design and application of these devices. First, we will discuss the cell and extracellular matrix issues for these devices. Second, we will address membrane formulation and properties with particular emphasis on diffusive transport characterization. Third, we will present a case study of the evaluation of these devices in small and large animal models of Alzheimer's disease.

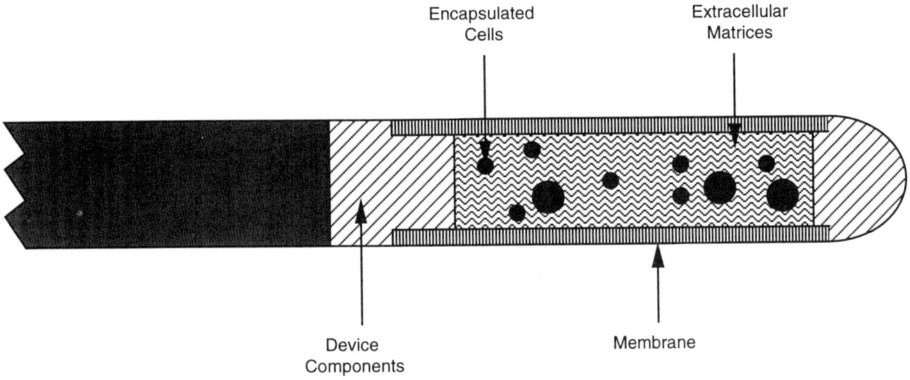


**Figure 1.** Schematic representation of a hollow fiber membrane based bioartificial organ for use in lumbar intrathecal, lateral ventricular, or parenchymal implantation.

## CELLS AND EXTRACELLULAR MATRICES USED IN ENCAPSULATION

### Cells

Cells used within these membrane based devices fall into one of three basic types: (1) primary post-mitotic cells such as porcine islets of Langerhans, bovine adrenal chromaffin cells or porcine hepatocytes; (2) immortalized (or dividing) cells such as PC12 (derived from a rat pheochromacytoma) or fibroblasts; and (3) cell lines that have been engineered to secrete a bioactive substance such as baby hamster kidney (BHK) cells which secrete human nerve growth factor (hNGF). Dividing tissue has advantages over post-mitotic tissue in that they can be banked and thus more easily tested for sterility and contaminants.

### Extracellular Matrices

*In vivo,* extracellular matrices (ECMs) provide control of cell function through the regulation of morphology, proliferation, differentiation, migration and metastasis.[26-29] Within a capsule, ECMs were originally employed simply to prevent aggregation of cells *(immobilization)* and resultant central necrosis, but have since been found beneficial to the viability and function of cells that require immobilization as well as a *scaffolding* for anchorage dependent cell lines. Currently employed matrices are derived from naturally occurring polysaccharides (sodium alginate, chitin, chitosan or hyaluronic acid) or biologically-derived materials (Matrigel, vitrogen, Type I and IV collagen). Sodium alginate has been used with a wide variety of cell types (e.g. rat, canine and porcine islets and bovine adrenal chromaffin cells)[7,12]. Chitosan is deacetylated chitin and has been used as a substrate for PC12 cells.[9,30] Matrigel is a complex matrix consisting of growth factors and basement membrane, which has been used to study attachment, migration and differentiation of many cell types including neuronal, epithelial and endothelial cells. Low concentrations of laminin, fibronectin, Type I and Type IV collagen have been reported to support attachment, but not cell spreading, of hepatocytes.[31,32] Collagen has also been reported as a matrix for baby hamster kidney cells genetically engineered to secrete human nerve growth factor.

## PREPARATION OF MEMBRANES USED FOR IMMUNOISOLATION

The majority of thermoplastic ultrafiltration (UF) and microfiltration (MF) membranes used to encapsulate cells are manufactured from homogenous polymer solutions by phase inversion. Ultrafiltration membranes have pore sizes ranging from 5 nm to 0.1 µm while microfiltration (or microporous) membranes have pores ranging from 0.1 µm to 3 µm. Phase inversion is a very versatile technique allowing for the formation of membranes with a wide variety of nominal molecular weight cutoffs, permeabilities and morphologies[33]. The morphology and membrane properties depend on thermodynamic parameters and kinetics of the process. The polymer is dissolved in an appropriate solvent. The solution is then cast as a flatsheet or extruded as a hollow fiber. As part of the casting or extrusion procedure, the polymer solution is precipitated by a phase transition which can be brought about by a change in temperature or solution composition. This process involves the transfer of a single phase liquid polymer solution into a two phase system that consists of a polymer-rich phase that forms the membrane structure and a second liquid polymer-poor phase that forms the membrane pores. Any polymer that will form a homogenous solution which under certain temperatures and compositions will separate into two phases can be used. Thermodynamic and kinetic parameters such as chemical potential of the components and the free energy of mixing of the components determine the manner in which the phase separation takes place[33]. The process can be described by a polymer/solvent/non-solvent ternary phase diagram.

If membrane strength is limiting for the overall device strength, then the membrane must be manufactured with certain considerations in mind. For example, the membrane dimensions, composition and structure may have to be altered to increase the strength. Choosing a material that is inherently stronger (i.e. more ordered), or has a higher molecular weight, with which to cast the membrane should increase the overall mechanical properties. UF or MF membranes can be fabricated with macrovoids within the wall or as an open cell foam where the microvoids are interconnected. By incorporating techniques that increase this isoreticulated structure within the membrane wall, the tensile strength can be increased generally at the same overall membrane porosity, thus maintaining the same overall diffusive transport. The strength can also be improved by increasing the cross-sectional area of the membrane by thickening the walls. Decreasing the overall membrane porosity will also serve to increase the overall membrane strength. Examples of both macrovoid containing and isoreticulated structures are shown in Figure 2.

The outer morphology of the membranes can be altered during fabrication or by a post treatment to improve the reaction required for a successful implant. Using various phase inversion techniques, the outer surface of the membrane can range from a rejecting skin to a structure that is large enough to allow cells to enter into the wall itself (approximately 10 μm in diameter). The combination of proper membrane transport and outer morphologies may also be achieved using composite membranes. Brauker[34] and Boggs *et al.*[35] have used such membranes for the treatment of Type I diabetes.

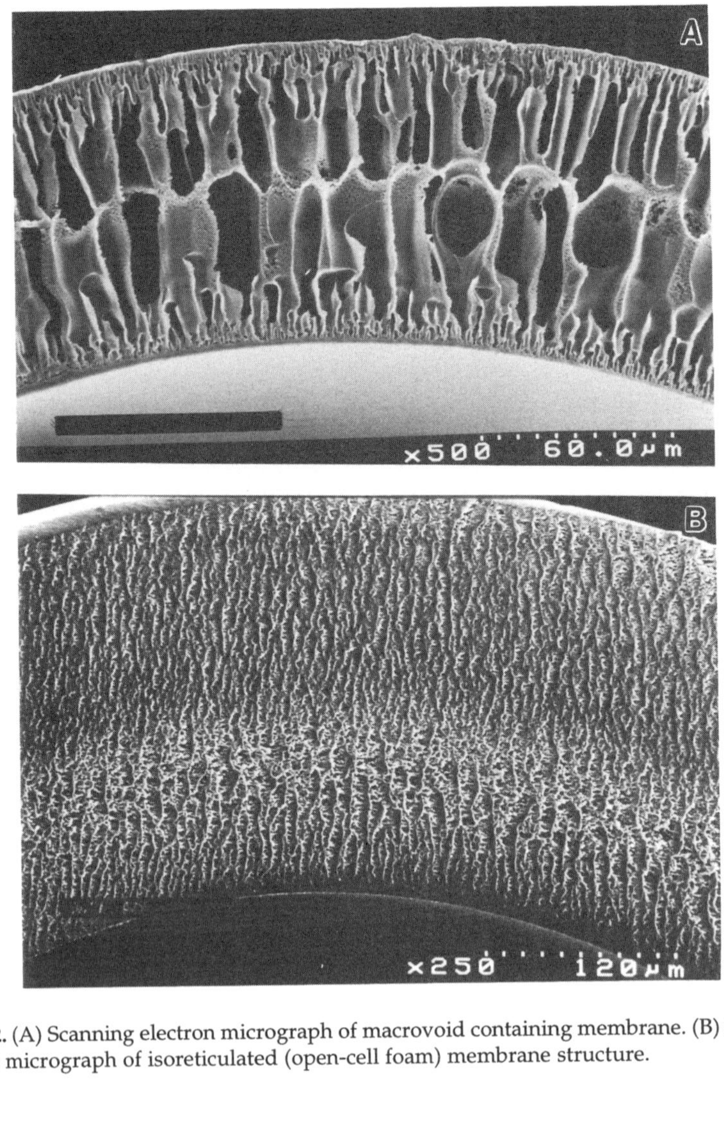


**Figure 2.** (A) Scanning electron micrograph of macrovoid containing membrane. (B) Scanning electron micrograph of isoreticulated (open-cell foam) membrane structure.

## MEMBRANE DIFFUSIVE TRANSPORT CHARACTERIZATION

Transport of these membranes is typically characterized in terms of their capability to retain marker molecules in convective sieving experiments that comparing which species are past or are retained by the membrane. Here the rejection coefficient, R, is defined as:

$$R = 1 - C_f/C_r \qquad (1)$$

where $C_f$ and $C_r$ are the concentrations of a marker in the filtrate and retentate, respectively.

Although standard convective measurements give an approximate idea of what size molecules can pass through a membrane, they are unable to predict the rate or selectivity in diffusion-based devices. Since all currently successful immunoisolation systems are primarily diffusion based[35-40], it is essential to know diffusive transport properties in order to understand the intra-capsular environment of encapsulated cells, optimize the passage of essential metabolic species and delivery of product, and assess the degree of immune exposure.

### General Theory

In the absence of electrochemical or buoyant effects, the one-dimensional flux, N [$g/cm^2$ s], of a dissolved solute, i, may be represented by the following equation:

$$N_i = (-D\, \partial C/\partial x)_i + (1-\sigma_i)\, \varpi_i \qquad (2)$$

where D is the diffusion coefficient [$cm^2/s$], C is the concentration [$g/cm^3$], x is the distance [cm], $\sigma$ is the Stavermen reflection coefficient [dimensionless], and $\varpi$ is the mass velocity per surface area [$g/cm^2$ s]. The concentration and pressure dependent terms above represent the diffusive and convective components of solute transport. Equation (2) can usually only be integrated for special cases where the interdependence of convection and diffusion is known or where the two components can be assumed to be independent and additive. However, in the case of no transmembrane convection (velocity, $\varpi_i = 0$), applicable for many immunoisolation devices, the second term drops out, and Equation (2) becomes Fick's first law which can be integrated assuming a homogenous membrane and a solute concentration independent diffusion coefficient[41], yielding:

$$N_i = [k_t\,(C_l - C_b)]_i = (k_t\,\Delta C)_i \qquad (3)$$

where $\Delta C$ is the bulk solution concentration difference between the fiber lumen (l) and the bath (b) [$g/cm^3$], and $k_t$ is the overall mass transfer coefficient [cm/s]. Although non-linearities in the concentration profile caused by chemical, charge, or other interactions of the solute with the membrane material may make the internal concentration impossible to predict, they can usually be taken into account by using overall "lumped-parameter" mass transport coefficients. Ideally, solute concentration is uniform in the liquid region on either side of the membrane and decreases linearly across the membrane. In actuality, such profiles are approximated only when the membrane resistance is large compared to the unstirred liquid resistances. Even with aggressive stirring, the transport of small, rapidly diffusing solutes can lead to gradients within the media which, if neglected, may cause large errors in calculated membrane mass transfer coefficients. The contribution of the unstirred liquid regions, often referred to as boundary layers, may be accounted for by defining the reciprocal overall mass transfer coefficient as an inverse sum of the coefficients for each region:

$$\frac{1}{k_t} = \frac{1}{k_b} + \frac{1}{k_m} + \frac{1}{k_l} \tag{4}$$

where the subscripts "b", "m", and "l" refer to the outside bath, membrane and fiber lumen respectively.

**Small Solutes**

To minimize the effect of boundary layers, the membrane diffusivity of rapidly diffusing small solutes was measured using the flowing-type system in which solute diffuses from a recirculating bath fluid through the membrane wall and is carried away by buffer that is slowly pumped through the fiber lumen at a known flow rate. The mixing cup concentration of solute "i", $(C_{mc})_i$, which is equal to the flux divided by the average velocity, was measured in the collected lumen fluid. Using Equation assuming the buffer entering the fiber lumen is solute-free, the overall device-averaged mass transfer coefficient can be expressed in terms of the following easily measured variables:

$$(k_t)_i = (C_{mc})_i \, Q_f \,/(A \, \Delta C_i) \tag{5}$$

where A is the log mean membrane surface area calculated for an annulus (= $2\pi(r_2\text{-}r_1)z/(\ln(r_2/r_1)$, where $r_2$ and $r_1$ are the outer and inner radii, respectively, z is the length), and $Q_f$ is the volumetric flowrate through the fiber lumen in μl/min.

*Membrane $(k_m)_i$:* The transport resistance of the membrane is a complicated function of bulk diffusivity, D; membrane thickness; tortuosity; equilibrium partition coefficient (which in turn is a function of porosity, pore size distribution and membrane/solute interactions); and reduced pore diffusivity. In most practical situations, $(k_m)_i$ is impossible to calculate from fundamental principles and is instead calculated from measurements of overall mass transport. Effectively, $(k_m)_i$ can be viewed as the proportionality constant relating the measured flux, $N_i$, to the concentration difference of solute "i" across the membrane if the solution concentration at the membrane/solution interface were known.

*Bath Side Boundary Layer $(k_b)_i$:* A boundary layer develops because of the mass transport resistance, $(1/k_b)_i$, of the fluid immediately adjacent to either side of the membrane. The result of this resistance is that a concentration gradient must exist between the surface solution and the bulk solution in order to drive the species across this region. Stirring or flow can reduce, but not eliminate the boundary layer resistance. As will be shown later, the boundary layer is numerically important only for rapidly diffusing small molecules where $R_b$, $R_m$ and $R_l$ are within an order of magnitude of each other.

The experimental system for the measurement of small solute diffusion is designed such that the bath solution flows tangential to the surface of the hollow fiber. (This provides a simpler method for estimating extra-lumenal flow than for most hollow fiber dialyzers). In this cross-flow configuration, the bath side mass transfer coefficient can be calculated using a Sherwood number (Sh) analogy for flow across a tube[42] such that:

$$Sh_{bi} = \{k_{bi}\, d_2/D\}_i = 0.911\, Re^{0.385}\, Sc^{0.333} \tag{6}$$

$Re = vd_2/\upsilon$ Reynolds number
$Sc = \upsilon/D_i$ Schmidt number

valid for $4 \leq Re \leq 40$; where $d_2$ is the outer fiber diameter [cm]; $D_i$ is the bulk solution diffusivity of solute "i" [$cm^2/s$]; v is the bath fluid velocity relative to the stationary fiber [cm/s]; and $\upsilon$ is the kinematic viscosity of the bath fluid [$cm^2/s$]. The value of v used in this analysis is defined as the volumetric flowrate through the bath divided by the cross-sectional area of the fiber.

*Fiber Lumen Boundary Layer* $(k_l)_i$: The transport resistance of a liquid flowing inside of a hollow fiber has been studied extensively in the field of dialysis. This resistance can be represented by a Sherwood number:

$$(k_l)_i = (Sh_l)_i D_i/d_l \tag{7}$$

where $d_l$ is the fiber inner diameter. Colton and Lowrie[43] present the graphical solution for $(Sh_l)_i$ for the case of axial laminar flow through a hollow tube, with no ultrafiltration, and neglecting axial diffusion. $(Sh_l)_i$ is shown to be a strong function of the dimensionless length, $\chi$:

$$\chi = zD/vd^2 = zD\pi/4Q_f \tag{8}$$

where z is the fiber length [cm], and v is the bulk average velocity of fluid in the fiber [cm/s], and $Q_f$ is the volumetric flowrate through the fiber lumen [$cm^3/s$]. $(Sh_l)_i$ is also a function of the wall Sherwood number, $Sh_w$, which is the sum of the membrane and bath side Sherwood numbers.

In most practical cases, for set geometry, flowrates, fiber diameter, and solute, "i": $(k_b)_i$ is a constant, and $(Sh_l)_i$ can often be taken as a pseudo constant (numerically equal to 4.0) because of the low dependence on $(Sh_w)_i$, in which case $k_l$ and $k_b$ can be explicitly calculated and $(k_m)_i$ can then be calculated in a single equation:

$$(k_m)_i = 1/\{(A\, \Delta C/Q_f\, C_{mc}) - (1/k_l + 1/k_b)\}_i \tag{9}$$

The log mean concentration difference, $\Delta C_{lm}$ [$g/cm^3$], is usually used in Equation (9):

$$(\Delta C_{lm})_i = \{(\Delta C_0 - \Delta C_z)/\ln (\Delta C_0 / \Delta C_z\}_i \tag{10}$$

where the subscripts 0 and z refer to the beginning and end of the fiber respectively. After steady state has been achieved, if the bath flow is run in excess and there is no solute in the starting lumen reservoir, then $C_{b0} = C_{bz} = C_b$, $C_0 = 0$, and $C_{mc}$ = mixing cup concentration. Thus,

$$(k_m)_i = 1/\{(A/Q_f \ln [C_b/(C_b - C_{mc}] - (1/k_l + 1/k_b)\}_i \tag{11}$$

**Large Solutes**

"Large solutes" are practically defined as solutes with a molecular weight or size equal to or greater than the nMWCO of the membrane (defined here as the molecular weight at 90% rejection in a convective measurement). These solutes are sterically excluded from all but the largest pores, creating a membrane transport resistance which is at least an order of magnitude greater than the surrounding bath and lumen resistances. Since $k_l$ and $k_b$ were calculated to be less than 10% of the membrane resistance, this contribution to the overall resistance was assumed to be insignificant and the elaborate calculations introduced for the smaller solutes to compensate for the boundary layer contributions were not necessary. The large solutes used in the current diffusion study included both proteins (Bovine Serum Albumin, Immunoglobulin-G and Apoferritin) and fluorescein-tagged dextrans (10K - 2000K). Because long measurement times were necessary to attain detectable quantities of these large solutes, the membrane mass transfer coefficient, $(k_m)_i$, was measured in the static system, wherein marker solution was loaded inside the fiber and allowed to diffuse through the membrane into the surrounding bath which was initially filled with solute-free buffer. Stirring was not required to reduce boundary layers as diffusion through the membrane was the rate limiting step in transport. Equating the flux $N_i$ from equation (3) with the flux across the membrane we obtain:

$$N_i = V_b(dC_b/dt)/A = \{k_t \Delta C\}_i \tag{12}$$

A mass balance equates the amount of solute leaving the fiber to the amount entering the bath:

$$V_b dC_b/dt = -V_l\, dC_l/dt \tag{13}$$

Assuming that $k_t \approx k_m$, then combining equation (13) with (12) and integrating we obtain the following pseudo-steady-state transport expression presented by Colton *et al.*[44]:

$$\ln\left[\frac{(C_l - C_b)_t}{(C_l - C_b)_0}\right] = -(k_m)_i At\left[\frac{1}{V_l} + \frac{1}{V_b}\right] \tag{14}$$

where the mass transfer coefficient is in [cm/s], $V_b$ is the bath volume [$cm^3$], $V_l$ is the lumen volume [$cm^3$], and t is the time [s]. For protein diffusion experiments, samples were removed from the bath at timed intervals and a replacement volume of solute-free buffer added to avoid depleting the bath reservoir. This dilution is accounted for by modifying equation (14) with the following[44]:

$$\ln\left[\frac{(C_l - C_b)_{n+1}}{(C_l - C_b')_n}\right] = -(k_m)_i A t_n \left[\frac{1}{V_l} + \frac{1}{V_b}\right] \tag{15}$$

where :

$$(C_b)'_n = (C_b)_n \,(1 - V_r/V_b) \tag{16}$$

$$(C_l)_{n+1} = (C_l)_n + (V_b/V_l)\,\{(C_b)'_n - (C_b)_{n+1})\} \tag{17}$$

$t_n$ is time of sampling of the $n^{th}$ sample, $(C_l)_n$ and $(C_b)_n$ are the lumen and bath concentrations of the $n^{th}$ sample respectively, and $V_r$ is the replacement volume. For the dextran diffusion experiments, it was not necessary to account for dilution of the bath since fewer samples were taken and smaller sample volumes were required due to the increased sensitivity in assaying of the fluorescently tagged dextrans.

## Results

All measurements were performed on a single formulation of a hollow fiber prepared from solutions of Poly(acrylonitrile-co-vinyl chloride) in an organic solvent via phase inversion. The results validating the system can be found in Dionne *et al.*[45] Here, we show an example of the final diffusive characterization of a single membrane type used in transplantation studies.

Figure 3 contains the membrane mass transfer coefficient and relative membrane diffusivity ($D_m/D_{H20}$) for all markers plotted versus molecular size. The convective sieving curve using protein markers is also shown.

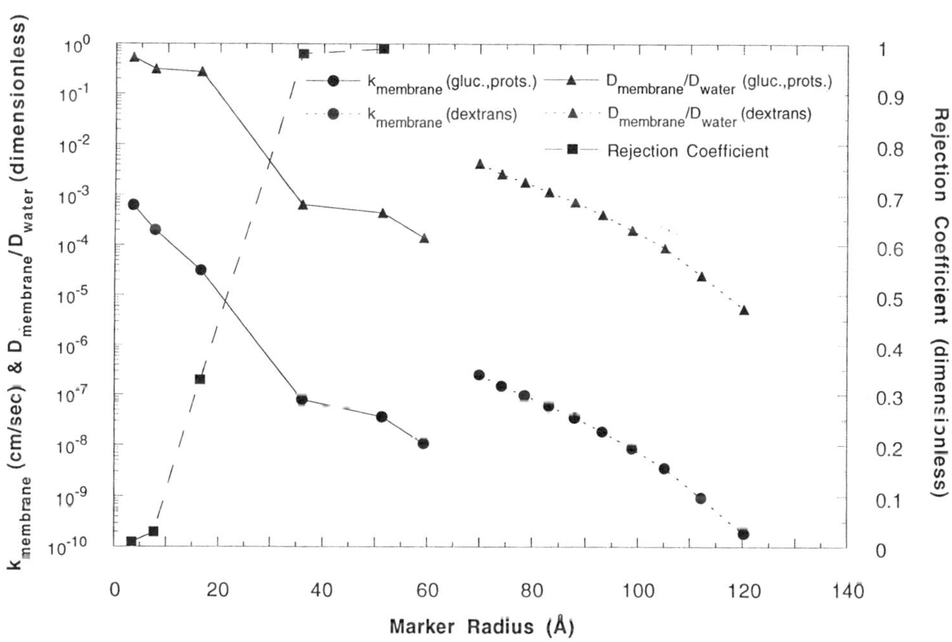


**Figure 3.** Convective rejection coefficient, membrane mass transfer coefficient and membrane diffusivity relative to water vs. molecular size at 37 °C for glucose and proteins and at 4 °C for dextrans.

## CASE STUDY: USE OF ENCAPSULATED CELLS IN THE CENTRAL NERVOUS SYSTEM

Much attention has been focused on cell transplantation in the Central Nervous System (CNS). The use of this approach in both the parenchymal and intrathecal spaces (lateral ventricle and lumbar intrathecal) has particularly strong advantages. First, it provides site specific delivery of various agents. Many of the factors of interest do not cross the blood-brain barrier (e.g. dopamine for Parkinson's patients) or have a relatively short half-life (e.g. dopamine and ciliary neurotrophic factor, CNTF, for amyotrophic lateral sclerosis, ALS, patients). Second, encapsulation provides the added safety benefit of being, in some forms, retrievable. Table 1 shows the patient population for some of the CNS disorders for which delivery of various factors may be possible through encapsulated cells.[46] The following is a case study for the use of encapsulated cells in the treatment of Alzheimer's disease.

**Table 1.** Prevalence of central nervous system disorders which may be treated via encapsulated cells.

| Indication | Prevalence (US only)* |
|---|---|
| Dementias | 4,000,000 - 5,000,000 |
| (Alzheimer's) | (3,000,000 - 4,000,000) |
| Parkinson's | 800,000 |
| Amyotrophic lateral sclerosis (ALS) | 25,000 |
| Huntington's disease | 12,000 - 25,000 |
| Brain tumor (all types) | 1,000,000 |
| Chronic pain | 600,000 |
| Multiple sclerosis | 200,000 - 500,000 |
| Epilepsy | 2,000,000 |

*Sources: National Parkinson Foundation, Parkinson's Disease Foundation, National Information Center for Orphan Drugs and Rare Diseases (ODPHP), National Health Information Center, and National Institute of Neurological Disorders and Statistics.

Alzheimer's disease (AD) affects 4-6% of the population over the age of 65 and is the most prevalent form of adult onset dementia. Current estimates project that its incidence will triple over the next 75 years. The institutional cost for AD patients is 3 billion dollars per year. There is no currently available treatment for AD patients. This disorder is characterized by a progressive deterioration of cognitive and mnemonic ability.

Basal forebrain cholinergic neurons degenerate in several human dementing illnesses including Alzheimer's disease (AD). The loss of cortical cholinergic neurotransmission resulting from the degeneration of these neurons has been suggested to play an integral role in the cognitive decline observed in AD. At present, treatments are ineffective for slowing or preventing the loss of cholinergic neurons or the associated memory deficits. Converging lines of evidence indicate that nerve growth factor (NGF) has potent target-derived trophic and tropic effects upon cholinergic basal forebrain neurons.

In various animal studies and in one human patient, NGF was delivered into the lateral ventricles via a chronically indwelling pump/cannula system.[47] Indeed, to date, long-term administration of NGF has been limited to intraventricular infusions using cannulae or pumps. Unfortunately, the need for multiple intraventricular injections of NGF over the long-term presents clinical as well as practical problems. These routes of administration are confounded by the need for repeated injections or refilling of pump reservoirs to maintain NGF levels and avoid the degradation of NGF in solution. In addition, chronic low dosage infusion of NGF is difficult to obtain using current pump technology. An alternative method of delivering NGF is the implantation of cells that have been genetically modified to produce NGF. This avoids the problem of degradation and repeated refilling while allowing a localized distribution of NGF within the CSF or parenchyma. Below, we summarize work directed at delivering human NGF into rodent and primate model of AD using polymer-encapsulated cells which have been genetically modified to secrete hNGF.

Prior to beginning human clinical trials with NGF, several criteria should be met including a consistent, reliable and well characterized source of hNGF. The hNGF should be biologically active and affect basal forebrain cholinergic neurons in experimental animals including nonhuman primates. Transplantation of cells which have been genetically modified to secrete high levels of hNGF is one means of meeting these criteria. Recently, we have modified baby hamster kidney (BHK) fibroblasts to produce high and stable levels of hNGF. These BHK cells have been encapsulated within ultrafiltration grade poly(acrylonitrile-co-vinyl chloride) membranes and have been shown to survive implantation for extended periods of time while continuing to release biologically-relevant levels of hNGF. Together with the inherent advantages of using a genetically engineered cell line, encapsulation permits molecules of interest to diffuse from the capsule into the surrounding host tissue.

To begin to evaluate the potential of encapsulated BHK-hNGF cells as a treatment for AD, we have looked at the ability of these cells to prevent the loss of cholinergic neurons following axotomy. Following aspirative lesions of the fimbria/fornix, the cholinergic neurons within the medial septum atrophy and die. This cell loss has been shown to be prevented by concurrent administration of NGF and provides an appropriate system for examining the neuroprotective effects of encapsulated cells. In our first studies[17], rats received lesions of the fimbria/fornix followed by intraventricular implants of BHK-hNGF or BHK-Control (non-transfected) cells. The BHK-control implanted animals had an extensive loss of ChAT-positive neurons ipsilateral to the lesion. In contrast, ChAT-positive neurons ipsilateral to the site of BHK-hNGF capsule insertion appeared normal with no apparent atrophy or loss of ChAT-immunoreactivity. Quantitation of ChAT-positive neurons for the two groups revealed that with the BHK-control capsules only 14±3% of the neurons remained ChAT-positive on the lesioned side of the brain compared to the non-lesioned side, while with the BHK-hNGF capsules, 88±10% of the cholinergic neurons were immunopositive for ChAT.

Following these results, we extended our studies to the primate order.[18] Cynomolgus primates received transections of the fornix followed by placement of encapsulated BHK-hNGF or BHK-Control cells into the lateral ventricle. In the control animals, a significant reduction was observed in the number of cholinergic neurons ipsilateral to the lesion. NGFr-positive neurons were decreased 54% and ChAT-immunoreactive neurons were decreased 53% within the medial septum ipsilateral to the lesion/implant relative to the con-

diagonal band. More importantly, these changes were most severe in those animals with the greatest cognitive impairments. Following training, animals received bilateral intraventricular implants of encapsulated BHK-hNGF or BHK-Control cells. The 18 and 24 month old animals receiving BHK-hNGF cells showed a significant improvement in cognitive function. No improvements or deleterious effects were observed in the young nonimpaired animals. Anatomically, the hNGF released from the encapsulated cells was sufficient to increase the size of the atrophied basal forebrain and striatal cholinergic neurons to the size of the neurons in the young healthy rats. Furthermore, there was no evidence that the BHK-hNGF cells produced changes in mortality, body weights, somatosensory thresholds, potential hyperalgesia, or activity levels, suggesting that the levels of hNGF produced were not toxic or harmful to the aged rats.

## REFERENCES

1. J.Sagen, *ASAIO Journal* 38:24 (1992).
2. O. Lindvall, et al. *Prog. Brain Res.* 82:729 (1990).
3. F. P. Sullivan *et al., Transplantation* (Baltimore) 44:465 (1987).
4. C. Ricordi, D. W. Scharp, P. E. Lacy, *Transplantation* (Baltimore) 45:994 (1988).
5. A. M. Posselt et al., *Science* 249:1293 (1990).
6. J. Sagen, H. Wang, P.A. Tresco, P. Aebischer, Transplants of immunologically isolated xenogeneic chromaffin cells provide a long-term source of pain-reducing neuroactive substances, *J. Neurosci.* 13(6):2415 (1993).
7. J.M. Joseph, M.B. Goddard, J. Mills, V. Padrun, A. Zurn, B. Zelinski, J. Favere, J.P. Gardaz, F. Mosimann, J. Sagan, L. Christenson, P. Aebischer, Transplantation of encapsulated bovine chromaffin cells in the sheep subarachnoid space: a preclinical study for the treatment of chronic pain, *Cell Transplantation* 3:355 (1994).
8. P. Aebischer, S.R. Winn, and P.M. Galletti, Transplantation of neural tissue in polymer capsules, *Brain Research* 448:364 (1988).
9. P. Aebischer, M. Goddard, P. Signore, and R. Timpson, Functional recovery in hemiparkinsonian primates transplanted with polymer encapsulated PC12 cells, *Exp. Neurology* 126:1 (1994).
10. W.L. Chick, J.J. Perna, V. Lauris, D. Law, P.M. Galletti, G. Panol, A.D. Whittemore, A.A. Like, C.K. Colton, M.J. Lysaght, Artificial pancreas using living beta cells: effects of glucose homeostasis in diabetic rats, *Science* 197:780 (1977).
11. D.W. Scharp, N.S. Mason, and R.E. Sparks, Islet immuno-isolation: the use of artificial organs to prevent tissue rejection, *World J. Surg.* 8:221 (1984).
12. P.E.Lacy, O.H. Hegre, A. Gerasimidi-Vazeou, F.T. Gentile, K.E. Dionne, Maintenance of normoglycemia in diabetic mice by subcutaneous xenografts of encapsulated islets, *Science*, 254:1782, (1991).
13. D.F. Emerich, J.P. Hammang, E.E. Baetge, S.R. Winn, Implantation of polymer-encapsulated human nerve growth factor-secreting fibroblasts attenuates the behavior and neuropathological consequences of quinolinic acid into rodent striatum, *Exp. Neurology* (in press) (1994).
14. Y. Sagot, S.A. Tan, E. Baetge, H. Schmalbruch, A.C. Kato, P. Aebischer, Polymer encapsulated cell lines genetically engineered to release ciliary neurotrophic factor can slow down progressive motor neuronopathy in the mouse, *Eur. Jour. of Neurosci.* 7:1313 (1995).

15. D. Hoffman, X.O. Breakfield, M.P. Short, and P. Aebischer, Transplantation of polymer-encapsulated cell line genetically engineered to release NGF, *Exp. Neurology* 122:100 (1993).
16. J.H. Kordower, S.R. Winn, Y.-T. Liu, E.J. Mufson, J.R. Sladek, J.P. Hammang, E.E. Baetge, and D.F. Emerich, The aged monkey basal forebrain: rescue and sprouting of axotomized basal forebrain neurons following grafts of encapsulated human NGF-secreting cells, *Proc. Natl. Acad. Sci.* 91:10898 (1994).
17. S. Winn, J. Hammang, D. F. Emerich, A. Lee, R. D. Palmiter, and E.E. Baetge, Polymer-encapsulated cells genetically modified to secrete human nerve growth factor to promote the survival of axotomized septal cholinergic neurons, *Proc. Natl. Acad. Sci.* 91: 2324 (1994).
18. D.F. Emerich, S.R. Winn, J. Harper, J. Hammang, E.E. Baetge, J.H. Kordower, Implants of polymer-encapsulated human NGF-secreting cells in the nonhuman primate: rescue and sprouting of degenerating cholinergic basal forebrain neurons, *Journal of Comparative Neurology* 349:148 (1994).
19. M.D. Lindner, C.E. Kearns, S.R. Winn, B.R. Frydel, D.F. Emerich, Effects of intraventricular encapsulated hNGF-secreting fibroblasts in aged rats. *Cell Transplantation* (in press) (1995).
20. P. Aebischer, E. Buschser, J.M. Joseph, J. Favre, N. deTribolet, M.J. Lysaght, S.A. Rudnick, M.B. Goddard, Transplantation in humans of encapsulated xenogeneic cells without immunosuppression - a preliminary report, *Transplantation* (in press) (1994).
21. N.A-M. Pochon, B. Heyd, N.Déglon, J-M. Joseph, Z. Zurn, E.E> Baetge, J.P. Hammang, M. Goddard, M. Lysaght, F. Kaplan, A.C. Kato, M. Schluep, L. Hirt, F. Regli, F. Prochet, N. deTribolet, P. Aebischer, Gene therapy for Amyotrophic Lateral Sclerosis (ALS) using a polymer encapsulated xenogeneic cell line engineered to secrete hCNTF, *Gene Therapy Protocols* (submitted) (1995).
22. D.W. Scharp, C.J. Swanson, B.J. Olack, P.P. Latta, O.D. Hegre, E.J. Doherty, F.T. Gentile, K.S. Flavin, M.F. Ansara, P.E. Lacy, Protection of encapsulated human islets without immunosuppression in patients with type i and type ii diabetes and in non-diabetic controls, *Diabetes* 43:1167 (1994).
23. S.J. Sullivan, T. Maki, K.M. Borland, M.D. Mahoney, B.A. Solomon, T.E. Muller, A.P. Monoco, W.L. Chick, Biohybrid artificial pancreas: long-term implantation studies in diabetic, pancreatectomized dogs, *Science* 252: 718 (1991).
24. P. Aebischer, J. Sagen, J. Joseph, M. Goddard, F. Mosimann, Xeno-transplantation of encapsulated bovine chromaffin cells in sheep subarachnoid for the treatment of pain: a preclinical study, *Cell Transplantation* in press (1995).
25. P. Soon-Shiong, E. Feldmen, R. Nelson, J. Komtebedde, O. Smidrod, G. Skjak-Braek, T. Espevik, R. Heintz, M. Lee, Successful reversal of spon taneous diabetes in dogs by intraperitoneal microencapsulated islets. *Transplantation* 54:769 (1992).
26. J. C. Y. Dunn, R. G. Tompkins, M. L. Yarmush, *Biotechnol. Prog.* 7:237 (1991).
27. D. J. Mooney *et al.*, *J . Cell Physiol.* 151:497 (1992).
28. A. Rotem, M. Toner, R. A. Tompkins, M. C. Yarmush, *Biotechnol. Prog.* 40:286 (1992).

29. 5. UyaTna, P.-M. Kaufmann, T. Takeda, J. P. Vacanti, *Transplantation* (in press) (1995).
30. D.F. Emerich, B.R. Frydel, T.R. Flanagan, M. Palmatier, S.R. Winn, and L. Christenson, Transplantation of polymer encapsulated PC12 cells: use of chitosan as an immobilization matrix, *Cell Transplantation*, 2:241 (1993).
31. M. Tachibana, G. R Nagamatsu, J. C. Addonizio, J. *Urol.* 133: 866 (1988).
32. A. Atala *et al., J. Urology* 148:658 (1992).
33. H. Strathmann, Production of microporous media by phase inversion processes,*in* "Material Science of Synthetic Membranes,"*American Chemical* Society 165 (1985).
34. J. Brauker, Bioarchitecture of polymers determines soft tissue response to implants, ACS Polymer Sci. and Engineering - Biomaterials for the 21st Century Meeting, Palm Springs, CA (1992).
35. D. Boggs, A. Khare, D. McLarty, R. Pauley, S.M. Sternberg, Membrane for immunoisolation, Proceedings of the North American Membrane Society, 6th Annual Meeting, Breckenridge, CO (1994).
36. C.K. Colton, E.S. Augoustiniatos, Bioengineering in development of the hybrid artificial pancreas, *J. Biomech. Eng. (Trans ASME)* 113:152 (1991).
37. L. Christenson, K.E. Dionne, M.J. Lysaght, Biomedical applications of immobilized cells, in "Fundamentals of animal cell encapsulation and immobilization", Goosen, M.F.A. ed., C.R.C. Press, Boca Raton, (1993)
38. R. Langer, J.P. Vacanti, Tissue engineering, *Science* 260:920 (1993).
39. C. Mullon, K. Dunleavy, A. Foley, J. O'Neil, J. Rudolph, C. Toscone, I. Otsu, T. Maki, A. Monoco, K. Gagnon, S. Naik, H. Santangini, D. Trenkler, H. Jauregui, B. Solomon, Use of synthetic membranes in cell based medical devices - artificial pancreas and liver assist device, *Poly. Mat. Sci. Eng.* 70:221 (1994).
40. M.J. Lysaght, B. Frydel, F.T. Gentile, D.F. Emerich, S.R. Winn, Recent progress in immunoisolated cell therapy, *J. Cell Biochem.* 56:196 (1994).
41. M.J. Lysaght, U. Baurmeister, Dialysis, *in* "Kirk-Othmer Encyclopedia of Chemical Technology, 4th Ed., 8", John Wiley & Sons, New York (1993).
42. K. Keith, W.Z. Block, "Basic heat transfer", Harper and Row, New York (1980).
43. C.K. Colton, E.G. Lowrie, Hemodialysis: physical principles and technical considerations, *in* "The kidney", Brenner, B.M. and Rector, F.C., eds., W.B. Saunders Co. ,Philadelphia (1981).
44. C.K. Colton, K.A. Smith, P. Stroeve, E.W. Merrill, Laminar flow mass transfer in a flat duct with permeable walls, *AICHE J.* 17:773 (1971).
45. K.E. Dionne, B.M. Cain, R.H. Li, E.J. Doherty, M.J. Lysaght, D.H. Rein, F.T. Gentile, Transport characterization of membranes for immunoisolation, *Biomaterials* (in press) (1995).
46. J.S. Swen, T.R. Flanagan, T.G. Wiggans, Assessing Commercial Potential Of Central Nervous System Delivery Approaches, *Methods in Neuroscience* 21:485 (1994).
47. L. Olson, A. Nordberg, H. von Holst , L. Backman, T. ebendal, I. Alafuzoff, K. Amberla, P. Hartvig, A. Herlitz, A. Lilja, H. Lundqvist, B. Langstrom, B. Meyerson, A. Persson, M. Vlitanen, B. Winblad, A. Seiger. Nerve growth factor affects $^{11}$C-nicotine binding, blood flow, EEG, and verbal episodic memory in an Alzheimer patient (case report). *J. Neuronal Transmission (P.-D. Sect.)* 4:79 (1992).

# STUDIES IN THE DEVELOPMENT OF MEMBRANE-BASED EXTRACORPOREAL DEVICES FOR THE THERAPEUTIC MANAGEMENT OF DISEASES WITH CIRCULATING PATHOMOLECULES

S.K. Gautam

Research Associate, Membrane Technology Lab,
Chemical Eng Dept and School of Biomedical Engg
Indian Institute Of Technology, Bombay-400076

## INTRODUCTION

Extracorporeal devices [ECDs] are medical appliances placed outside the body which permit the processing of blood under controlled conditions. The extracorporeal treatment of blood can assist and support a failing organ in line with its style of functioning and thereby abort the life-threatening consequences of the failed organ. Organ failure can be the long term consequence of circulating pathomolecules lodging in the capillary network of organs leading to a number of biological events. Circulating immune complexes (CIC), one such species of pathomolecule leading to organ failure, have been detected in the blood of patients of a number of malignancies and autoimmune diseases (Theofilopoulos & Dixon, 1979). Due to the continuous release of tumor antigens in these diseases, immune complexes are formed which later act as "blocking factors" and mask, modulate, or cause shedding of tumor associated antigens so that sensitized effector lymphocytes no longer recognise their antigenicity. Because of the critical role of CIC in the pathogenesis of lesions leading to organ failure, considerable efforts have been directed towards the development of techniques to artificially remove them from the blood, a recommended modality of therapeutic management of many of these diseases (WHO, 1977). Extracorporeal treatment of blood based on crossflow membrane filtration of blood has opened the possibility of physically manipulating the immune response (the so called `immunosurgery') by removal of such CIC at the bedside thereby forming what can be termed `therapeutic artificial organs'.

At present, there exists a strong competition between membrane filtration and other techniques in the area of extracorporeal manipulation of blood. Membrane filtration principle has the advantage of being simple, rapid and is recognized as a safe and efficient blood purification modality. Membranes form the heart of most extracorporeal devices because of the gentleness and speed required in the on-line bedside processing of blood. However, membranes cannot achieve the specificity of immunoadsorption systems. Tailoring membranes for specific applications is still an unfulfilled dream for membrane scientists (Cheryan, 1986). There is the need to improvise the current membrane-based blood filtration technology to make it more selective, more predictable and more controlled.

## PREVIOUS STUDIES AND LACUNAE

The classical studies of Zydney & Colton (1982) in the prediction of flux indicated that the permeate flux in crossflow membrane plasmapheresis is directly proportional to the shear rate. The incorporation of the shear-enhanced diffusivity term in the equation for the prediction of flux gave good agreement with most experimental results. However, issues of membrane fouling and the time dependency of flux have not been adequately addressed (Davis, 1992).

It is clear from earlier studies that multiple routes exist to achieve the desired amount of selectivity in membrane filtration. In a study on plasmapheresis, Zborowski and Malchesky (1990) pointed out that temperature-dependent blood filtration can achieve selectivity in a limited manner. However, the effect of temperature on the sieving characteristics in crossflow membrane plasmapheresis needs to be investigated further to exploit the possibility of tailoring selectivity by manipulating the temperature of operation.

The feed viscosity is critical in filtration systems but unfortunately, has not received its due consideration in most previous studies with blood. The whole blood viscosity not only depends on the temperature and haematocrit but also on a number of other parameters such as red blood cell rigidity, plasma protein concentration, shear rate, shear history etc. Also, the whole blood/plasma viscosity varies between health and disease conditions (Dientenfass, 1976). Earlier studies on extracorporeal devices have not taken into consideration the changed fluid profile in diseases.

Most efforts to develop viable therapeutic artificial organs have floundered at the point of blood incompatibility. As such there is a strong need to develop reliable screening procedures for the qualitative and quantitative estimation of hemocompatibility especially in extracorporeal circuits. The material, the hemodynamics (nature of blood flow) and the biological environment are the key components which interact in varying ways and to varying degrees leading to the process of compatibility of foreign interfaces. Hemocompatibility, a multiparameteric function cannot be reliably predicted by the presently available in vitro tests since most of the tests ignore considerations of hemorheological parameters (Bruck, 1980).

Since most of the available ECDs have complex geometries, the experimental performance evaluation is of much importance in the proper control of the extracorporeal processing of blood. A more "universal" relationship can be obtained if models of the dimensionless form are used. Unfortunately data in the dimensionless form is rare in literature [Cheryan, 1986].

## FOCUS OF THE WORK

The present work is a multi-disciplinary attempt at a comprehensive study of specific bioengineering problems related to the development of highly selective disease-specific membrane-based therapeutic extracorporeal devices in diseases with no conventional cures such as drugs or surgery.

## APPROACH AND METHOD OF THE INVESTIGATION

### Size Characterization Of Immune Complexes

Since sieving is the mechanism of separation in membrane filtration, size characterization of the target biomolecule species is critical. Systemic lupus erythematosus (SLE) is a typical immune-complex disease characterized by manifestations such as dermatitis, nephritis, vasculitis and arthritis due to immune complex deposits. Earlier studies have given no conclusive idea regarding the size of the CIC in SLE. Results of earlier studies range from predominantly small ie., <7S, intermediate ie., 7 to 19 S (Ooi et al, 1977), large ie., >19S (Nydegger et al, 1974) or variable (Tung et al, 1981 & Cairns et al, 1980) in size. This is probably because most of the techniques used in these studies such as size exclusion gel permeation chromatography, sucrose density gradient ultracentrifugation, cryoelectron microcopy, ultracentrifugation, fluorescence quenching, antigen binding etc can be subject to artifacts in that the CIC undergoes substantial dissociation during the measurement process.

Quasi-elastic light scattering (QELS) is a new, non-invasive and accurate technique to determine the size of immune complexes (Yarmush et al, 1988). In this study, QELS particle size characterization of the CIC aggregates has been done with an aim to evolve an appropriate membrane separation strategy to improvise the current blood filtration technology and make it highly selective.

Size Characterization Experiments: Blood samples were collected by venipuncture from overnight fasting patients of systemic lupus erythematosus (SLE) at the King Edward Memorial Hospital, Bombay. This was allowed to clot in sterile glass tubes for 2 hours, then centrifuged at 1000 x g for 10 minutes at room temperature. The obtained aliquots of serum combined with equal volumes of 6000 M Polyethylene glycol was dissolved at concentrations ranging from 2% to 12% in phosphate-buffered saline (150 ul of serum and 30 µl of PEG-EDTA-saline solution) and kept at 4°C overnight before again centrifuging at 10,000 x g for 20 minutes at 4° C and the supernant carefully discarded. The sedimented precipitate was resuspended in 2 ml of PEG in 0.01 M EDTA and saline, recentrifuged, supernants removed and the precipitate dialysed to remove the PEG. Particle size was evaluated by quasi-elastic light scattering with the help of BI-90 (Brookhaven Instrument Corpn, USA) which uses a 5 MW He Ne laser, 0.6328 µm light wavelength and a scattering angle of 90°. Measurements were taken at 40°C, 35°C, 30°C, 25°C, 20°C , 15°C, 10°C and 5°C to investigate the cryoaggregation phenomena in SLE. To find out the growth rate constants the following model of Smoulechousky was used (Jossang et al 1985)

$$R/Ro = (1+I_R t)^R \qquad \text{..........} \qquad (1)$$

where R = final radius of particle
Ro = initial radius (before aggregation)
I = growth rate constant
= constants

**Plasmapheresis Experiments To Study Filtration Characteristics**

The parallel plate geometry is a simple and preferred format for mass transfer studies. The Minitan-S crossflow system used in most of our plasmapheresis experiments is a parallel plate thin channel filtration cell with a channel height of 0.045 mm. Polyvinylidene fluoride (hydrophillic Durapore) membranes of pore diameters 0.2µm, 0.45µm and 0.65µm and effective filtration area of 32 $cm^2$ were used in our studies. The choice of hydrophillic polyvinylidene fluoride was made because these membranes have low-protein-binding [1$\mu g/cm^2$] and remain less affected by the plasma protein adsorption. Domestic goat's blood has been chosen as the appropriate working fluid because it represents mammalian blood and has basic properties such as the proportion of plasma proteins, the reaction of the coagulation system to foreign surfaces and trauma.

Continuous-flow flatsheet membrane plasmapheresis experiments have been carried out using thin-channel crossflow filtration device (Minitan-S, Millipore Corpn, U.S.A.) with PVDF Durapore membranes 0.65µm pore size and 30 cm2 effective filtration area [Fig 1]. Experiments were performed, using anticoagulated (sodium citrate 3.8 gm% in proportion of 10:1) fresh goats' blood of haematocrit 34% pumped by a peristaltic pump at a controlled rate from a reservoir placed in a temperature controlled water bath maintaining minimal changes in temperature (±1°C) in the circuit during the perfusion. The inlet pressure was monitored by a sensitive pressure transducer connected to a wheatstone bridge with an amplifier (Polyrite, Medicare, India) and the filtrate and retentate collected at atmospheric pressure. Runs lasting 15 mins each were performed and various filtration parameters measured and samples of the feed and permeate collected during steady state

and biochemical (total proteins, albumin, fibrinogen, plasma free haemoglobin), hemorheological (plasma/whole blood viscosity profiles, platelet aggregation, red cell intrinsic viscosity), hematological & surface morphological (SEM & EDAX) parameters evaluated. The change in the intrinsic membrane properties was evaluated before and after each run by measuring the normal saline permeability (NSP) at room temperature. Thorough cleaning of the membrane was performed after the completion of each run using a detergent (sodium lauryl sulphate) and Hemazyme (Miles India, Baroda) as per our cleaning protocol. Total proteins was estimated by the Biuret method, albumin by the Bromocresol green method, fibrinogen by the modified Ratnoff & Menzie method with the help of a electronic autoanalyser (Polimak, Italy). Immunoglobulin values have been calculated by subtracting albumin and fibrinogen values from total protein (with an error of ± 5%). The relative plasma filtration ratio defined as plasma filtration rate at given temperature at time t/normal saline flux at 27°C at time t, was also evaluated.

**Temperature-related Plasmapheresis Experiments**

Temperature effects on crossflow membrane plasmapheresis have been investigated with the help of hydrophillic polyvinylidene fluoride (PVDF) Durapore membranes and goat's blood as the working fluid. Runs were performed at 6±1, 10±1, 20±1, 30±1, & 40±1 °C, the various filtration parameters measured and samples of the feed and permeate collected during steady state. The filtration and sieving properties have been characterized by evaluating the normal saline (0.9 gm%) flux and the sieving coefficients of albumin, immunoglobulins & fibrinogen, respectively.

**Tailoring Membranes for Specific Blood Processing Applications**

Hollow fibre membranes are best suited for blood applications since they mimic physiological blood flow in the capillary network, the natural body filters. Hollow fibre membranes are spun using a simple, inexpensive laboratory fabricated spinneret by phase-inversion technique (Fig 12). Cellulose acetate (CA), N,N Dimethylformamide (DMF) and acetone were used as a three component model system and membranes of both flat sheet type and hollow fiber type were cast in the laboratory for specific blood processing applications. A thorough characterization of the cast membranes was done by a simple modifications of the usual membrane evaluating tests such as bubble point test, normal saline permeability (NSP) test, blood passage test and scanning electron microscopy (SEM).

**Evaluating the Fluid Profile**

Hemorrheological abnormalities should be expected in Organ failure for different reasons such as altered acid-base balance which decreases red blood cell deformability and altered hormone levels (elevated parathormone levels in case of renal failure) which increases osmotic fragility of erythrocytes. Hemorheology determines blood fluidity and regulates the onset of kinetics of the thrombotic process, a major threat to patients undergoing extracorporeal processing support. Rheological monitoring in organ failure is thus a key factor in the successful long term extracorporeal support. It seems of interest to study the haemorrheological parameters in acute and chronic renal failure since a large proportion of renal failure cases do undergo extracorporeal manipulation of blood such as haemofiltration to remove the small and middle molecular nitrogenous wastes. This will also help establish whether the organ (renal) failure is always associated with significant changes in rheological parameters. If so, then the extracorporeal blood purification schemes should take into account the changed fluid profile because any membrane-based process is significantly affected by rheological status of the feed.

**Method of Evaluating Rheological Profile of Blood in Renal failure:** The study was performed on selected cases of non-dialysed patients newly admitted with renal failure, both acute (15 patients) and chronic (24 patients) renal failure admitted to the King Edward Memorial Hospital, Bombay and another 50 normal controls. None had undergone recent blood transfusion or dialysis. None had palpable spleens, concurrent infections nor were on catabolic drugs and all were normoglycaemic. Patients under anticoagulant, anti-inflammatory agents, -blockers and calcium inhibitors were excluded. Blood samples were collected (about 8 ml) from the antecubital vein in the morning before the patients had eaten or drunk anything. The samples were collected in a EDTA bulb (800 μl) and analysed within two hours for haematological, biochemical and rheological profiles. The haemorheological work-up was performed with a low shear coaxial viscometer (Contraves-LS 30) at shear rates from 0.512 $sec^{-1}$ to 128.5 $sec^{-1}$, at a temperature of 37 °C. Results are expressed in mPa.S. The haematological work-up consisted of blood cell counts and evaluation of haemoglobin, haematocrit, mean corpuscular volume, mean corpuscular haemoglobin concentration etc. Statistical analysis of the results was performed as per the Student's t test. Results are given in mean values + standard deviation.

### 'Hemorheological' Screening of Blood-Contacting Materials in Extracorporeal Circulation

The decisive events which control haemocompatibility occur at the molecular level and affect the various subphases of blood rheology (Fig 2). The extracorporeal manipulatory process subjects blood to various fluid dynamic alterations such as varying turbulence, shear rates etc. The complexity and extreme shear-sensitivity of human blood rheology which is influenced by a series of simultaneous processes can be used to our advantage in the screening of blood-contacting surfaces in extracorporeal circuits. Haemorheology thus offers a new dimension for obtaining predictive information on the outcome of the compatibility process in extracorporeal circuits.

Materials known to be haemocompatible (medical grade silicon and polyvinylchloride of blood bags) and materials known to be extremely haemo-incompatible (pyrex glass, copper, cotton fabric, all commercial grade) were chosen along with other materials to be tested (polyvinylidene fluoride membranes from Millipore, USA) for haemocompatibility. Samples of equal surface area were mixed with 10 ml each of freshly drawn anticoagulated (with EDTA) whole human blood (from healthy volunteers) in plastic vials and incubated at 37 °C along with a control blood sample (without any material) for two hours and the haemorheological parameters hemorheological parameters evaluated as per the guidelines of the International Committee for Standardization in Hematology [ICSH] within three hours after collection of the samples. Contraves LS-30 viscometer [Contraves AG] was used to evaluate the whole blood viscosity, intrinsic viscosity of the red blood cells and the plasma viscosity. Chronolog [model 707] Dual channel recorder was used to measure the platelet aggregation. Biochemical analysis has been done by the Poli-mak M-10 microprocessor controlled fully automatic microanalyser which measures the values of chemical substances in the blood by spectrophotometry. Total protein estimation was done by the classical Biuret reaction, albumin by the Bromocresol green method, fibrinogen by the modified Ratnoff & Menzie method. Estimation of the plasma free haemoglobin was carried out based on the measurement of oxyhemoglobin at 415 nm, the soret band of maximal absorbance of oxyhemoglobin.

### Dimensional Analysis to Evaluate Performance of the ECD

The technique of dimensional analysis is used to understand the complex correlations between different variables involved in the filtration process.

For steady-state flow, the dependent variable J (permeate flux) across the device is a function of independent variables such as:

| | | |
|---|---|---|
| Fluid properties | : | density, viscosity. |
| Membrane properties | : | porosity, pore size distribution, material properties. |
| Geometrical properties | : | length, characteristic diameter and number of the channels. |
| Flow properties | : | system pressure, inlet flow rate. |

For non-steady state, time should be included in the above list

Assumptions:

1) The system is isothermal
2) Viscosity and density of blood remains constant
3) Membrane material properties do not change with respect to system pressure

Under these assumptions

J = F [Q, d, p, μ, time, membrane properties, length & no: of channels] (1)

where

J = Filtration flux
Q = Inlet flow rate
P = Pressure differential across the membrane
p = density of feed
v = transmembrane filtration velocity
Re = Reynolds number
t = time
l = length of the membrane
V = feed velocity

Under steady-state, for a particular geometry and membrane, equation (1) can be rewritten in non-dimensional form as

$$J/Q = F_1 [\, p\,(4Q/d\, d\,)\, d/\mu] = F_1 [Re] \qquad .......(2)$$

where

Re = Reynolds number

Similarly, the pressure differential P, depends upon membrane properties, device geometry, fluid properties and system pressure. The functional form for P can be written as

$$P = F_2 (\, p, d, \mu, v) \qquad .....(3)$$

or, in non-dimensional form as

$$\Delta P/1/2\, pv^2 = F_2\, (pvd/\mu) \qquad .....(4)$$

where

v = characteristic transmembrane velocity

For a particular membrane system and at steady state

$$J/Q = a\, [Re]^{n1} \qquad ...........(A)$$

$$\Delta P/\, p\, v = b\, [Re]^{n2} \qquad ...........(B)$$

$$J/Q = c\, [Vt/l]^{n3}\text{, if Re is constant}.....(C)$$

Here n1, n2, n3, a, b,c, are constants which we should obtain by plotting JQ vs Re, $\Delta P/1/2\ rv^2$ vs RE, J/Q vs vt/L

## EXPERIMENTAL RESULTS AND DISCUSSION

### Size Characterization Of Immune Complexes Aggregates In SLE

Fig 3 is a plot of cryoaggregate size [in nm] as a function of temperature [in kelvin] as estimated by QELS in cases of SLE. It can be observed that the size of the cryoaggregates increases as the temperature is lowered below the physiologic and reaches upto 2.5 μm at 7° C. Fig 4 shows the histograms of the results of QELS studies over a range of CIC concentrations in patients of SLE at body temperature of 37.4°C. To investigate the growth rate of the aggregates, time-related experiments were carried out. Growth rate constants were calculated using the Smoulchouski model and plotted in Fig 5 as a function of the CIC concentration at 20° C. The slope of the straight line is = 9.1 x $10^{-3}$ /min PEG unit and a linear relationship between the CIC concentration and growth rate constant can be observed.

It can be observed that the size of the aggregates formed is higher in the plasma samples of patients with higher CIC levels confirming that the cryoaggregate size is plasma CIC level dependent. At the body temperature of 37.4° C, the aggregates are dissociated into immune complexes of nanometers size range. This gives us an effective mode of `tailoring' the pathoaggregate to shift it towards one extreme end of the plasma protein size spectrum. Such technique can help in the selective separation of the pathoaggregate.

It has been demonstrated that on cooling the serum of patients suffering from certain immune complex diseases, the immune complexes aggregate out (Saha et al, 1970). With increasing concentrations, molecular interactions become important. Also, as the temperature decreases, the conformational structure of the Fab fragment of one complex, changes such that it is in a favourable position to interact with the Fc fragment of the other resulting in the formation of large, insoluble aggregates which further decreases their collective solubility and makes it temperature sensitive.

### Plasmapheresis Experiments

Fig 6 is a plot of plasma filtration rate as a function of time for four runs with 0.45 μm PVDF Durapore membrane. It is observed that flushing helps increase the plasma flux but cannot restore it to its original values. Fig 7 is a comparative plot of plasma filtration rates in cc/min as a function of time of operation for the two membrane systems 0.22 μm and 0.45 μm. It is observed that the 0.45 μm membrane starts with higher plasma filtration rate, experiences more blockage and gives lower yield than the 0.22 μm membrane after a short of operation. Fig 8 shows a comparison of plasma free haemoglobin (PF Hb) in the retentate and the resistance due to plugged cells during CFMP. Fig 9, a EDAX plot of minerals deposited on the membrane surface after 30 hrs of plasmapheresis shows iron, calcium & chlorine deposited in excess to their serum levels. The EDAX studies show that iron, calcium & chlorine are deposited in excess on the membrane surface. Following haemolysis iron deposits in its trivalent form and can be a marker to quantify haemolysis. Salts of calcium can be implicated in the blockage mechanism since it is deposited in excess of the amount that can be explained away by the serum calcium levels. This has special relevance in blood processing systems since deposited calcium not only causes blockage but further triggers the coagulation mechanism leading to formation of clot with disastrous iatrogenic consequences.

### Temperature-Dependent Filtration Experiments

Fig 10 shows the sieving coefficients (defined as Cp/Cf) of the three major plasma proteins as a function of temperature in the plasmapheresis experiments. It is seen that the sieving of fibrinogen is very low at temperatures < 20°C even though the nominal pore size is 0.65 μm. This shows that at reduced temperatures fibrinogen can be separated from the albumin though conventionally it is believed that fractionation of solutes cannot be accomplished easily unless there is atleast a tenfold difference in molecular weights (Cheryan, 1986)

Fig 11 is a plot of rejection coefficients [defined as 1-Cp/Cf] of the representative solutes against the logarithm of their Stokes diameters. A linear regression line has been drawn, the slope of which closely relates to the membrane selectivity. The diameter at which the rejection is 50% is taken to be the mean `effective' filtration pore size of the membrane (Michealis, 1980). It is seen that the mean `effective' sieving pore size increases with temperature increase from 0.02 μm at 10°C to 1.1 μm at 40°C. High sieving of the plasma proteins at 40°C suggests that donor plasmapheresis can be more efficient at higher than physiological temperatures. Recently, we have evidence that such a scheme, if applied for donor plasmapheresis, would have a preventive role against accidental HIV contamination which is rapidly inactivated at 42 -45°C (Gordon et al, 1988). It is seen that the sieving of fibrinogen is very low at temperatures less than 20 °C even though the nominal pore size is 0.65 μm. This is probably because fibrinogen has an elongated molecular structure which increases the `effective molecular size' and gets retained. At 6 °C, it is observed that the `effective sieving curves' of fibrinogen and albumin are far apart from each other. This shows that at lower temperatures one could simply separate fibrinogen from the albumin though conventionally it is beleived that fractionation of solutes cannot be accomplished easily unless there is atleast a tenfold difference in molecular weights (Cheryan, 1986). By exploitation of this differential sieving effect at lower temperatures, sharper selectivity can be achieved in plasma fractionation schemes. The resultant of the membrane transport mechanism and associated membrane-protein interactions phenomena such as membrane fouling/plugging determine the final `effective' sieving curves. Since temperature can modify both the protein size and the occurence of associated phenomena in membrane filtration, one could exploit temperature effects to improve selectivities in plasmapheresis.

## Tailoring Membranes for Specific Blood Processing Applications

Fig 12 shows the components and the design of the laboratory spinneret. Table V shows the effect of casting solution composition on the resulting hollow fibre properties.
At dope composition of 17.5-42.5-40 and 18.5-30-50 (CA:A:DMF) the viscosity of the dope was too low and so hollow fibres could not be obtained. Fig 13 shows the effective pore size as determined by Michaelis co-relationship. Table VI sums up the process condition variables for the spinning of hollow fibres as done in our laboratory. Fig 14 shows the optimized casting dope composition for the various blood processing applications.

Spinning hollow fibres for specific blood processing applications can be accomplished by exercising a meticulous control over the movement of the composition pathway between the original compositions and the final membrane. Hollow fibres require more viscous dopes than flatsheet membranes of similar properties probably because the unsupported thread line of the nascent hollow-fibre membrane must establish mechanical integrity in a very short time. The thumb rule is that the dope solution which falls continuously without breaking up from a glass rod is good for hollow fibre making. The optimum air gap distance h (higher and lower air gaps lead to snapping and entangled of the fibre respectively), humidity, flow rate of bore fluid and the velocity of dope extrusion control the gelation process and are important in determining the final membrane properties. At higher relative humidity, water enters the membrane at a rate higher than acetone-DMF removal and highly porous membranes suitable for cryofiltration and plasma separation result. Compact hemofiltration membranes with pore size in dalton range result at low humidity and long evaporation times. Higher the percentage of cellulose acetate, lower the permeability and the sieving character moves towards the lower side of the blood sieving spectrum.

**Characterization of the Fluid Profile in Renal Failure**

The results of the study are shown in tables I to IV. It can be clearly inferred that the haematocrit, red cell count, haemoglobin, mean corpuscular haemoglobin and mean corpuscular haemoglobin concentration in chronic renal failure are grossly reduced as compared with normals. This might be due to the occurrence of anaemia due to the reduced secretion of erythropoietin from the diseased kidney. There is a similar but milder change in acute renal failure. Platelet counts are increased in chronic renal failure showing a tendency towards excess thrombogenesis. The leucocytes are increased in chronic renal failure probably suggesting an increased susceptibility to infections. The plasma viscosity is increased in chronic renal failure. The intrinsic viscosity of red blood cell is also increased in both chronic and acute renal failure showing that red blood cells become rigid and less deformable in renal failure and any membrane-based extracorporeal manipulative process such as haemofiltration should take into account this changed rheological properties.

Our study on feed characterization confirms the existence of haemorheological abnormalities in patients suffering with acute and chronic renal failure. Haemorheological `fingerprints' have to be taken into consideration when designing extracorporeal therapies since they affect the effectiveness and safety of the therapy.

**`Hemorheological' Screening of Blood-Contacting Materials**

The shear profiles of blood exposed to the materials and control (blood unexposed to any material) are plotted as the log of apparent viscosity to log of shear rate in Figs 15a & $15_b$. The N value (calculated from the slope of the curve, as per the Power-law model) gives the degree of Non-Newtonian behaviour of the fluid. It is observed that both the `n' and `k' values are significantly deviated from that of control in case of glass (n= 0.68, k = 144) , cotton fabric (n =0.68 k = 166) and copper (n =0.73 k =92) thus breaching the limits of haemo-rheocompatibility. All others show values closer to the controls' (n = 0.63 ± 0.02, k = 120 ± 20). It can be seen that blood-incompatible materials such as glass, cotton fabric and commercial grade copper show significantly altered N & K whereas compatible materials show minimal variation. Such shear profiles give us information about the overall effect of individual material-blood interactions. The N & K values of the blood after contact with the material gives us a birds eyeview of the entire gamut of haemorheological effects specific to that substance and helps qualify the individual blood-material interactions.

**Dimensional Analysis to Evaluate Performance of the ECD**

Fig $16_a$, $16_b$, & $16_c$ are plots characterizing our Minitan-S thin channel crossflow membrane system. New non-dimensional numbers are evolved which represent grouping of relevant filtration parameters and which contribute to the global characterization, optimal utilization and to find the border between stable & unstable operating conditions in the CFMF extracorporeal device. P/ pv , the `dimensionless filtration pressure coefficient' incorporates the pressure differential across the membrane and the transmembrane plasma filtration velocity. This dimensionless number `focusses and magnifies' the hindrance at the pore interface, quantifies filtration-related phenomena such as fouling and polarization and gives a good estimate of the efficiency of the membrane system.

Also, dimensionless saline flux [NSP after flushing/NSP initial] a dimensionless quantity, gives a good estimate of the effective reduction in the pore size/number and quantifies the blockage of the membrane system. The relative filtration ratio [NSP before the run/1st minute PFR] is a useful parameter in quantifying flux particularly in notoriously fouling systems known to remain at `quasi-steady-state'.

## SUMMARY & SCOPE OF WORK

Results of the present study have brought to fore the fact that membrane filtration offers significant flexibility as a haemoseparation unit operation. Selectivity in membrane processing of blood can be approached at three levels:

* tailor-casting the membrane to the requirements of the specific application
* `tailoring' the target molecule to the membrane type
* `tailoring' the sieving spectrum by exploiting the operating conditions and fluid management techniques.

Application-specific ECDs are the need of the present day transfusion medicine centres the world over. Availability of such devices will lead to improved management both in donor hemapheresis and in therapeutic hemapheresis particularly with regards to those diseases which have no `exact cures'.

**Table 1.** Red blood cell studies in normals, CRF & Ac GN

| Parameter | Normals | CRF | Ac GN |
|---|---|---|---|
| Haematocrit (%) | 42 + 5.0 | 22.8 + 6.7 | 26.8 + 6.2 |
| RBC count (x $10^6$/cc) | 5 + 0.8 | 3.8 + 2.0 | 3.9 + 1.1 |
| Mean Corpuscular Volume ( $U^3$) | 87 + 5.0 | 88.7 + 9.7 | 85.7 + 7.4 |
| Mean corpuscular haemoglobin (Uug) | 29 + 2.0 | 24.7 + 6.2 | 25.5 + 3.1 |
| Mean corpuscular haemoglobin concn (%) | 34 + 2.0 | 27.3 + 6.4 | 29.4 + 4.26 |
| Haemoglobin (g/100cc) | 14 + 2.0 | 10.6 + 7.1 | 9.6 + 3.33 |
| Red cell distribution width (U) | 90 +20.0 | 90.0 +12.0 | 91.4 +12.61 |

**Table 2.** Average whole blood viscosities corrected to 45% Hct in cases of Acute Glomerulonephritis, SLE & CRF

| Shear rate | Ac.GN | SLE | CRF [other causes] |
|---|---|---|---|
| 0.51 | 25.82 + 11.12 | 63.43 + 20.00 | 33.41 + 21.70 |
| 1.28 | 17.10 + 8.02 | 20.55 + 7.82 | 21.07 + 10.50 |
| 3.23 | 9.35 + 5.66 | 10.06 + 1.97 | 10.90 + 8.04 |
| 4.39 | 6.99 + 1.94 | 8.26 + 1.94 | 7.84 + 2.36 |
| 8.11 | 5.40 + 1.93 | 5.86 + 1.48 | 6.12 + 1.92 |
| 20.40 | 3.93 + 1.10 | 4.37 + 0.82 | 4.23 + 1.26 |
| 51.20 | 3.25 + 0.82 | 3.49 + 0.66 | 3.85 + 0.96 |
| 94.50 | 2.90 + 0.66 | 3.49 + 0.50 | 3.11 + 0.57 |

**Table 3.** Platelet studies in normals, CRF & Ac GN

| Parameter | Normals | CRF | Ac GN |
|---|---|---|---|
| Platelets (x $10^3$/cc) | 150-400 | 450-120 | 285 + 115 |
| Mean platelet volume ($U^3$) | 8-12 | 11+ 1.5 | 10.7 + 1.2 |
| Plateletocrit (%) | 0.43 +0.5 | 0.43 +0.5 | 0.33 + 0.01 |
| Platelet distribution width ($U^3$) | 13.5 + 5 | 13.7 + 5 | 13.3 + 4 |

**Table 4.** Haemorheological fingerprints normals, CRF & Ac GN

| Parameters | Normals | CRF | Ac GN |
|---|---|---|---|
| Plasma viscosity (mPa.S) at 94.5/sec | 1.27 + 0.09 | 1.93 + 0.18 | 1.42 +0.46 |
| Whole blood viscosity | | | |
| 94.5/sec | 4.66 + 1.2 | 3.08 + 1.0 | 2.89 + 1.0 |
| 1.26/sec | 15.24 + 3.0 | 20.05 + 13.0 | 16.22 + 10 |
| Low shear viscosity/ high shear viscosity ratio | 3.27 | 6.50 | 5.61 |
| Relative viscosity (WBV/PV) as LS | 12 | 15.78 | 12.77 |
| Intrinsic viscosity of red blood cell | 2.73 + 0.68 | 3.76 + 0.75 | 3.35 + 0.7 |
| Albumin/fibrinogen ratio | 14 -20.5 | 13.68 | 11.63 |
| Albumin/globulin ratio | 1.4 | 1.69 | 1.4 |
| Haematocrit (%) | 42.4 +4.4 | 22..8 + 6.7 | 26.8 + 6.2 |

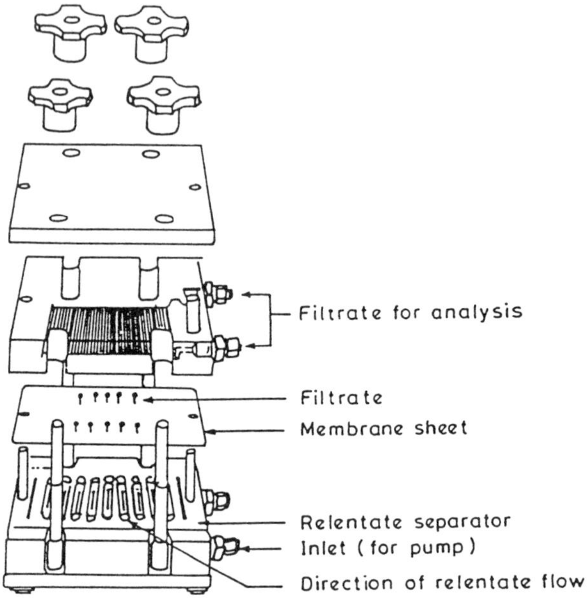


Fig 1. The crossflow filtration device (Minitan-S, Millipore Corpn, U.S.A.) with PVDF Durapore membranes 0.65 μm pore size and 30 cm2 effective filtration area.

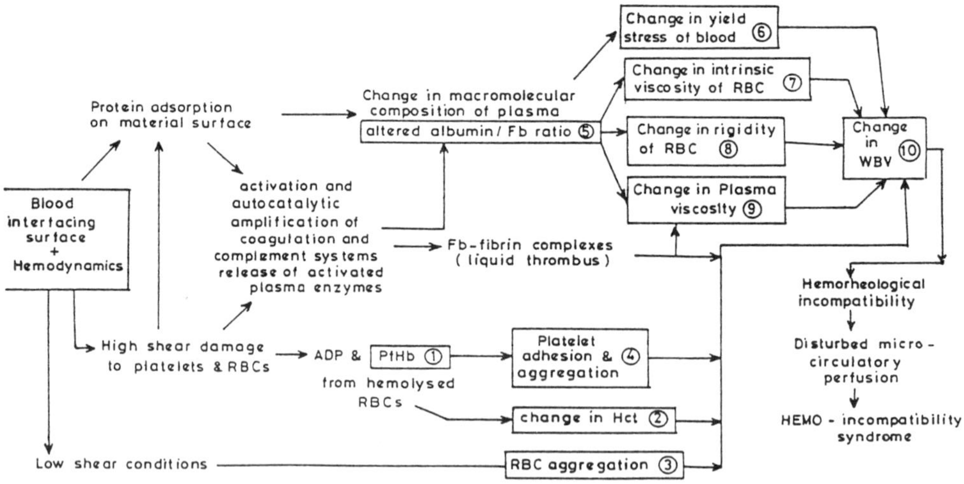


Fig 2. A Conceptual schematic of the hemorheological pathway during blood-material interactions.

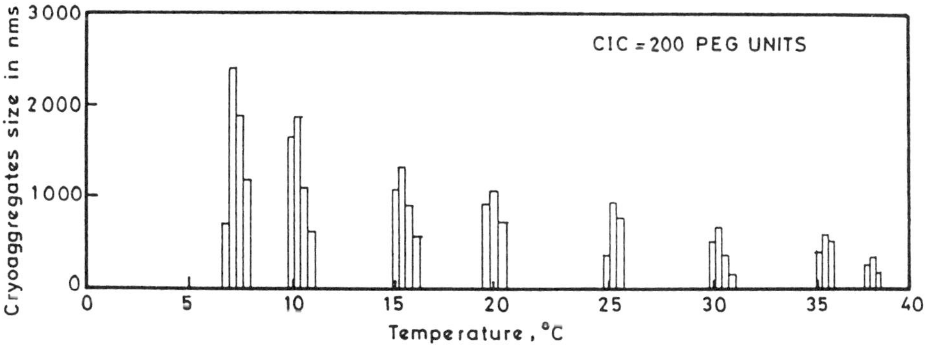


Fig 3 is a plot of cryoaggregate size [in nm] as a function of temperature [in kelvin] as estimated by QELS in cases of SLE.

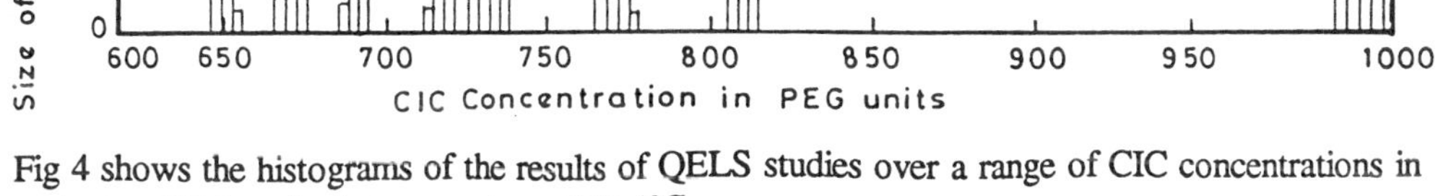


Fig 4 shows the histograms of the results of QELS studies over a range of CIC concentrations in patients of SLE at body temperature of 37.4°C.

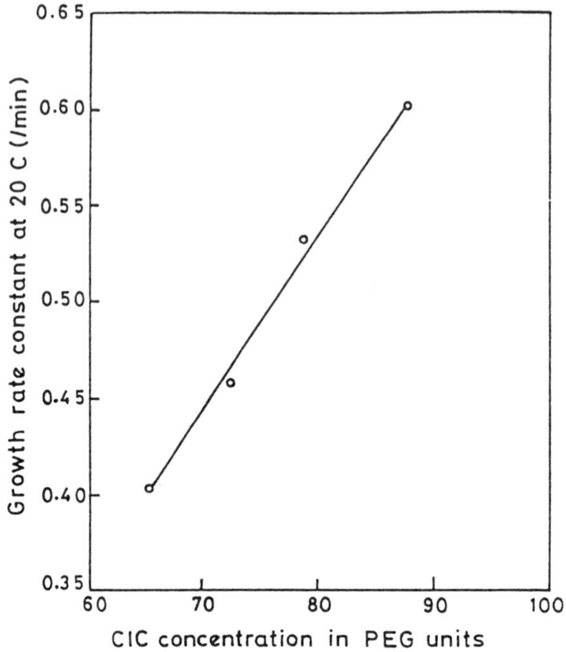


Fig 5  Growth rate constants were calculated using the Smoulchouski model and plotted as a function of the CIC concentration at 20° C.

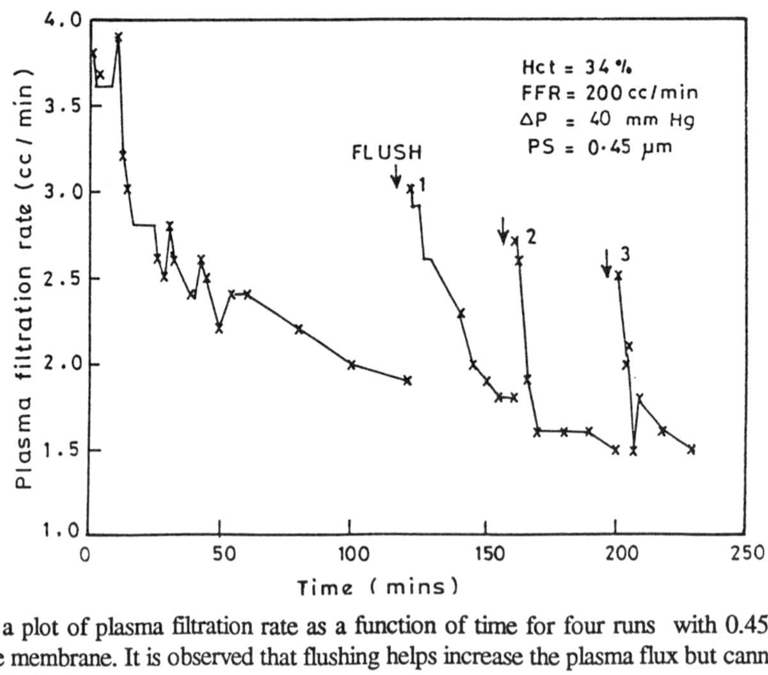


Fig 6  is a plot of plasma filtration rate as a function of time for four runs with 0.45 μm PVDF Durapore membrane. It is observed that flushing helps increase the plasma flux but cannot restore it to its original values.

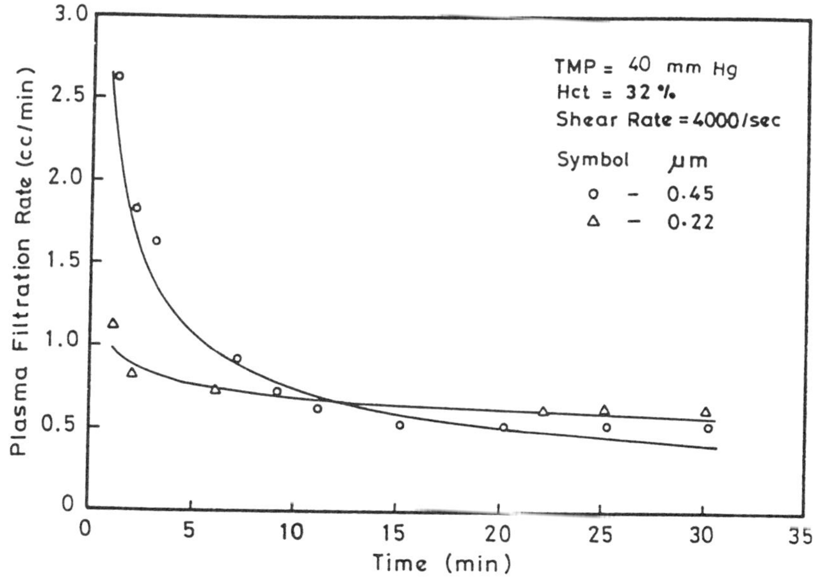


Fig 7 is a comparative plot of plasma filtration rates in cc/min as a function of time of operation for the two membrane systems 0.22 μm and 0.45 μm.

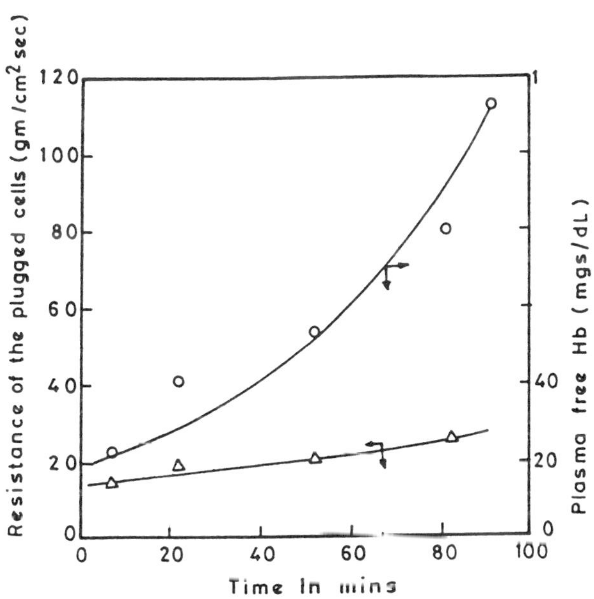


Fig 8 shows a comparison of plasma free haemoglobin (PF Hb) in the retentate and the resistance due to plugged cells during CFMP.

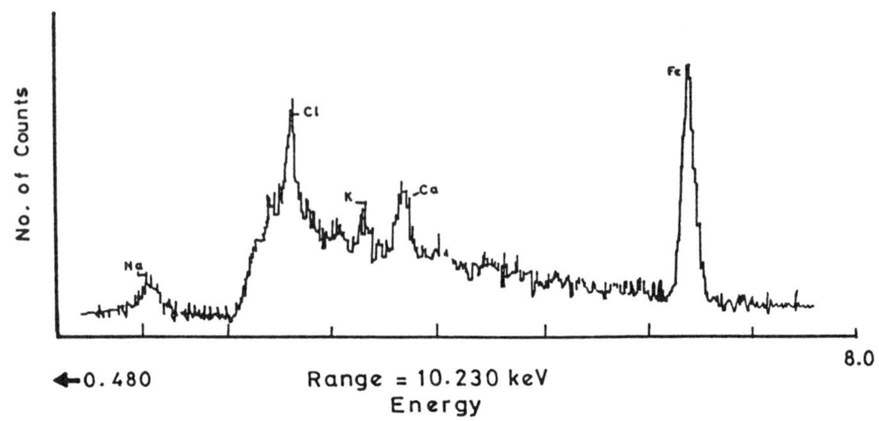


Fig 9, a EDAX plot of minerals deposited on the membrane surface after 30 hrs of plasmaphresis shows iron, calcium & chlorine deposited in excess to their serum levels.

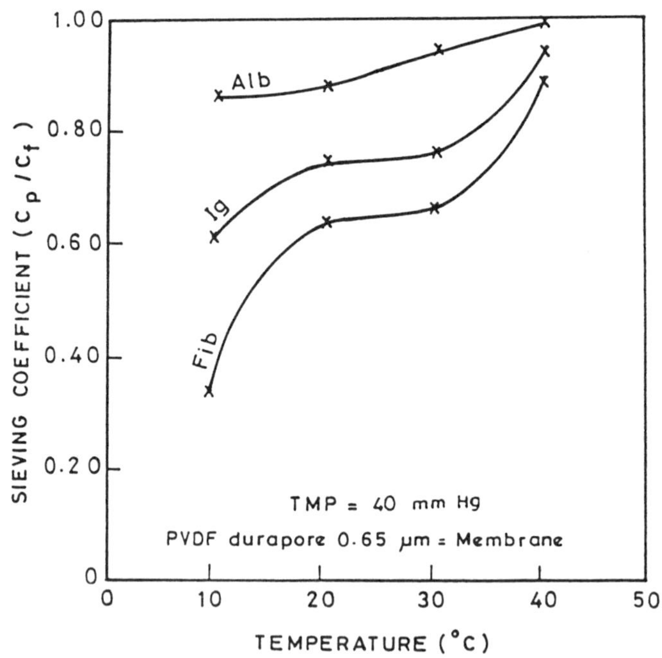


Fig 10 shows the sieving coefficients (defined as Cp/Cf) of the three major plasma proteins as a function of temperature in the plasmapheresis experiments.

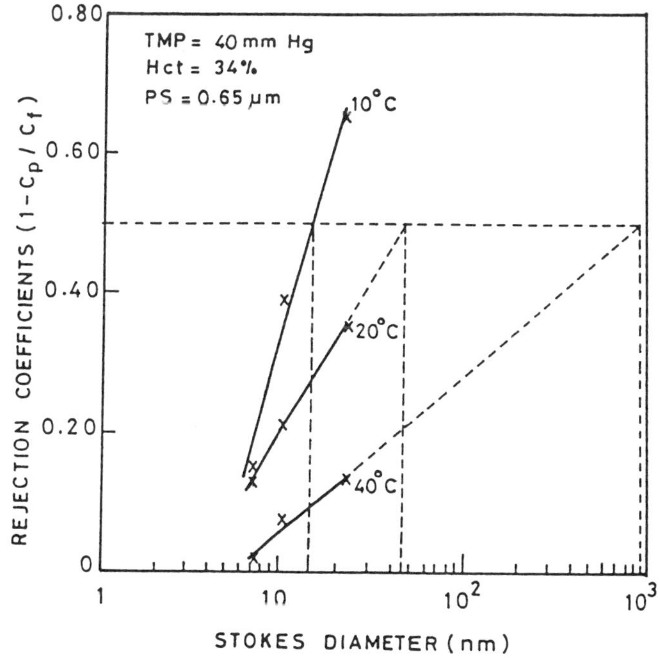


Fig 11 is a plot of rejection coefficients [defined as 1-Cp/Cf] of the representative solutes against the logarithm of their Stokes diameters. A linear regression line has been drawn, the slope of which closely relates to the membrane selectivity.

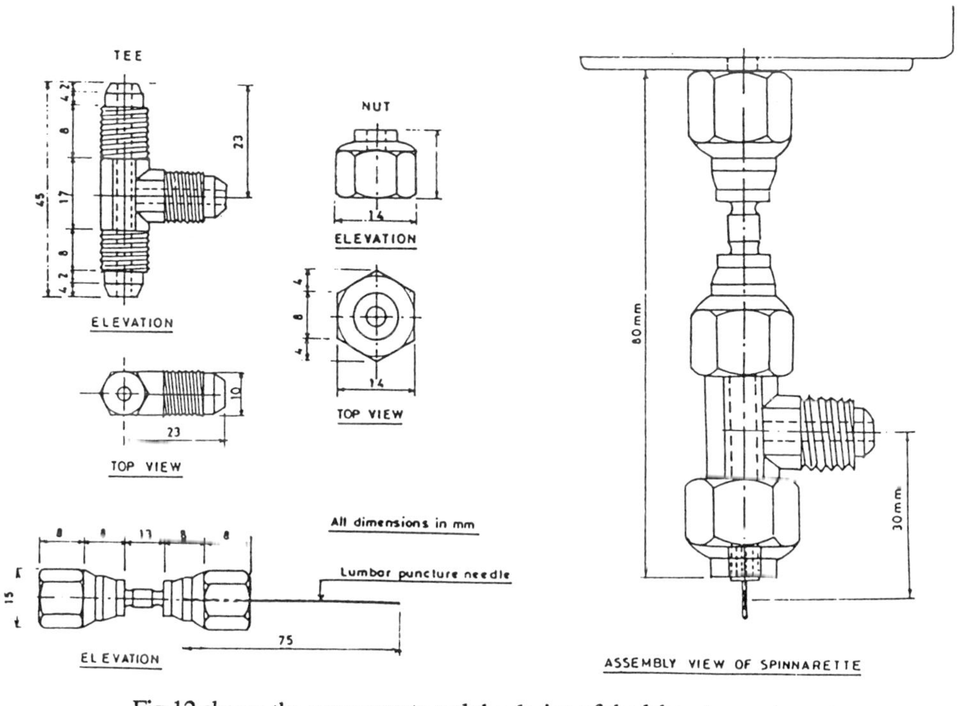


Fig 12 shows the components and the design of the laboratory spinneret.

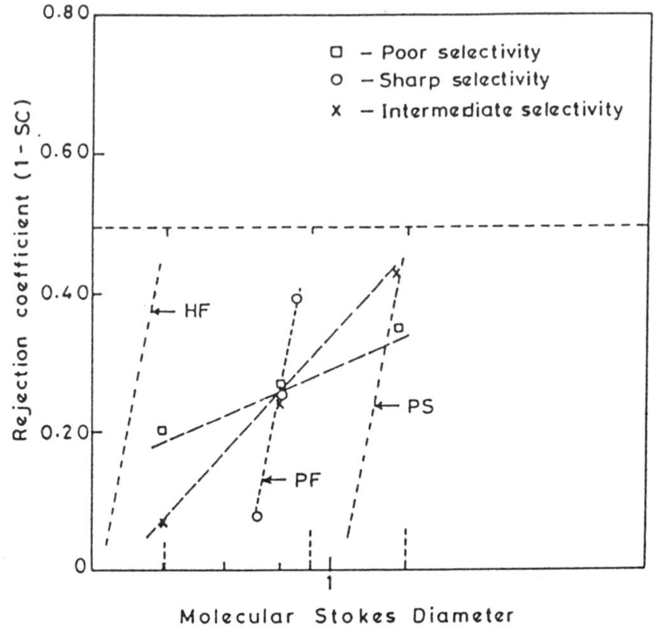


Fig 13 shows the effective pore size as determined by Michaelis co-relationship, for blood processing applications.

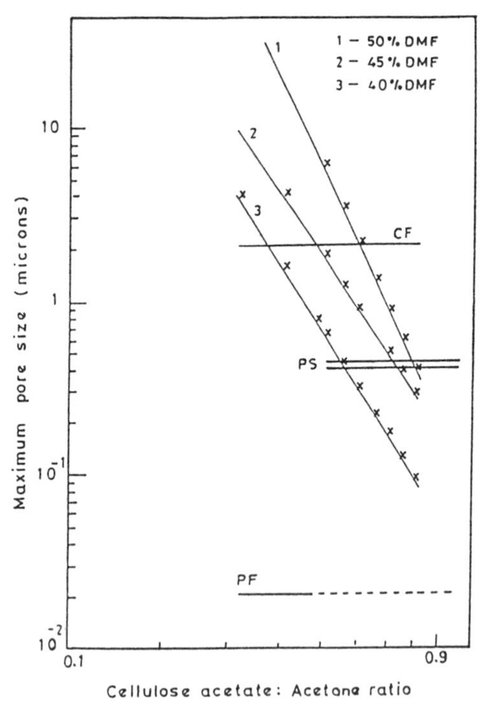


Fig 14 shows the optimized casting dope composition for the various blood processing applications.

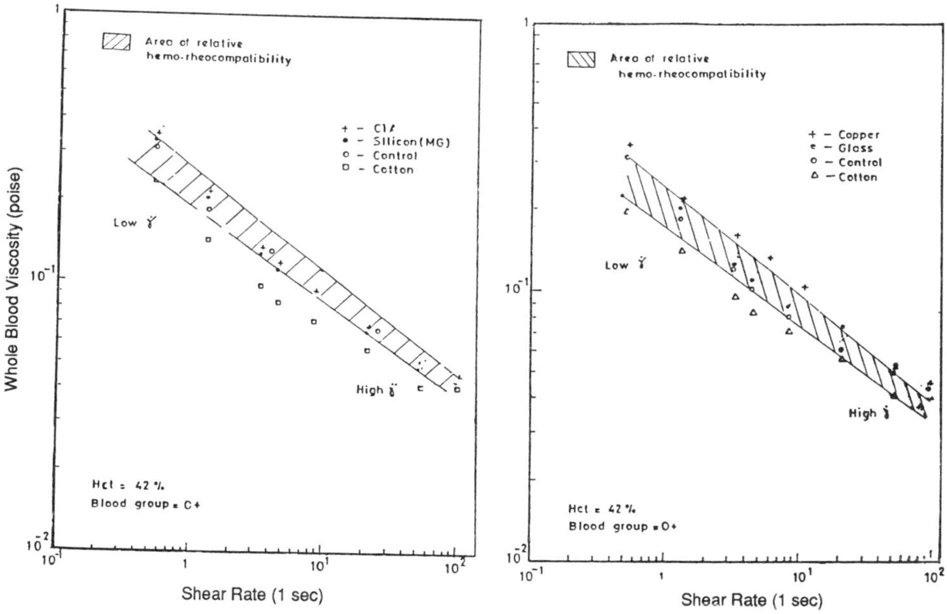


Fig 15 a & b shows profiles of blood exposed to materials and control plottes of log of apperent visionity to log of sheer exctre.

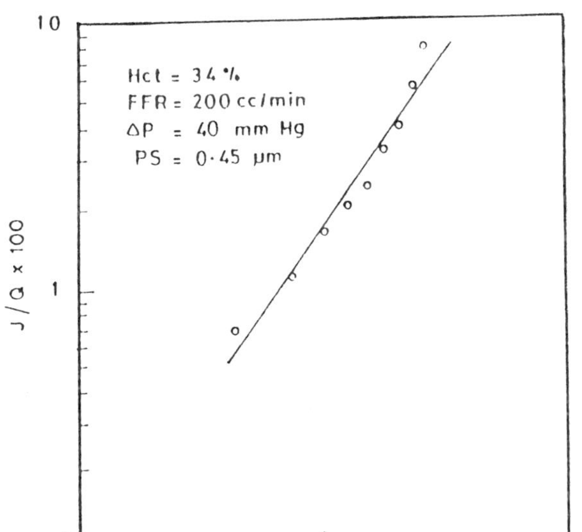


Fig 15 c shows ECD system characterization by dimensionless numbers.
(Here J/Q $= 62 [Re]^{0.4}$)

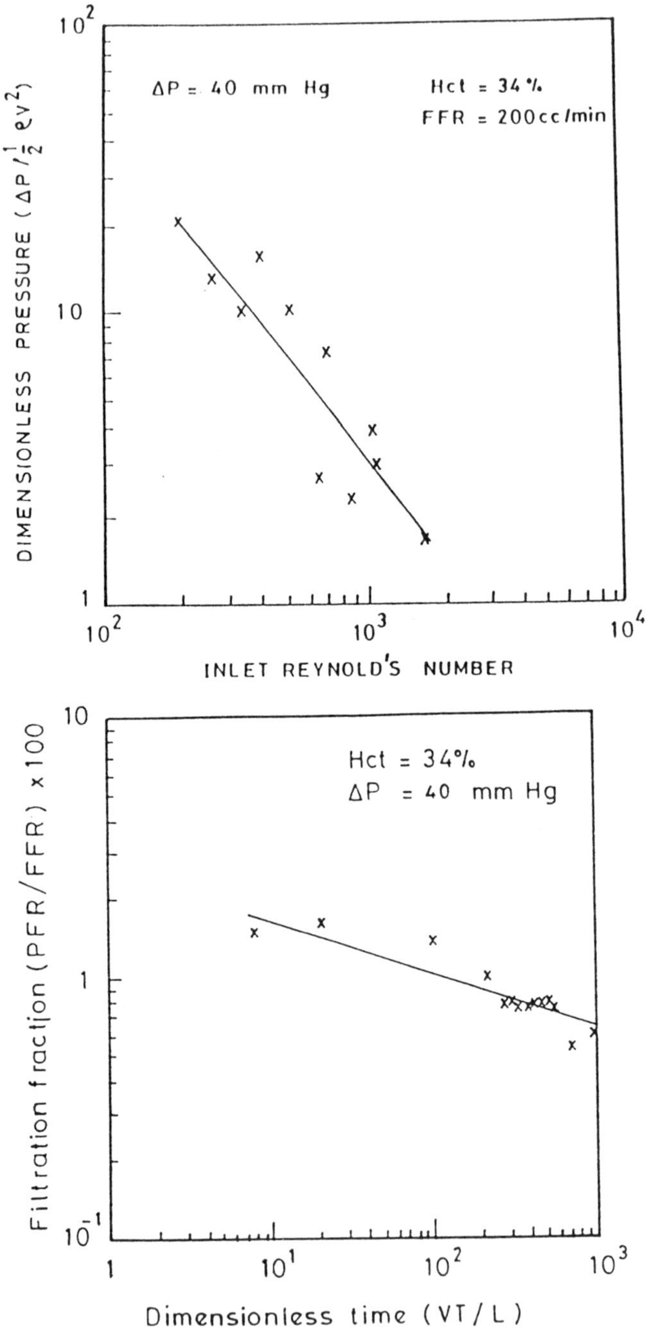


Fig 16 a, b. Characterization of membrane-based blood filtration system

REFERENCES

Bruck, S. (1980) in Evaluation of Biomaterials, Winen G.D, Leray, J.L. & Groot K. L. [Eds] 1980, John Wiley & Sons Ltd

Davis, R.,H. (1992) `Theory of crossflow microfiltration', in The Membrane handbook, Ho, W.S.W & Sirkar, K.K. [Eds] London, pp 484-497

Dientenfass, L. (1976) in Rheology of blood in diagnostic and preventive medicine, Dintenfass,L [Ed], Butterworths, London.

Cairns, S.A., London, A., Mallick, N.P.`The value of three immune complex assays in the mangaement of systemic lupus erythematosus: An assessment of immune complex levels, size, and immunochemical properties in relation to disease activity and manifestations' Clin Exp Immunol. v 40 p 273 (1980)

Cheryan, M. (1986) in Ultrafiltration handbook, Munir Cheryan [Ed], Technomic, Lancaster (U.S.A.) pp 67, 89.

Gordon, I.M., Jenson, I.C., Curtam, C.C.,Mobley, P.M. & Alora, R.C. (1988) "Thermotropic lipid phase separation in the human immunodeficiency virus" Biochim Biophys Acta 943 pp 331-342

Hardin, J (1981) `Antigen-antibody complexes' in Adv Immunology,Wari, D.M. [Ed], Longman Pvt Ltd, Singapore pp 366

Jossang, T., Feder, J., and Rosenqvist, E. (1985) `Heat Aggregation Kinetics of Human IgG', J.Chem.Phys., 82 (1), pp 574-594.

Michealis, L.(1980) `Analysis and prediction of sieving coefficients for ultrafiltration membranes- a universal correlation?' Sep Sci Technol v 15 pp 1305-1322

Munir Cheryan (1986) `Models to predict limiting flux' in Ultrafiltration handbook, Munir Cheryan [Ed], Technomic, Lancaster (U.S.A) pp 89

Nydegger, U.E., Lambert, P.H., Gerber, H., Miescher, P.A. `Circulating immune complexes in the serum in systemic lupus erythematosus and in carriers of hepatitis B antigen: Quantitation by binding to radiolabelled C1q'. J Clin Invest v 54 p 297 (1974)

Ooi, Y.M., Valllota, E.H., West, C.D. `Serum immune complexes in membranoproliferative and other glomerulonephritides' Kidney Int v 11 p 275 (1977)

Saha, A., Chowdhury, P., Sambury, S., Smart, K. & Rose, B (1970) `Studies on Cryoprecipitation', J.Biol.Chem., v 245 pp 2730-2736.

Tung, K.S..K., DeHoratius, R.J., Williams, R.C. Jr. `Study of circulating immune complex size in systemic lupus erythematosus' Clin. Exp. Immunol. v 43 p 615 (1981)

Theofilopoulos, A. & Dixon, F.J. `The biology and detection of immune complexes', Adv. Immun., v 28 p 89 (1979)

WHO Technical report (1977) `The role of immune complexes in diseases', Geneva.

Yarmush, D.M., Murphy, R.M., Colton, C.K., Fisch, M., & Yarmush, M.L. (1988) `Quasi-elastic light scattering of antigen-antibody complexes', Molec. Immun., v 25, pp 17-25

Zborowski, M. & Malchesky, P.S. (1990) "Pore size and temperature effects in membrane separation of albumin from immunoglobulin", Trans. Am. Soc. Artif. Intern. Organs v 36: pp 730-733

Zydney, A.L. & Colton, C.K. [1982] "Continuous flow membrane plasmapheresis: Theoritical models for flux and hemolysis prediction" Trans. Am. Soc. Artif. Intern. Organs v 28 pp 408-412.

# CHARACTERIZATION OF LIPOSOMAL SYSTEMS ENTRAPPING BORON-CONTAINING COMPOUNDS IN RESPONSE TO pH GRADIENTS

Angela M. Moraes[1], Maria Helena A. Santana[2], and Ruben G. Carbonell[3]

[1] On leave from: Dept. Proc. Químicos/FEQ/State University of Campinas
Campinas, SP 13083-000, Brazil
[2] Dept. Proc. Químicos/FEQ/State University of Campinas
Campinas, SP 13083-000, Brazil
[3] Dept. of Chemical Engineering, North Carolina State University
Raleigh, NC 27695-7905

## ABSTRACT

Boron neutron capture therapy (BNCT) is based on the nuclear reaction that occurs when a stable isotope, Boron-10, is irradiated with low-energy thermal neutrons to yield ionizing Helium and Lithium ions that are highly damaging and usually lethal to cells. The successful treatment of cancer by BNCT requires the selective concentration of Boron-10 within malignant tumors. Liposomes have been used as drug delivery vehicles for *in vivo* application, including several anticancer agents. The ability of two Boron-containing compounds, l-p-borono-phenylalanine (BPA) HCl and o-carboranylpropylamine chloride (CPA) to accumulate within unilamellar liposomes passively and in response to a transmembrane pH gradient are compared. Characterization of the obtained systems is performed for conventional and polyethylene glycol (PEG)-modified (stealth) liposomes, in terms of lipid and drug contents, vesicle size and stability. The results indicate that BPA can be successfully encapsulated in conventional liposomes by passive loading, while the active loading approach is more suitable for the entrapment of CPA.

## INTRODUCTION

Although modern standard treatments have been successful in curing some types of cancer, various forms are still incurable, such as glioblastoma multiforme, a class of brain tumor. An ideal therapeutic method would consist of a regimen that selectively destroys malignant cells without seriously damaging normal surrounding tissue. The selective destruction of tumor cells is theoretically possible with the use of a binary approach based on the neutron capture reaction observed with the $^{10}B$ nucleus and a thermal neutron. The reaction involved in this therapy, named Boron Neutron Capture Therapy (BNCT), is shown in Figure 1.

$$^{10}B + {}^{1}n_{th} \rightarrow [^{11}B] \begin{cases} \nearrow\ {}^{4}He + {}^{7}Li + 2.79\ MeV\ \ (6\%) \\ \searrow\ {}^{4}He + {}^{7}Li + 0.48\ MeV\ \gamma + 2.31\ MeV\ (94\%) \end{cases}$$

**Figure 1** - The two parallel nuclear fission reactions that occur upon capture of a thermal neutron by a $^{10}B$ nucleus (Hawthorne, 1993).

During their passage through the interior of a cell, the energetic products resulting from the reaction can cause extensive damage to organic materials through ionization processes, generating cytotoxic compounds and destroying biologically active molecules (Barth et al., 1990). If the $^{10}B$ is selectively accumulated in tumor cells, theoretically the radiation should not affect normal tissue, since the ions generated can travel approximately only one cell diameter (10 to 14 μm).

The development of boron-containing compounds that show some affinity to tumors and of effective targeting strategies for their selective transport have been the objects of study of several research groups. In the past few years, the search for suitable tumor-localizing boron compounds has been intense, and among the many compounds that have been screened in animals, some have presented favorable properties, such as [$^{10}B$]Boronophenylalanine (BPA), whose structure is shown in Figure 2(a). Several investigators have noticed significant BPA accumulation in rat intracerebral melanomas (Matalka et al., 1993), rat intracerebral gliomas (Matalka et al., 1994), mouse mammary carcinomas and rat brain gliosarcomas (Coderre et al., 1990). In addition, BPA has not been found to be toxic in animal studies (Coderre et al., 1990; Saris et al., 1992). This compound was used in the treatment of melanoma patients, resulting in good local control of the tumors and tolerable skin reaction (Fukuda et al., 1994).

Another chemically attractive class of compounds that can be used as agents for BNCT are carboranes containing several atoms of boron in their structure, such as the carboranylpropylamine (CPA) illustrated in Figure 2(b). This class of compounds is known to be resistant to metabolic degradation and can exhibit significant lipophilicity, permitting easier penetration into the tumor cell membrane (Barth et al., 1992). Some of these carboranes have been synthesized, but the evaluation of their selective uptake by malignant cells has proceeded slowly. Good reviews of other compounds used for BNCT have been published by Barth et al., 1990; 1992 and Hawthorne, 1993.

A potential strategy to improve the delivery of boron-containing compounds to tumors is the use of liposomes, vesicular structures consisting of hydrated lipid membranes (Shelly et al., 1992). The selection of the lipid composition and of the method used for drug incorporation in liposomes are important elements in the design of delivery systems, since they determine the final characteristics of the preparations. Liposomes sterically stabilized by the inclusion of lipid derivatives of polyethylene glycol (PEG), for instance, present prolonged circulation *in vivo*, ultimately enhancing the probability of their accumulation in tumors (Lasic and Martin, 1995).

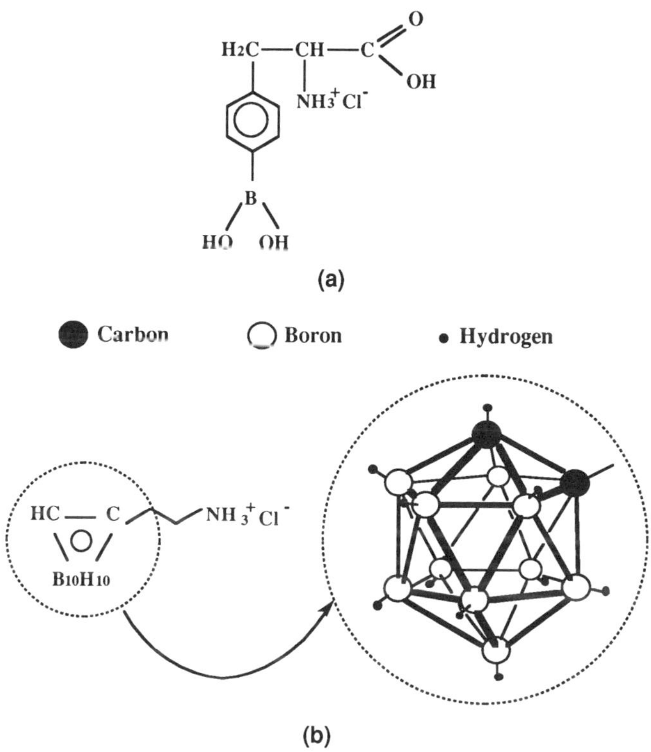


Figure 2 - Chemical structure of L-p-boronophenylalanine HCl (a) and o-carboranylpropylamine HCl (b).

The encapsulation of drugs in liposomes can involve either passive or active loading procedures (Mayer et al. 1986). In passive loading, drug and lipids are codispersed and the approach relies on the ability of liposomes to capture a certain aqueous volume (containing dissolved solutes) during vesicle formation. Trapping efficiencies for small unilamellar vesicles (less than 100 nm diameters) vary from less than 1% to 5%, due to their small internal volume. In active loading, the drug (normally a lipophilic amine) is added to preformed liposomes with an imposed pH gradient (the pH in the aqueous core is normally lower). The drug redistributes across the membrane in response to the pH gradient, resulting in trapping efficiencies that can approach 100% (Mayer et al., 1990; Madden et al., 1990; Mayer et al., 1993).

Liposomes have proven to be useful tools for the delivery of numerous substances to tumor cells, including some boron-containing compounds. Yanagië et al. (1991) observed that antidody-targeted multilamellar liposomes passively encapsulating $Cs_2B_{12}H_{11}SH$ selectively bound pancreatic carcinoma cells *in vitro*. In *in vivo* applications, however, these immunoliposomes would probably not be as effective, since multilamellar vesicles are often rapidly removed from the blood stream by the reticuloendothelial system cells.

The hydrophilic salts $Na_2B_{10}H_{10}$, $Na_2B_{12}H_{11}SH$, $Na_4B_{20}H_{17}OH$, $Na_3B_{20}H_{19}$, $Na_2B_{20}H_{18}$ and its photoisomer were also passively encapsulated in liposomes (Shelly et al., 1992). The results of this study indicated that some of the liposomal preparations tested selectively delivered the borane anions to EMT6 tumor cells implanted in rats, although the boron compounds used did not exhibit affinity for tumors.

Other boron compounds successfully encapsulated in the aqueous core of liposomes are $NaB_{10}H_9NH_3$ and the isomers of $Na_3B_{20}H_{17}NH_3$ (Feakes et al., 1994). In this study, both sterically stabilized (liposomes containing 5 % PEG) and conventional lipid vesicles (without PEG) were used in biodistribution experiments, resulting in high initial tumor uptake followed by an increase in tumor boron concentration for both $Na_3B_{20}H_{17}NH_3$ isomers. In another study, Feakes et al. (1995) demonstrated that significantly enhanced boron delivery to murine tumors can be obtained with liposomes encapsulating simultaneosly the lipophilic *nido*-carborane species K[*nido*-7-$CH_3$ $(CH_2)_{15}$-$C_2B_9H_{11}$] incorporated in the liposomal membrane and the hydrophilic compound $Na_3$[1-(2'-$B_{10}H_9$)-2-$NH_3B_{10}H_8$] entrapped in the aqueous core of liposomes.

Several of the preparations described in the studies referred to above delivered the boron compounds to the target cells in concentrations high enough to achieve the necessary therapeutic range. Although detailed *in vitro* and *in vivo* tests were reported in these studies, few aspects concerning the optimization of the entrapment of the boron compounds and the physical characterization of the liposomal sytems were described.

In this paper, the preparation and characterization of conventional and stealth liposomes containing the boronated compounds o-carboranylpropyl amine chloride and l-p-boronophenylalanine HCl encapsulated are described. Since CPA exhibits relatively high water solubility (around 16 mg/ml) and has a primary amine group in its molecule (meaning that it is a natural candidate to accumulation within liposomes in response to pH gradients), this boron compound should be successfully entrapped into liposomes by using either passive or active loading techniques. BPA should be able to accumulate into lipossomes by passive loading provided that its low solubility in water is enhanced, while the entrapment by active loading will basically depend on its charge at the pH used. To the best of our knowledge, no boron-containing compounds have ever been encapsulated in liposomes by using the active loading approach. None of the substances examined here has previously been encapsulated in liposomes. The effects of entrapment conditions, such as temperature, incubation time, boron compound to lipid molar ratio and liposome composition are evaluated. In addition, since *in vivo* the integrity of liposomes is strongly affected by interactions between vesicles and amphiphilic molecules, the stability of some of the liposomal systems in non-ionic detergent solutions is described.

## MATERIALS AND METHODS

### Materials

Cholesterol (Chol), dimyristoyl phosphatidyl ethanolamine (DMPE), distearoyl phosphatidyl choline (DSPC), N-[2-hydroxyethyl]piperazine-N'-[2-ethanesulfonic acid]) (HEPES), citric acid and ß-D(-)fructose were obtained from Sigma Chemical Company (St. Louis, MO). 1,2-Dimyristoyl-*sn*-glycero-3-phosphoethanolamine-N-[Poly(ethylene glycol) 2000] (DMPE-PEG) was obtained from Avanti Polar Lipids Inc. (Alabaster, AL). 2,4,6-trinitro benzenesulfonic acid (TNBS) was obtained from Pierce (Rockford, IL). The non-ionic surfactant penta ethylene glycol mono-*n*-dodecyl ether ($C_{12}E_5$) was obtained from Nikko Chemical Co. (Tokyo, Japan). Polystyrene 96-well round-botton microtiter plates were obtained from Corning Inc. (Corning, NY). The boron containing compounds l-p-boronophenylalanine HCl (BPA) and o-carboranylpropylamine HCl (CPA) were kindly supplied by Boron Biologicals, Inc. (Raleigh, NC) and used without further purification.

### Methods

**Vesicle Preparation.** Small and large unilamellar vesicles with composition DSPC/Chol (60:40 mol %), DSPC/Chol/DMPE (40:40:20 mol %) and DSPC/Chol/DMPE/ DMPE-PEG (40:40:8:12 mol%) were prepared by standard sonication procedures (Szoka and Papahadjopoulos, 1980) or by extrusion (MacDonald et al., 1991). Briefly, the lipids were dissolved in chloroform:methanol (9:1 v/v) and dried in a rotatory evaporator to form a lipid film that was then hydrated with 300 mM citric acid solution at pH 4.0 (for active loading of the drug) or with an aqueous solution containing either BPA or CPA (for passive loading) for 15 minutes at 65°C. The resulting multilamellar vesicles were subjected to five freeze-thaw cycles (-37°C and 80°C, respectively), to enhance solute distribution and trapped liquid volume (Mayer et al., 1985). Small unilamellar vesicles were obtained by sonication at 80°C for 2.5 hours at 60% power using a sapphire-tipped sonicator (Heat Systems Ultrasonics W385, Farmingdale, NY). Non-dispersed lipid was removed by centrifugation at 4200 g for 20 minutes. Large unilamellar liposomes were obtained by extruding the suspension 31 times through two stacked 100 nm Nuclepore polycarbonate filters (Costar, Cambridge, MA), using a LiposoFast extruder (Avestin Inc., Ottawa, Canada).

**Generation of Transmembrane pH Gradients.** Transmembrane pH gradients for boron compound active loading were obtained by gel permeation chromatography, passing the vesicles suspended in citric acid buffer at pH 4.0 through a Bio-Gel P-6 column (bed volume equal to 10 ml) pre-equilibrated with 20 mM HEPES, 150 mM NaCl (pH 7.5).

**Entrapment of Boron-Containing Compounds.** For active loading, the liposomal samples were diluted to suitable lipid concentrations with 20 mM HEPES, 150 mM NaCl (pH 7.5) and then mixed with the appropriate amount of boron compound dissolved in the same buffer. The resultant suspensions were incubated for various periods of time and temperature at 200 rpm in a rotatory evaporator (vacuum disconnected). For passive loading, the boron compound encapsulation was carried out during the liposome preparation through the hydration of the lipid film with the boron compound suspended in the appropriate aqueous solution. The non-entrapped boron compound was separated from the liposomes by gel permeation chromatography, passing 3 ml of the sample through a Bio-Gel P-6 column (bed volume equal to 10 ml) pre-equilibrated with 20 mM HEPES, 150 mM NaCl (pH 7.5). Trapping efficiencies were calculated as the boron compound to lipid molar ratio after separation of free boron compound divided by the boron compound to lipid molar ratio before separation. The number of boron compound molecules per conventional vesicle was estimated from liposome diameter and the correspondent boron compound to lipid molar ratio.

**Vesicle Characterization.** The vesicle-containing eluants from both active and passive loading procedures were assayed for lipid concentration by phosphate detection using the technique of Chen et al. (1956), for boron compound concentration by reaction of the primary amines with 2,4,6-trinitrobenzenesulfonic acid (method modified from New, 1990), and for vesicle diameter by quasi-elastic light scattering (Ford, 1985).

**Evaluation of Liposome Stability in Non-ionic Surfactant Solution.** The stability of liposomes was evaluated by adding aliquots of liposomes suspended in 10 mM HEPES buffer, pH 7.4, to solutions containing $C_{12}E_5$ solubilized in the same buffer, in 96-well polystyrene microtiter plates, according to the technique used by Steffen (1993). The final concentration of total lipid was 1 mM and the final concentration of detergent ranged from 0 to 50 mM. The evaluation of the interactions between vesicles and $C_{12}E_5$ was performed 2 hours after surfactant action, by monitoring the overall solution absorbance at 340 nm, using a microtiter plate reader model EL 340, from Bio-Tek Instruments. The plate was incubated at room temperature without shaking, but the samples were shaken vigorously in the plate reader for 3 seconds prior to absorbance measurement. The normalized mean absorbance at equilibrium, defined as (A-Ao)/Ao, where A is the absorbance at different detergent concentrations and Ao is the absorbance at zero detergent concentration, is plotted against detergent concentration, defined as the surfactant molar concentration divided by the sum of the surfactant and total lipid molar concentrations. According to Steffen (1993), this procedure smooths differences in initial absorbances due to vesicle preparation conditions, providing equitable comparisons.

## RESULTS AND DISCUSSION

### Entrapment of BPA in Liposomes

**Passive Loading.** The passive loading of BPA either suspended as a slurry in 10 mM HEPES buffer or as a more soluble fructose complex was evaluated using sonicated and extruded liposomes. The results can be observed in Table 1. No accumulation of the compound was observed for conventional DSPC/DMPE/ Chol sonicated vesicles (results not shown), possibily due to the extremely low solubilityof BPA in the tested aqueous solution (10 mM HEPES buffer). However, when using sonicated PEG-containing liposomes, the BPA solubility was improved, possibly by interactions of this boron compound with the polymer chains, leading to BPA accumulation into the vesicles. The low trapping efficiency obtained for passive loading is typical for sonication procedures, since the capture volume is normally small for vesicles with diameters less than 100 nm (New, 1990).

The solubility of BPA in aqueous solutions can be enhanced to up to 60 mg/ml by the addition of fructose at a 1:1 fructose/BPA molar ratio, followed by correction of the pH to 8.8, as reported by Matalka et al. (1993). Higher amounts of BPA were encapsulated in extruded liposomes with the use of this procedure to enhance BPA solubility, as illustrated in Table 1. DSPC/Chol liposomes passively entrapped the fructose/BPA complex, although at low efficiency, giving an internal BPA concentration equivalent to the one obtained for sonicated PEG-containing liposomes. Therefore, since the DSPC/Chol vesicles have larger average diameters, the final number of boron atoms per vesicle is 5 times higher, resulting in systems that would be more adequate for the delivery of this agent for BNCT. The alternative entrapment of the BPA/fructose complex in sonicated liposomes is not possible, due to degradation of fructose by oxidation (caused by prolonged exposure to a temperature of 80°C).

Table 1 - Passive loading of BPA in liposomes, performed at initial BPA to lipid molar ratios of 50:5.

| System Characterization | Vesicle Composition | |
|---|---|---|
| | DSPC/Chol/DMPE/ 12%DMPE-PEG | DSPC/Chol |
| BPA Suspension Media | 10 mM HEPES | 50 mM fructose |
| Vesicle Preparation Method | Sonication | Extrusion |
| Final BPA to Lipid Molar Ratio | 0.0256 | 0.0543 |
| Vesicle Average Diameter (nm) | 92.3 | 134.4 |
| Trapping Efficiency (%) | 0.3 | 0.5 |
| BPA Molecules per Vesicle | 2051 | 10221 |
| BPA Concentration in Vesicle Aqueous Core (mM) | 21.5 | 24.5 |

**Active Loading.** The encapsulation of BPA by active loading was performed using sonicated liposomes. As liposomes containing DMPE in their composition can be easily converted to immunoliposomes by the covalent attachment of antibody glycolipids to the primary amino groups of liposomal phosphatidylethanolamine, only vesicles composed by DSPC, DMPE and cholesterol were used for the evaluation of BPA active loading, at initial BPA to lipid molar ratios of 0.4:2. This value was chosen such that approximately 50% of the boron compound would accumulate inside the vesicles, according to the Henderson-Hasselbach equation (Madden et al., 1990). The vesicles were incubated with BPA in different conditions of time and temperature and the results are illustrated on Figures 3 and 4.

The transmembrane proton gradient procedure did not yield large amounts of encapsulated BPA in liposomes. Increases in the incubation time and temperature did not result in more effective entrapment, as indicated by Figure 3. Only after 120 minutes of incubation at 65°C larger amounts of BPA were encapsulated, although the overall concentration was still low and some aggregation was noticed, as indicated by the increase in vesicle average diameters illustrated in Figure 4.

This boron compound possibly did not accumulate into liposomes by using the active loading approach due to its low ability to cross lipid bilayers, as indicated by its low partition coefficient of 0.007 in octanol/water at 25°C. In addition, at pH 7.5, BPA is probably a neutral species that cannot permeate through the lipid bilayer. The presence of hydroxyl groups in this compound, which would limit its permeability to lipid bilayer membranes, could also have prevented BPA accumulation within liposomes, as reported for the active loading of catecholamines (Nichols and Deamer, 1976).

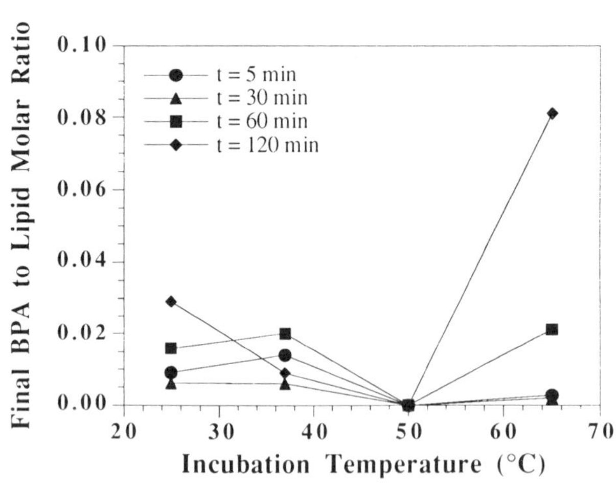


Figure 3 - Effect of the variation of time and temperature of incubation of the mixture BPA/liposomes (DSPC/Chol/DMPE) on the final BPA to lipid molar ratio. Initial BPA to lipid molar ratio = 0.4/2.

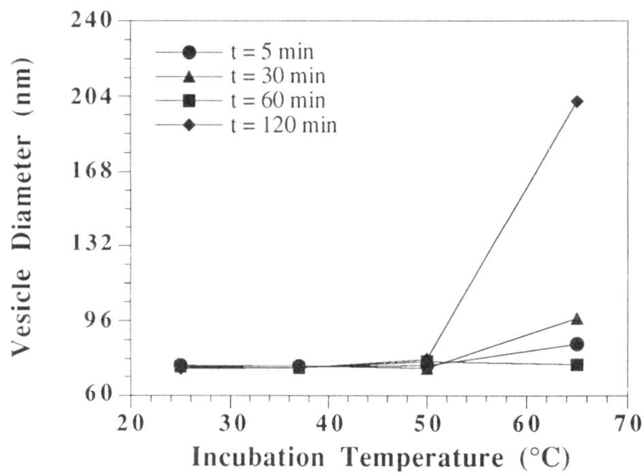


Figure 4 - Effect of the variation of time and temperature of incubation of the mixture BPA/liposomes (DSPC/Chol/DMPE) on final vesicle diameters. Initial BPA to lipid molar ratio = 0.4/2.

## Entrapment of CPA in Liposomes

**Passive Loading.** The passive encapsulation of CPA in extruded liposomes composed by DSPC/Chol/DMPE was tested using an initial CPA to lipid molar ratio equal to 50:7.5. After the removal of the non-encapsulated boron compound, a final CPA to lipid molar ratio of 1.58:3.80 was achieved, in vesicles with an average diameter of 97.7 nm. The encapsulation efficiency of this preparation was equal to only 6.5 %, however, the number of boron atoms per vesicle was estimated to be 470,064. If it is assumed that all the CPA is in a soluble form in the aqueous core of the liposomes, the approximate CPA concentration would be equal to 294 mM, 6 times higher than the CPA concentration in the hydration step. This indicates that a large fraction of the CPA is associated with the lipid membrane. However, this vesicle preparation was not stable, resulting in aggregation and fast increase in sample turbidity after removal of the non-encapsulated CPA. Upon three days of storage at 5°C, a large amount of precipitated material was observed in the botton of the flask, resulting in a release of 23% of the originally entrapped CPA. These results suggest that other encapsulation techniques for CPA should be used to prepare stable systems suitable for BNCT.

**Active Loading.** Figures 5 and 6 illustrate the effects of liposome incubation conditions on CPA uptake and vesicle diameter using an initial CPA to lipid molar ratio of 0.4:2, employing the active loading approach as an alternative method for the entrapment.

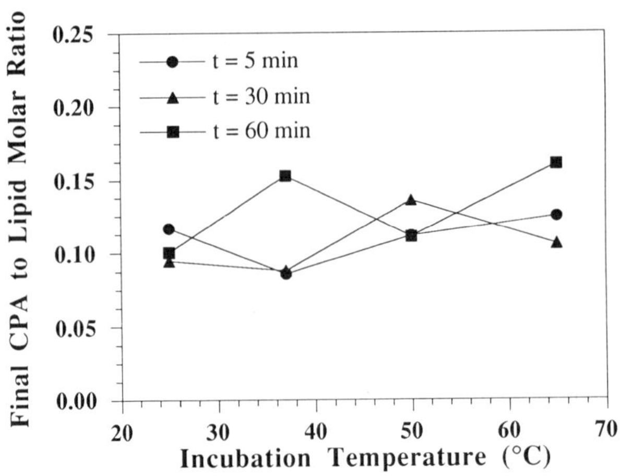


Figure 5 - Effect of the variation of time and temperature of incubation of the mixture CPA/liposomes (DSPC/Chol/DMPE) on the final CPA to lipid molar ratio. Initial CPA to lipid molar ratio = 0.4/2.

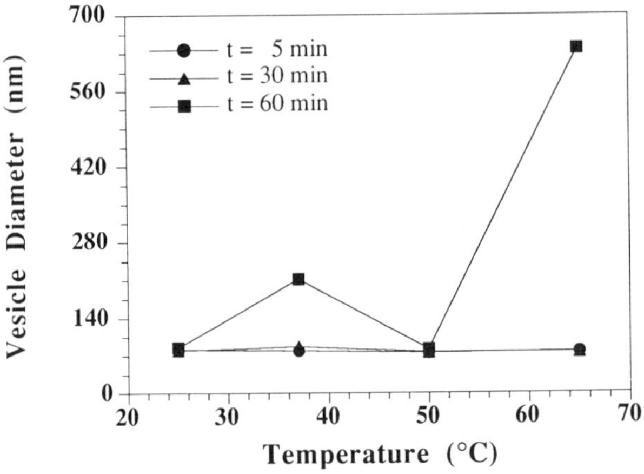


Figure 6 - Effect of the variation of time and temperature of incubation of the mixture CPA/liposomes (DSPC/Chol/DMPE) on final vesicle diameters. Initial CPA to lipid molar ratio = 0.4/2.

As indicated in Figure 5, CPA entrapment characteristics are relatively insensitive to changes in incubation conditions. The final CPA to lipid molar ratios varied from 0.086 to 0.161 (equivalent to trapping efficiencies of 43 to 80.5%, respectively). The larger CPA to lipid molar ratios were obtained for longer incubation times, although vesicle aggregation was noticed after 1 hour of incubation at 37 and 65°C (Figure 6). Extensive vesicle aggregation was observed after 2 hours of incubation (data not shown), when large visible flakes could be easily detected. Therefore, 50°C and 30 minutes were selected as the incubation conditions for further experiments due to thc achicvcmcnt of high trapping efficiency without vesicle aggregation.

Once the incubation conditions were selected, the CPA encapsulation efficiencies were measured at various initial CPA to lipid molar ratios, using conventional and stealth liposomes. The results are shown in Figures 7 and 8. The active entrapment of CPA in conventional vesicles is more sensitive to changes in the boron compound to lipid molar ratio of the initial incubation mixture than to changes in time and temperature of incubation. Increasing CPA to lipid ratios from 0.11 to 0.35 results in an increase of the uptake levels, up to 181 mmol of CPA per mol of lipid. However, when the initial CPA to lipid ratio is increased above 0.35, the CPA loading decreases slightly, stabilizing around 160 mmol of CPA per mol of lipid, apparently achieving an equilibrium. These results indicate the existence of a maximum CPA loading in these vesicles. Vesicle diameters increased from 3 to 32 % as a result of their incubation with the boron compound at the different CPA/lipid molar ratios tested (Figure 8).

Vesicles containing 12%PEG showed much lower final CPA to lipid molar ratios. The maximum load of CPA in these vesicles was around 50 mmol of CPA per mol of lipid, about 3.6 times lower than for conventional vesicles. 12%PEG-vesicle diameters before incubation were 98.8 nm (28.9 nm larger than conventional liposomes). After incubation, the 12%PEG-vesicles increased at most 7% in diameter, consistent with the fact that less CPA could be incorporated when compared to the other vesicle systems studied.

Figure 9 illustrates an estimation of the correspondent number of boron atoms entrapped per vesicle. In both liposomal systems, relatively large amounts of boron were encapsulated. For initial CPA to lipid molar ratios above 0.11, conventional liposomes can accomodate approximately twice more boron atoms per vesicle than 12%PEG-vesicles.

## Stability of Liposome Preparations in Non-Ionic Surfactant Solutions

The addition of low concentrations of surface active agents to liposome dispersions can have a profound effect on the vesicle membrane, causing significant changes in thc suspension turbidity. Variations in turbidity are related to changes in vesicle diameter. As a result, measurements of the absorbance of the solution after surfactant addition is a controllable measure of the stability of a liposomal preparation. Conventional and stealth liposomes with or without CPA actively entrapped exhibit turbidity changes when exposed to $C_{12}E_5$ solutions, as illustrated on Figure 10. In these experiments, the final CPA to lipid molar ratios were 0.08 and 0.04, for conventional and 12%PEG-vesicles respectively.

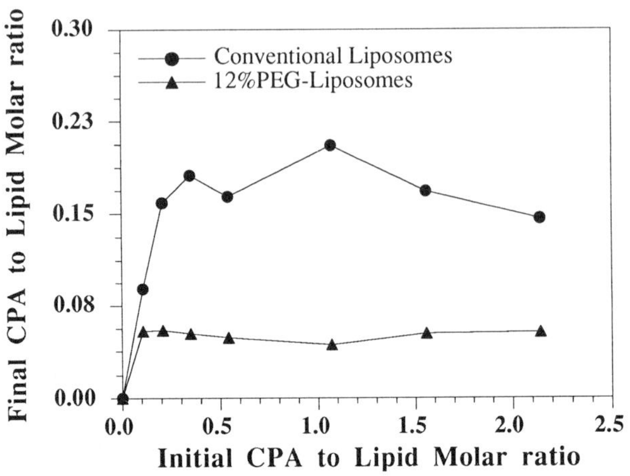


Figure 7 - Influence of the initial CPA to lipid molar ratio on entrapped CPA to lipid molar ratio. Conventional liposomes (DSPC/DMPE/Chol, 40:20:40 mol%, 69.9 nm) and 12%PEG-liposomes (DSPC/DMPE/Chol/DMPE-PEG, 40:8:40:12 mol%, 98.8 nm) having a pH gradient were incubated at 50°C for 30 minutes in the presence of CPA at the indicated drug to lipid ratios. Initial lipid concentration: 2 mM.

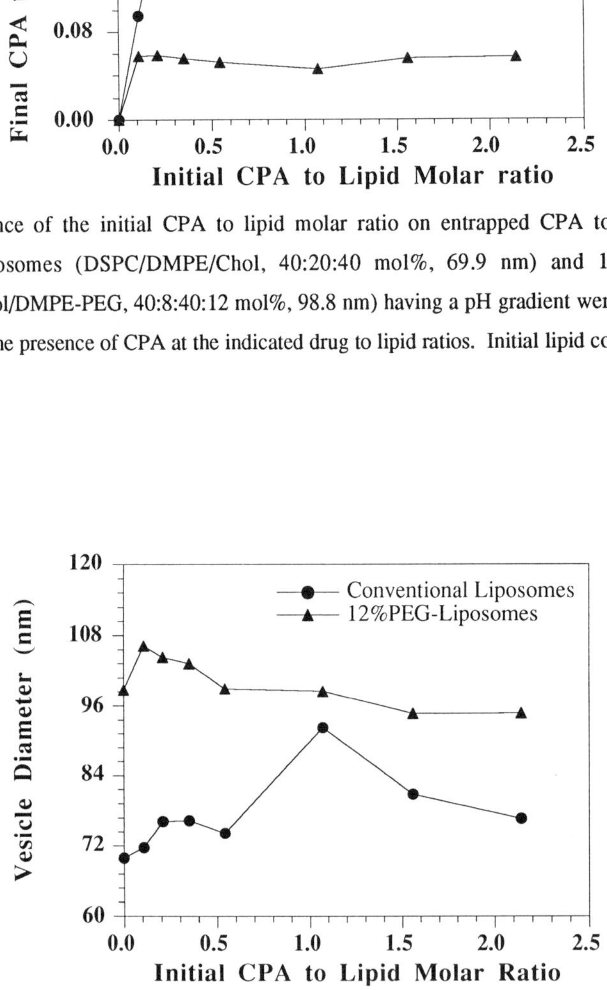


Figure 8 - Influence of the initial CPA to lipid molar ratio on vesicle diameter after removal of non-entrapped CPA. Conventional liposomes (DSPC/DMPE/Chol, 40:20:40 mol%, 69.9 nm) and 12%PEG-liposomes (DSPC/DMPE/Chol/ DMPE-PEG, 40:8:40:12 mol%, 98.8 nm) having a pH gradient were incubated at 50°C for 30 minutes in the presence of CPA at the indicated drug to lipid ratios. Initial lipid concentration: 2 mM.

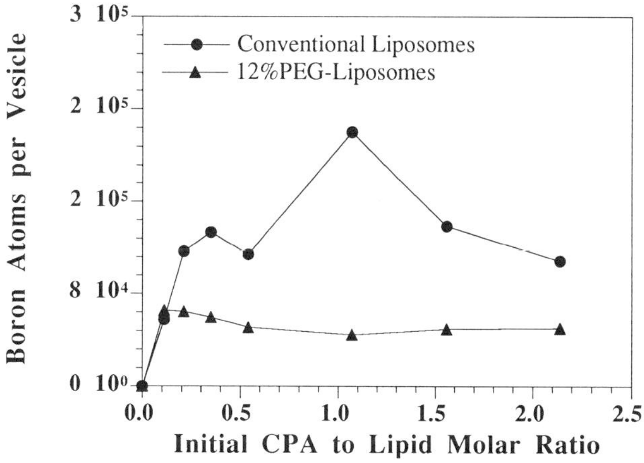


Figure 9 - Number of boron atoms per liposome obtained at different initial CPA to lipid molar ratios. conventional liposomes (DSPC/DMPE/Chol, 40:20:40 mol%, 69.9 nm) and 12%PEG-liposomes (DSPC/DMPE/Cho/DMPE-PEG, 40:8:40:12, 98.2 nm) presenting a pH gradient were incubated (2mM) at 50°C for 30 minutes in the presence of CPA at the indicated drug to lipid ratios.

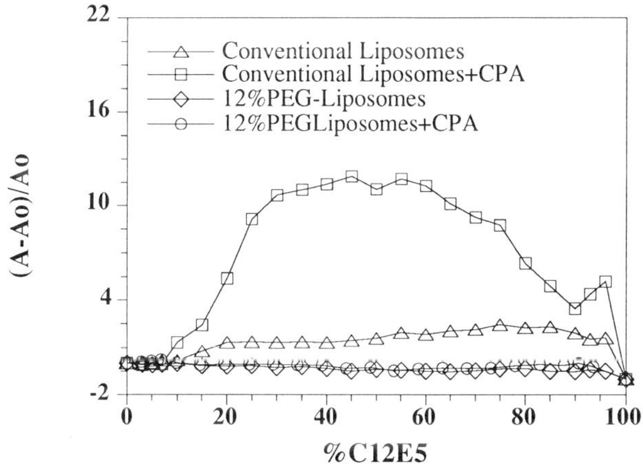


Figure 10 - Effect of different concentrations of $C_{12}E_5$ on the absorbance of conventional and stealth liposomes (conventional liposomes: DSPC/DMPE/Chol, 40:20:40 mol% and 12%PEG-liposomes: DSPC/DMPE/Chol/DMPE-PEG, 40:8:40:12 mol%). Incubation at room temperature, for 2 hours. Lipid concentration: 1mM.

For conventional liposomes, the measured absorbance followed the classic 3-stage pattern for vesicle solubilization by detergents (Nilsson et al., 1987; Edwards et al., 1989; Edwards and Almgren, 1990). For conventional liposomes with no entrapped CPA, the increases in absorbance began around 10 mol%, until a plateau of about 1.4 was reached at 25 mol % $C_{12}E_5$, continuing until 50%, where another rise is seen until the maximum normalized absorbance is reached at about 75% detergent. This rise is followed by a steady decrease in absorbance at larger surfactant concentrations. The behavior of this system can be explained by the following sequence of events. Low concentrations of detergent (below its critical micelle concentration of $6.5*10^{-5}$ M) do not cause any major variation of vesicles structure, however, above a critical surfactant concentration (25 mol %), the liposomes probably change from small to large unilamellar vesicles. Above 50 mol % $C_{12}E_5$, mixed micelles are formed and detergent concentrations higher than 75 mol % possibly induce complete lipid solubilization.

In the detergent concentration range of 0 to 10%, conventional vesicles with encapsulated CPA presented similar behavior as plain liposomes. However, normalized absorbances increased markedly until a maximum value of about 12 was reached at 40 mol% $C_{12}E_5$, continuing until about 60% surfactant. For detergent concentrations larger than 60%, the absorbance decreased with the addition of $C_{12}E_5$.

The interaction pattern between empty stealth liposomes and $C_{12}E_5$ dispersions is different from the one obtained for conventional liposomes. The changes in absorbance are much smaller for PEG-containing liposomes, indicating higher vesicle stability. CPA loaded liposomes containing 12%PEG behave similarly to stealth liposomes without CPA.

These results suggest that during the active loading process, the CPA is also incorporated into the liposomal bilayer. The accomodation of CPA molecules in the lipid membrane probably disturbes the uniformity of the bilayer, resulting in less stable vesicles. The presence of the PEG chains causes an increase in vesicle stability and a reduction in CPA loading within the membrane. Therefore, when conventional or stealth liposomes loaded with CPA are exposed to $C_{12}E_5$, the presence of the CPA in the bilayer probably changes the pattern of lipid packing. At low detergent concentrations, the surfactant molecules are unable to penetrate significantly into the bilayer. At sufficiently high detergent concentrations, the accommodation of a few $C_{12}E_5$ molecules in the bilayer causes massive instability, either by the creation of holes in the liposomal membrane, that would provide more space to detergent accommodation, or by vesicle aggregation. Further increases in detergent concentration could lead to the coexistence of fused vesicles and mixed micelles, with the ultimate disruption of the liposomes.

The solubility of CPA in the lipid membrane is possibly affected by the geometrical restrictions originated from the limited space available in the bilayer and the fixed orientation of the bilayer constituents. If the CPA were predominantly localized in the vesicle aqueous core, increases in detergent concentration would lead to CPA leakage from the vesicles. This event would not impose further changes in absorbance when compared to empty vesicles at the same lipid composition.

The consideration that CPA is predominantly incorporated into the liposomal bilayer is also justified when evaluating the data obtained from the incubation of CPA with the vesicles at different initial molar ratios. In all situations, it seems that a maximum loading is obtained. These values could result from the accommodation of the maximum number of CPA molecules inside the bilayer, depending both on the lipid composition and on the presence of external PEG chains, that would reduce the permeability of the membrane to the CPA by steric hindrance.

## CONCLUDING REMARKS

The experimental results obtained suggest that its possible to succesfully encapsulate both boron compounds tested in liposomes.

BPA can be entrapped in PEG-liposomes using the passive loading approach, while the addition of fructose to BPA solutions improves the encapsulation of this compound in conventional liposomes. Transmembrane proton gradients trapping procedures, however, did not yield large amounts of encapsulated BPA in conventional liposomes.

The passive loading of CPA in conventional liposomes resulted in unstable preparations and in vesicle aggregation immediately after the entrapment. Transmembrane proton gradients (active loading) can be used to accumulate CPA within conventional and stealth liposomes. For conventional liposomes, the CPA entrapment characteristics are more sensitive to the drug-to-lipid molar ratio than to the incubation conditions. CPA could be more efficiently accumulated in conventional liposomes than in 12%PEG-liposomes, indicating that the presence of PEG chains reduces the incorporation of this compound in the liposomes.

PEG-modified vesicles with and without CPA exhibit larger stability in $C_{12}E_5$ solutions than conventional liposomes. CPA-loaded conventional liposomes are less stable than conventional vesicles without the boron compound, suggesting that CPA is predominantly located in the lipid bilayer.

In summary, this paper demonstrates that the undestanding of the effects that control the entrapment of therapeutic compounds in lipid vesicles is essential to produce liposomes with specific requirements, such as size and drug to lipid ratio. The preparation protocols can also control the final stability of the resulting liposomal system.

## ACKNOWLEDGEMENTS

The authors gratefully acknowledge the support of this work by the William R. Kenan, Jr. Institute for Engineering, Technology & Science, at North Carolina State University, and The Conselho Nacional de Desenvolvimento Científico e Tecnológico (CNPq), Brasília, DF,

Brazil. The authors also thank Boron Biologicals Inc., Raleigh, NC, U.S.A. for the donation of the boron compounds BPA and CPA.

## REFERENCES

Barth, R. F.; Soloway, A. H. and Fairchild, R. G. - Perspectives in cancer research: boron neutron capture therapy of cancer - *Cancer Research*, vol. 50, pp. 1061-1070, 1990.

Barth, R. F.; Soloway, A. H.; Fairchild, R. G. and Brugger, R. M. - Boron neutron capture therapy for cancer: realities and prospects - *Cancer*, vol. 70, No. 12, pp. 2995-3007, 1992.

Chen Jr., P. S.; Toribara, T. Y. and Warner, H. - Microdetermination of phosphorus - *Analytical Chem.*, vol. 28, No. 11, pp. 1756-1758, 1956.

Coderre, J. A.; Glass, J. D.; Fairchild, R. G.; Micca, P. L.; Fand, I. and Joel, D. D. - Selective delivery of boron by the melanin precursor analogue p-boronophenylalanine to tumors other than melanoma - *Cancer Research*, vol. 50, pp. 138-141, 1990.

Edwards, K.; Almgren, M.; Bellare, J. and Brown, W. - Effects of Triton X-100 on sonicated lecithin vesicles - *Langmuir*, vol. 5, pp. 473-478, 1989.

Edwards, K. and Almgren, M. - Kinetics of surfactant-induced leakage and growth of unilamellar vesicles - *Progress in Colloid & Polymer Science*, vol. 82, pp. 190-197, 1990.

Feakes, D. A.; Shelly, K.; Knobler, C. B. and Hawthorne, F. - $Na_3[B_{20}H_{17}NH_3]$: Synthesis and liposomal delivery to murine tumors - *Proc. Natl. Acad. Sci. USA*, vol. 91, pp. 3029-3033, 1994.

Feakes, D. A.; Shelly, K. and Hawthorne, M. F. - Selective boron delivery to murine tumors by lipophilic species incorporated in the membranes of unilamellar liposomes - *Proc. Natl. Acad. Sci. USA*, vol. 92, pp. 1367-1370, 1995.

Ford, N. C. Jr. - In Dynamic Light Scattering: Applications of Photon Correlation Spectroscopy - R. Pecora (Ed.), Plenum Press, New York, pp.7-57, 1985.

Fukuda, H.; Ichihashi, M. and Mishima, Y. - [Neutron capture therapy for cancer] - *Gan To Kagaku Ryoho*, vol. 21, No. 3, pp. 374-378, 1994.

Hawthorne, F. M. - The role of chemistry in the development of boron neutron capture therapy of cancer - *Angewandte Chemie* (International Edition in English), vol. 32, pp. 950-984, 1993.

Lasic, D. and Martin, F. (Editors) - Stealth Liposomes, CRC Press, Inc., Boca Raton, 1995.

MacDonald, R. C.; MacDonald, R. I.; Menco, B. P.M.; Takeshita, K.; Subbarao, N. K. and Hu, L. - Small-volume extrusion apparatus for preparation of large, unilamellar vesicles - *Biochimica et Biophysica Acta*, vol. 1061, pp. 297-303, 1991.

Madden, T. D.; Harrigan, P. R.; Tai, L. C. L.; Bally, M. B.; Mayer, L. D.; Redelmeier, T. E.; Loughrey, H. C.; Tilcock, C. P. S.; Reinish, L. W. and Cullis, P. R. - The accumulation of drugs within large unilamellar vesicles exhibiting a proton gradient: a survey - *Chemistry and Physics of Lipids*, vol. 53, pp. 37-46, 1990.

Matalka, K. Z.; Bailey, M. Q.; Barth, R. F.; Staubus, A. E.; Soloway, A H.; Moeschberger, M. L.; Coderre, J. A. and Rofstad, E. K. - Boron neutron capture therapy of intracerebral melanoma using boronophenylalanine as a capture agent - *Cancer Research*, vol. 53, pp. 3308-3313, 1993.

Matalka, K. Z.; Barth, R. F.; Staubus, A. E.; Moeschberger, M. L. and Coderre, J. A. - Neutron capture therapy of a rat glioma using boronophenylalanine as a capture agent - *Radiation Research*, vol. 137, pp. 44-51, 1994.

Mayer, L. D.; Hope, M. J.; Cullis, P. R. and Janoff, A. S. - Solute distributions and trapping efficiencies observed in freeze-thawed multilamellar vesicles - *Biochimica et Biophysica Acta*, vol. 817, pp. 193-196, 1985.

Mayer, L. D.; Bally, M. B.; Hope, M. J. and Cullis, P. R. - Techniques for encapsulating bioactive agents into liposomes - *Chemistry and Physics of Lipids*, vol. 40, pp. 333- 345, 1986.

Mayer, L. D.; Tai, L. C. L.; Bally, M. B.; Mitilenes, G. N.; Ginsberg, R. S. and Cullis, P. R. - Characterization of liposomal systems containing doxorubicin entrapped in response to pH gradients - *Biochimica et Biophysica Acta*, vol. 1025, pp. 143-151, 1990.

Mayer, L. D.; Madden, T. D.; Bally, M. B. and Cullis, P. R. - pH Gradient-mediated drug entrapment in liposomes, in Liposome Technology, edited by Gregoriadis, G., 2$^{nd}$ edition, vol.II, pp. 27-44, CRC Press Inc., Boca Raton, 1993.

New, R. R. C. (Editor) - Liposomes: A Practical Approach, Oxford University Press, Oxford, 1990.

Nichols, J. W. and Deamer, D. W. - Catecholamine uptake and concentration by liposomes maintaining pH gradients - *Biochimica et Biophysica Acta*, vol. 455, pp. 269-271, 1976.

Nilsson, K.; Almgren, M.; Brown, W. and Jansson, M. - Effects of a non-ionic surfactant on small unilamellar lecithin vesicles - *Mol. Cryst. Liq. Cryst.*, vol. 152, pp. 181 - 203, 1987.

Saris, S. C.; Solares, G. R.; Wazer, D. E.; Cano, G.; Kerley, S. E.; Joyce, M. A.; Adelman, L. S.; Harling, O. K.; Madoc-Jones, H. and Zamenhof, R. G. - *Cancer Research*, vol. 52, pp. 4672-4677, 1992.

Shelly, K.; Feakes, D. A.; Hawthorne, M. F.; Schmidt, P. G.; Krisch, T. A and Bauer, W. F. - Model studies directed toward the boron neutron-capture therapy of cancer: boron delivery to murine tumors with liposomes - *Proc. Natl. Acad. Sci. USA*, vol. 89, pp. 9039-9043, 1992.

Szoka , F. and Papahadjopoulos, D. - Comparative properties and methods of preparation of lipid vesicles (liposomes) - *Ann. Rev. Biophys. Bioeng.*, vol. 9, pp. 467-508, 1980.

Steffen, D. M. - The effect of cholesterol, phosphatidylethanolamine and immobilized horseradish peroxidase on the stability of phosphatidylcholine vesicles when treated with nonionic surfactant - Master Thesis, Department of Chemical Engineering, NCSU, Raleigh, NC, 1993.

Yanagië, H.; Tomita, T.; Kobayashi, H.; Fujii, Y.; Takahashi, T.; Hasumi, K.; Nariuchi, H. and Sekiguchi, M. - Application of boronated anti-CEA immunoliposome to tumor cell growth inhibition in *in vitro* boron neutron capture therapy model - *The British Journal of Cancer*, vol. 63, No. 4, pp. 522 526, 1991.

# SECTION VI.

# MEMBRANE FOULING

# NON-SPECIFIC PROTEIN-MEMBRANE INTERACTIONS: ADSORPTION AND FOULING

Andrew L. Zydney

Department of Chemical Engineering
University of Delaware
Newark, DE 19716

## INTRODUCTION

One of the major factors limiting the development and application of biofunctional membranes is membrane fouling, the irreversible alteration in the membrane caused by specific physical and/or chemical interactions with various components present in the process stream. Fouling can cause dramatic changes in the rate of solute transport through the membrane, potentially destroying the membrane's ability to function in the particular biomedical device, biosensor, or bioreactor.

Proteins tend to have a particularly high affinity for many membrane surfaces. This is due in large part to the heterogeneity in the substituent amino acid residues which allows proteins to interact with the membrane surface via a wide range of electrostatic, hydrogen bonding, and van der Waals forces. In addition, protein desorption is highly unfavorable since it requires the simultaneous disruption of a large number of distinct interactions between the protein and the membrane.

The rate and extent of protein fouling in any given system will be determined by the strength of the intermolecular interactions between the proteins and the membrane in combination with the various hydrodynamic forces acting on the protein. These hydrodynamic forces not only determine the rate of protein transport towards the membrane (and thus the local protein concentration in the immediate vicinity of the membrane surface), they can also directly contribute to membrane fouling through the physical deposition of proteins onto the upper surface of the membrane. It is thus useful to consider two distinct types of fouling phenomena: protein adsorption, which refers to the specific intermolecular interactions between the protein and the membrane that occur in the absence of any ultrafiltration (i.e. in the absence of these hydrodynamic and body forces), and protein deposition, which refers to any additional protein that becomes associated with the membrane during filtration that is over and above that which would be adsorbed to the membrane in a static (non-flowing) system.

This chapter will examine some of our recent work on both protein adsorption and deposition, with particular emphasis on the effects of the membrane pore size on these fouling phenomena. In each case, we will try to highlight our current understanding of the underlying causes of protein adsorption and deposition in these systems as well as the effects that these phenomena ultimately have on membrane transport.

*Biofunctional Membranes*, Edited by
D.A. Butterfield, Plenum Press, New York, 1996

## PROTEIN ADSORPTION

Previous studies of protein adsorption to porous membranes are often quite contradictory in terms of both the actual amount of adsorbed protein as well as the effects of this adsorption on membrane transport (1, 2). Much of this discrepancy has to do with the failure to distinguish between protein adsorption and deposition, with the amount of protein deposited during filtration often being considerably greater than that due to simple adsorption. Experimental results for protein adsorption can also be quite sensitive to the specific procedures used to evaluate the protein uptake. For example, many investigators have used either radiolabelled or fluorescent-labelled proteins to directly measure the amount of protein adsorbed to the membrane surface, but the presence of the label can actually alter the adsorption characteristics, leading to potentially erroneous conclusions about the adsorption behavior (2). In addition, many investigators have reported protein adsorption levels based on the external (upper) surface area of the membrane, which ignores the adsorption that occurs throughout the internal pore area (with this internal area being as much as several orders of magnitude greater than the external surface area). It is also critically important to distinguish between protein adsorption within, or on, the permselective skin from that which occurs in the porous support structure or matrix of the asymmetric membranes used in many applications. This latter type of adsorption will generally have minimal effect on the overall transport characteristics of the membrane due to the very large pore size in the support structure. However, the large surface area available in the porous support can cause significant loss of valuable products in some applications, and it can significantly alter the overall device dynamics due to its effect on solute/protein transport. Several studies have shown that protein adsorption can occur in this porous support structure even with membranes that appear to be completely retentive to the protein of interest (3), with this effect generally attributed to the presence of a small number of large pores or defects in the membrane skin.

The effects of protein adsorption on membrane transport are very different depending upon the relative size of the protein and the membrane pores. The following discussion is thus divided into three subsections: adsorption to microfiltration membranes with pores that are much larger than the size of the proteins, adsorption to semipermeable ultrafiltration membranes with pores that are comparable in size to the proteins, and adsorption to fully retentive membranes which have pores that are considerably smaller than the proteins.

### Microfiltration Membranes

Most experimental studies of protein adsorption on microfiltration membranes have reported monolayer adsorption throughout the internal pore area (4). This includes data for a variety of proteins (bovine serum albumin, hemoglobin, yeast alcohol dehydrogenase, and β-lactoglobulin) on a number of different types of membranes including hydrophobic poly(vinylidene fluoride), polyethersulfone, aluminum oxide, nylon, and cellulose acetate. For example, Bowen and Hughes (5) showed that BSA adsorption on aluminum oxide (ceramic) membranes could be effectively described by a Langmuir type isotherm with a maximum adsorption level of about 2.5 $mg/m^2$, with the membrane pore area determined using nitrogen BET adsorption. There was also some evidence for a weakly bound multilayer adsorption on these membranes at fairly high protein concentrations, which is similar to the behavior seen in a number of studies of protein adsorption to non-porous surfaces. This weakly bound protein can often be removed by the washing/rinsing steps employed in many experimental systems. Protein adsorption on highly hydrophilic microfiltration membranes appears to be significantly less than a protein monolayer (4), although this reduced adsorption may be due, at least in part, to the desorption of loosely bound protein during the rinsing steps used prior to evaluating the amount of protein adsorbed to these microfiltration membranes.

The effect of monolayer protein adsorption on the solvent flux through a microfiltration membrane will generally be quite small. This can be most easily seen using a uniform cylindrical pore model with Poiseuille flow to evaluate the solvent velocity through the membrane:

$$V = \frac{N\pi r_p^4 \Delta P_{TM}}{8\mu\delta_m} \tag{1}$$

where N is the membrane pore density, $r_p$ is the pore radius, $\Delta P_{TM}$ is the transmembrane pressure drop, $\mu$ is the fluid viscosity, and $\delta_m$ is the membrane thickness. For a typical microfiltration membrane (with $r_p = 0.2$ µm), the flux reduction associated with monolayer adsorption of a protein with a radius of 30 Å is only about 6%. This small reduction in pore radius would only have a relatively small effect on the rate of solute transport through these microfiltration membranes.

## Semi-Permeable Ultrafiltration Membranes

Protein adsorption to ultrafiltration membranes with pores that are only slightly larger than the protein can be effected by "steric" interactions in the narrow pores. These steric effects arise from the loss of accessible surface area due to simple geometric constraints as well as the increase in free energy of the adsorbed protein arising from unfavorable intermolecular interactions associated with the proximity of the surrounding membrane surface. For example, a negatively charged protein may be able to adsorb strongly on a negatively charged surface by adopting an orientation in which the negatively charged amino acids are extended out into the bulk solution so that they have relatively little interaction with the membrane surface. This type of energy minimization will be much less effective in a narrow pore since these negatively charged amino acids will now be in close proximity to the negatively charged groups on the opposite surface of the pore wall.

We have examined the magnitude of these steric effects by measuring the uptake of radioactively labelled bovine serum albumin to a series of semipermeable polyethersulfone ultrafiltration membranes with different pore sizes (2). Equilibrium adsorption was not attained until almost 12 hrs due to the slow diffusive transport of BSA through the porous structure of the membrane. Thus, some care must be taken in interpreting literature data for protein adsorption to porous membranes to insure that the results are truly representative of the equilibrium behavior. The amount of protein adsorbed in the permselective skin of these asymmetric membranes was evaluated by subtracting off the contributions in the substructure and support matrix, both of which were evaluated from independent experimental measurements (2). It was also necessary to account for the preferential adsorption of the radiolabelled protein (relative to the unlabelled BSA), which was apparently due to the presence of the iodine on the amino acid tyrosine. The data are summarized in Table 1 along with the results for an 0.16 µm microfiltration membrane cast from the same base polymer (6). The amount of protein adsorbed per external area increased with increasing pore size for the smaller pore membranes due to the increase in the accessibility of the pores. Protein adsorption was maximum on the 300,000 molecular weight cut-off (300K) membrane before decreasing for the two largest pore membranes due to the reduction in the total pore area associated with the increase in pore size.

The internal pore area for these different membranes was estimated from data for the membrane hydraulic permeability as described by Robertson and Zydney (2). The mass adsorbed per total pore area was very similar for the three largest pore membranes, with the surface coverage in good agreement with that obtained for monolayer albumin adsorption on a variety of non-porous surfaces. Albumin adsorption to the 100K membrane was only about 50% of that in the larger pore membranes, reflecting the steric interactions that occur in the highly constricted pores of this membrane. There was almost no measurable albumin adsorption in the porous skin of the 50K membrane, although there was still considerable adsorption throughout the porous substructure and matrix (2).

**Table 1:** Protein Adsorption on Semi-Permeable Ultrafiltration Membranes

| Membrane | Mass Adsorbed per External Area ($mg/m^2$) | Mass Adsorbed per Pore Area ($mg/m^2$) | Hydraulic Permeability Ratio ($L_p/L_{po}$) |
|---|---|---|---|
| 50,000 | 20 ± 50 | 0.1 ± 0.2 | 0.68 ± 0.14 |
| 100,000 | 230 ± 90 | 2.0 ± 0.8 | 0.54 ± 0.11 |
| 300,000 | 440 ± 150 | 4.5 ± 1.5 | 0.20 ± 0.30 |
| 1,000,000 | 240 ± 150 | 4.0 ± 2.5 | 0.06 ± 0.03 |
| 0.16 μm | 130 ± 30 | 3.0 ± 0.4 | 0.79 ± 0.10 |

The final column in Table 1 shows the ratio of the membrane hydraulic permeability (or solvent flux) after albumin adsorption to that determined for the clean membrane. The reduction in permeability for the 1M membrane was more than a factor of 10, while the change in permeability for the 50K membrane was only 32 % and that for the 0.16 μm pore size was only 21 %. These solvent flux data can be used to evaluate the effective reduction in the membrane pore size using the simple cylindrical pore model (Eq. 1). The results for the 0.16 μm, 1M, and 300K membranes were all consistent with the adsorption of a protein monolayer, with the calculated thickness of the adsorbed layer being on the order of 50 Å. In contrast, the permeability data for the 50K and 100K membranes correspond to reductions in pore radius of only 3 and 7 Å, respectively, reflecting the much more complex adsorption behavior in the highly constricted pores of these smaller pore size membranes.

This type of protein adsorption will also have a significant effect on the rate of solute transport through these semipermeable ultrafiltration membranes. We have evaluated the convective (sieving) and diffusive transport properties of the polyethersulfone membranes shown in Table 1 both before and after protein adsorption using polydisperse dextrans, with the molecular weight distribution in the dextran samples determined using gel permeation chromatography (7). Additional details on these experiments and the associated data analysis are provided by Mochizuki and Zydney (7). The solute flux through the membrane ($N_s$) has contributions from both convection and diffusion:

$$N_s = S_\infty VC - D_{eff}\left(\frac{dC}{dz}\right) \tag{2}$$

where $S_\infty$, the asymptotic membrane sieving coefficient, provides a measure of the hindrance to convective transport, while $D_{eff}$ is the effective solute diffusion coefficient through the membrane. The dextran sieving coefficients and hindered diffusion coefficients after protein adsorption were uniformly smaller than those for the clean membranes. Experimental data obtained with the 1M membrane were again consistent with monolayer adsorption, with the effective pore size reduced by the size of an adsorbed monolayer. The effects of protein adsorption on the transport characteristics of the smaller molecular weight cut-off membranes were considerably more complex, reflecting the highly constricted geometry in the porous structure of these membranes (7). Experimental data for the 100K membrane showed more than a factor of 2.5 reduction in the sieving coefficient for a 20,000 molecular weight dextran (from $S_\infty = 0.55$ for the clean membrane to $S_\infty = 0.22$ after adsorption of BSA). A similar reduction was seen in the hindered diffusion coefficient, although the reduction in the hydraulic permeability (solvent flow) for this membrane was less than a factor of 2. Mochizuki and Zydney (7) used these changes in solvent flow, hindered diffusion, and sieving to conclude that protein adsorption occurs preferentially in the larger pores in the membrane pore size distribution, with the radius of these larger pores being effectively reduced by the protein adsorption.

## Fully Retentive Ultrafiltration Membranes

Protein adsorption to fully retentive ultrafiltration membranes is primarily a surface phenomenon since the membrane pores are largely inaccessible to the proteins. Most previous studies of adsorption to fully retentive membranes have implicitly assumed that the alteration in membrane transport characteristics was due to some type of pore blockage phenomenon. However, it must be recognized that pore blockage of an asymmetric ultrafiltration membrane is itself an extremely complex phenomena due to the very tortuous and interconnected pore structure of these asymmetric membranes. For example, even if part of the membrane surface were completely "blocked" by adsorbed protein, the solute and solvent that was transported around these blocked pores could potentially "percolate" down and through the area beneath the blocked pores due to the interconnected structure of the porous membrane.

Recent work in our laboratory has provided a very different physical picture of protein adsorption on these fully retentive membranes (8, 9). Typical data for the effect of BSA adsorption on the flux of an 8,000 molecular weight dextran through an AN69 polyacrylonitrile membrane are shown in Figure 1. The solute flux, scaled by the concentration driving force across the membrane, clearly decreases after adsorption. However, the magnitude of this effect is considerably greater at large velocities (i.e. in the regime dominated by convection) than at small velocities (i.e. in the regime dominated by diffusion). We were able to quantify this behavior using a two-layer membrane model, in which the adsorbed protein exists in a "porous" layer on the upper surface of the membrane that provides an additional resistance to both solute and solvent transport in series with that provided by the membrane. In this case, the adsorbed protein has a much greater effect on convection than diffusion because the protein layer is relatively thin (about 1 μm) compared to the overall thickness of the AN69 membrane (about 25 μm). A very different behavior was seen for asymmetric ultrafiltration membranes which have very thin skin layers. For example, the effective diffusion coefficient for a 16K dextran through a 30,000 molecular weight cut-off polyethersulfone membrane was reduced by more than a factor of 7 after a 3 hr exposure to blood plasma, while the sieving coefficient of this membrane decreased by only about a factor of 2 under the same conditions (8).

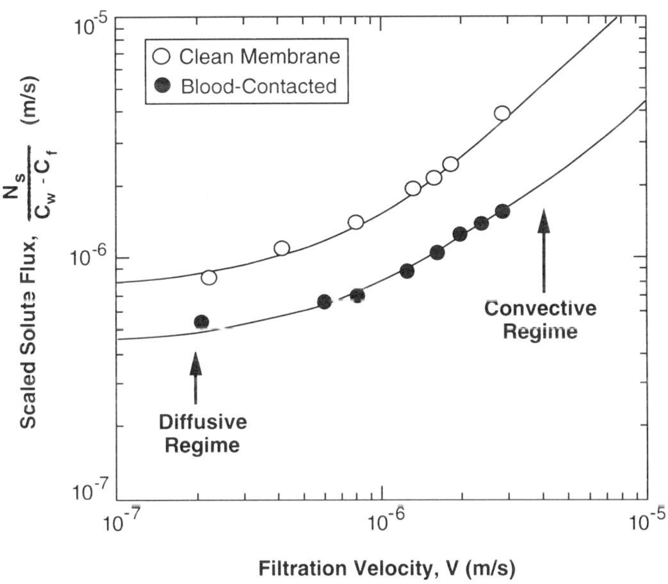


**Figure 1:** Scaled flux of an 8,000 molecular weight dextran through an AN69 polyacrylonitrile membrane both before and after exposure to whole blood.

One of the interesting implications of this two-layer membrane model is that the membrane sieving characteristics (i.e. convective solute flux) depend upon the flow direction, with the sieving coefficient determined primarily by the properties of the upstream layer of the membrane. In contrast, the diffusive solute flux and the membrane hydraulic permeability are determined by the weighted average of the properties of the protein layer and the membrane and are thus independent of the flow direction (9). This general behavior was verified experimentally for an AN69 polyacrylonitrile membrane exposed to whole blood for a period of approximately 3 hrs (9). In this case, the solute flux with the convective flow in the normal direction (i.e. with the blood-contacted layer on the upstream surface of the membrane) was more than a factor of two smaller than that obtained with the flow in the reverse direction (i.e. with the blood-contacted layer on the downstream surface) due to the development of an internal concentration polarization at the interface between the two "layers" of this blood-contacted membrane.

## PROTEIN DEPOSITION

Protein adsorption is an equilibrium process governed by the nature of the intermolecular interactions between the solute and the membrane. In contrast, protein deposition is a non-equilibrium process governed primarily by the hydrodynamic and body forces acting upon the macrosolute in the particular membrane device. The important hydrodynamic forces in membrane systems include the viscous drag force exerted on the particle by the flowing fluid, the hydrodynamic lift force arising from inertial interactions between the particle and solid boundary, and the diffusional forces associated with Brownian motion and shear-induced diffusion.

The effects of BSA adsorption and deposition on the hydraulic resistance of several different pore size membranes are shown in Table 2 (6). The total hydraulic resistance to the filtrate flow is shown for the clean membrane, for the membrane after equilibrium adsorption of bovine serum albumin in a static system, and for the same membrane after filtration of a 5 g/l BSA solution at 10 psi for 4 hrs in a stirred cell. In each case, the hydraulic resistance was evaluated from data for the pure saline flux obtained after thoroughly rinsing the stirred cell and membrane, thereby eliminating any interference from concentration polarization (bulk mass transfer) effects. Protein adsorption had the greatest effect on the partially permeable 1M membrane as discussed previously. Protein deposition caused a significant additional increase in the resistance of the 1M and 0.16 μm membrane, with the effect of deposition on the 0.16 μm membrane being considerably greater than that due to simple protein adsorption. Interestingly, the total hydraulic resistance to filtrate flow for the three membranes after BSA deposition were actually quite similar, despite the nearly two orders of magnitude difference in the resistance of the clean membranes.

**Table 2:** Effect of Protein Adsorption and Deposition on the Total Hydraulic Resistance to Flow

| | Hydraulic Resistance ($m^{-1} \times 10^{-11}$) | | |
|---|---|---|---|
| **Membrane** | **Clean** | **After Adsorption** | **After Filtration** |
| 30 K | 11 | 13 | 23 |
| 1 M | 0.62 | 9.6 | 30 |
| 0.16 μm | 0.33 | 0.41 | 23 |

The large increase in hydraulic resistance of these microfiltration membranes after protein filtration is directly attributable to the formation of a protein deposit on the upper surface of the membrane. The BSA deposits appear to consist of a number of relatively large protein aggregates embedded in a more amorphous protein matrix (10). Lee and Merson (11) showed that BSA and β-lactoglobulin formed sheet-like deposits on both ultrafiltration and microfiltration membranes, while immunoglobulin G formed deposits that were composed primarily of large protein granules.

The hydraulic resistance provided by these protein deposits has been shown to be a function of both the applied (transmembrane) pressure and the solution environment (6). The dependence on applied pressure reflects the compressibility of the protein deposit arising from the deformation and/or re-organization of the proteins within this tightly packed layer (6). Thus, the resistance of a BSA deposit increased by about a factor of 2 as the applied pressure increased from 3 to 10 psi. The dependence on the solution environment arises from a number of different phenomena including: changes in the protein packing density caused by the intermolecular electrostatic repulsion between charged proteins within the deposit, alterations in the conformation of individual proteins associated with intramolecular electrostatic effects, and/or development of an electroosmotic counterflow through these charged protein layers (12). The hydraulic resistance is generally greatest at the protein isoelectric point due to the increase in protein packing density caused by the reduction in intermolecular electrostatic repulsion as the pH approaches the pI (12). The hydraulic resistance tends to increase with increasing salt concentration at pH $\neq$ pI due to the increase in electrostatic shielding between the charged proteins.

These deposited protein layers can also have a dramatic effect on the rate of solute transport through these membranes. We have examined the sieving characteristics of BSA deposits formed on 0.16 μm microfiltration membranes using polydisperse dextrans (13). The microfiltration membrane itself had no measurable retention for any of these dextrans, but the fouled membrane had sieving characteristics similar to those of a 100,000 molecular weight cut-off ultrafiltration membrane. The dextran sieving coefficients were also found to be a function of the transmembrane pressure drop and the solution environment, with the general behavior being qualitatively consistent with that described above for the hydraulic resistance of the fouled membrane (13).

Several recent studies have demonstrated that the protein layers formed on the upper surface of these relatively large pore microfiltration membranes are directly associated with the convective deposition of small amounts of aggregated and/or denatured protein (4, 10). These protein aggregates can be formed during the initial preparation or fractionation of the proteins or during the various processing operations used in the particular membrane system (particularly those involving elevated temperatures or high shear stresses). Thus, very different results can be obtained using essentially the same protein due to small differences in their physicochemical characteristics (14). For example, different commercial preparations of BSA can have dramatically different fouling characteristics, with a heat-shock precipitated BSA preparation causing considerably more fouling than a cold-alcohol precipitated BSA due to the greater amount of aggregated and denatured protein in the heat-shock BSA. The fouling obtained in these systems could be almost completely eliminated by pre-filtering the BSA solutions through 100,000 molecular weight cut-off membranes prior to use which effectively removed the protein aggregates and oligomers from the bulk solution.

In order to effectively control protein fouling in these membrane systems, it is necessary to develop a more detailed understanding of the underlying physical and chemical mechanisms by which this protein aggregation actually occurs. We have demonstrated that BSA aggregation occurs through the formation of intermolecular thiol-disulfide linkages between BSA molecules as shown schematically in Figure 2. BSA has thirty-five cysteine amino acid residues, thirty-four of which are linked pairwise via seventeen intramolecular disulfide bonds. The remaining cysteine provides a free sulfhydryl (-SH) group near the amino terminus of the protein. This free sulfhydryl can serve as a nucleophile and attack an existing internal disulfide linkage in a separate BSA molecule as shown in Figure 2. The hydrogen on the free sulfhydryl is first extracted by a hydroxyl, with this ionized sulfhydryl reacting with a second BSA molecule. The reaction product is a BSA dimer, which has two free sulfhydryl groups that can participate in further thiol-disulfide interchange reactions leading to the formation of large protein aggregates. Kelly and

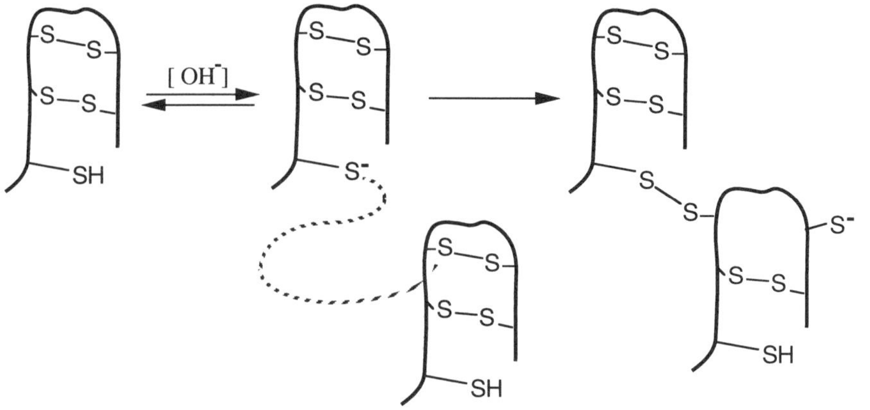


**Figure 2**: Schematic of the thiol-disulfide interchange reaction.

Zydney (10) have shown that capping this free sulfhydryl by a cysteinyl or carboxymethyl group completely eliminates the fouling seen during BSA microfiltration by blocking these thiol-disulfide interchange reactions. The extent of BSA aggregation, and in turn BSA fouling, can also be substantially reduced by the addition of metal chelators like EDTA or citrate, both of which reduce the catalytic activity of any divalent metal cations for these sulfhydryl-mediated reactions (10).

The rate of protein deposition in these membrane systems has generally been described using one of the classical models for particle deposition (4): pore blockage, pore constriction, or cake formation. Several recent studies have shown that the initial flux decline during both batch and cross-flow filtration appears to be best described using a pore blockage model in which the aggregates are assumed to reduce the number of open (accessible) pores on the membrane surface (4).

We have recently developed a more detailed mechanistic model for the flux decline during protein microfiltration which accounts for two distinct mechanisms of protein fouling: pore blockage associated with the deposition of large protein aggregates and the chemical attachment of native BSA to the growing deposit via the formation of inter-molecular disulfide linkages (15). The rate of change in the number of open pores ($N_{open}$) was expressed as:

$$\frac{dN_{open}}{dt} = -\alpha C_b A J_{v,open} - \beta C_b A J_{v,open}(N_o - N_{open}) \tag{3}$$

where $C_b$ is the bulk protein concentration, $J_{v,open}$ is the filtrate flux through the open pores, $N_o$ is the initial number of open pores, and A is the upper surface area of the membrane. The parameter $\alpha$ describes the efficiency of pore blockage, which was assumed to be proportional to the fraction of the total BSA present as aggregated protein. The second term in Eq. (3) describes the rate of native protein addition to the existing deposit and is modelled as being proportional to both the convective transport of native protein towards the membrane and to the number of aggregates already on the membrane surface (where the number of aggregates on the membrane is proportional to $N_o$ - $N_{open}$). Since the protein deposit is permeable to the filtrate, it is also necessary to account for the fluid flow through the "blocked" pores (15). One of the interesting predictions of this model is that the rate of flux decline initially increases during filtration due to the increase in the rate at which native protein is able to add to the growing deposit. This model has been shown to be in excellent agreement with experimental data for both the filtrate flux and the instantaneous rate of flux decline during the stirred cell filtration of BSA (15) and several other globular proteins. However, it is not currently possible to obtain a priori estimates of the two key fouling parameters for different proteins nor is it possible to quantify the effects of membrane pore size, morphology, and surface chemistry on these fouling parameters. In addition, this model completely neglects the complex changes in cake resistance and protein polarization that can occur over the course of the protein filtration.

The chemical attachment of native protein to these aggregates has also been studied in a stagnant system, i.e. in the absence of any continued filtration. Typical results are shown in Figure 3 for the adsorption of BSA to two separate 0.16 μm microfiltration membranes, both of which had been used to filter approximately 10 ml of BSA to deposit a small number of aggregates on the membrane surface. Results are shown for the percent increase in the hydraulic resistance of the membrane as a function of adsorption time. The membrane that was exposed to the standard BSA solution showed a considerable increase in resistance over time due to the chemical attachment of the native BSA to the existing deposit. In contrast, the cysteinylated BSA showed essentially no increase in resistance even after as much as 100 hrs of adsorption; the capping of the free sulfhydryl group completely eliminated the chemical addition of cysteinylated BSA to the existing deposit.

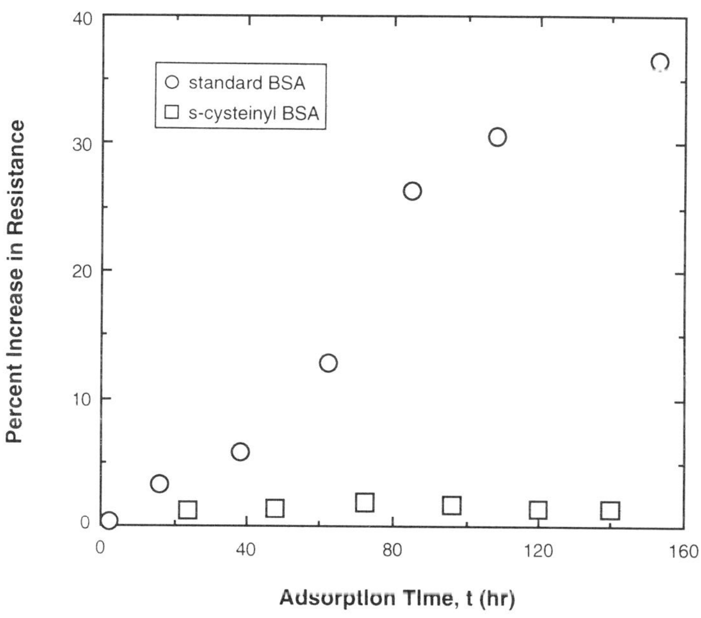


**Figure 3**: Increase in resistance of two membranes with initial protein deposits during the static adsorption of standard or cysteinylated BSA solutions.

## SUMMARY

Protein fouling remains an extremely complex subject. Recent work in our laboratory and in others around the world have begun to shed some important light on the underlying physical phenomena governing the rate and extent of fouling in different membrane systems. This work has very clearly demonstrated the important differences between protein adsorption, which occurs in both static and flowing systems, and protein deposition, which is inherently linked to the convective (filtrate) flow towards and through the membrane. The relative importance of these two phenomena depends upon the pore size and surface chemistry/morphology of the membrane as well as the detailed physico-chemical characteristics of the particular proteins involved in fouling. One of the primary factors determining the rate of protein deposition is the presence of protein aggregates or oligomers in these systems. Studies with bovine serum albumin have demonstrated that the chemical mechanism underlying the aggregation process is the thiol-disulfide interchange reaction, with the rate of protein fouling directly related to the kinetics of these sulfhydryl-mediated reactions. These aggregates are then able to serve as nucleation sites for the chemical attachment of native protein to the growing deposit. These results have provided important insights and guidelines for controlling membrane fouling in different systems, although much additional work needs to be done to develop more complete descriptions of the different fouling processes and to determine the generality of these results for other proteins and other membranes and membrane systems.

## ACKNOWLEDGEMENTS

This manuscript was prepared while the author was a Visiting Professor at the University of Colorado in Boulder.

## REFERENCES


1. A.D. Marshall, P.A. Munro, and G. Tragardh, The effect of protein fouling in microfiltration and ultrafiltration on permeate flux, protein retention and selectivity: a literature review, *Desalination* 91:65 (1993).
2. B.C. Robertson and A.L. Zydney, Protein adsorption in asymmetric ultrafiltration membranes with highly constricted pores, *J. Colloid Interface Sci.* 134:563 (1990).
3. J.M. Sheldon, I.M. Reed, and C.R. Hawes, The fine-structure of ultrafiltration membranes. II. Protein fouled membranes, *J. Membrane Sci.* 62:87 (1991).
4. G. Belfort, R.H. Davis, and A.L. Zydney, The behavior of suspensions and macromolecular solutions in crossflow microfiltration, *J. Membrane Sci.* 96:1 (1994).
5. W.R. Bowen and D.T. Hughes, Properties of microfiltration membranes. Adsorption of bovine serum albumin at aluminum oxide membranes, *J. Membrane Sci.* 51:189 (1990).
6. W.S. Opong and A.L. Zydney, Hydraulic permeability of deposited protein layers formed during ultrafiltration, *J. Colloid Interface Sci.* 142:41 (1991).
7. S. Mochizuki and A.L. Zydney, Effect of protein adsorption on the transport characteristics of asymmetric ultrafiltration membranes, *Biotech. Prog.* 8:553 (1992).
8. R. Boyd, L.J. Langsdorf, and A.L. Zydney, A two-layer model for the effects of blood contact on membrane transport in artificial organs, *Trans. Am. Soc. Artif. Intern. Organs* 40:M864 (1994).
9. L.J. Langsdorf and A.L. Zydney, Effect of blood contact on the transport properties of hemodialysis membranes: a two-layer membrane model, *Blood Purification* (1995).
10. S.T. Kelly and A.L. Zydney, Effect of thiol-disulfide interchange reactions on albumin fouling during membrane microfiltration, *Biotech. Bioeng.* 44:972 (1994).
11. D.N. Lee and R.L. Merson, Examination of cottage cheese whey by scanning electron microscopy: relationship to membrane fouling during ultrafiltration, *J. Dairy Sci.* 58:1423 (1974).
12. S.P. Palecek and A.L. Zydney, Hydraulic permeability of protein deposits formed during microfiltration: effect of solution pH and ionic strength, *J. Membrane Sci.* 95:71 (1994).
13. S. Mochizuki and A.L. Zydney, Sieving characteristics of albumin deposits formed during microfiltration, *J. Colloid Interface Sci.* 159:136 (1993).
14. S.T. Kelly, W.S. Opong, and A.L. Zydney, The influence of protein aggregates on the fouling of microfiltration membranes during stirred cell filtration, *J. Membrane Sci.* 80:175 (1993).
15. S.T. Kelly and A.L. Zydney, Fouling mechanisms during BSA microfiltration, *J. Membrane Sci.* (1995).

## CONTRIBUTORS


Ulrike Artelt
Institut fur Technische Chemie
Universitat Essen
D-45117 Essen
Germany

Leonidas G. Bachas
Department of Chemistry and
Center of Membrane Sciences
207 Chemistry/Physics Building
University of Kentucky
Lexington, Kentucky 40506-0055
U.S.A.

Anita Bhardwaj
Department of Chemistry
University of Kentucky
Lexington, Kentucky 40506-0057
U.S.A.

Dibakar Bhattacharyya
Department of Chemical and Materials
Engineering and
Center of Membrane Sciences
169A Anderson Hall
University of Kentucky
Lexington, Kentucky 40506-0046
U.S.A.

D. Allan Butterfield
Department of Chemistry and
Center of Membrane Sciences
409 Kinkead Hall
University of Kentucky
Lexington, Kentucky 40506-0057
U.S.A.

James P. Byers
Dept. Of Chemical Engineering
The University of Toledo
Toledo, Ohio 43606
U.S.A.

Ruben G. Carbonell
Department of Chemical Engineering
North Carolina State University
Raleigh, North Carolina 27695
U.S.A.

Dennis Chapman
Department of Protein and
Molecular Biology
Royal Free Hospital School of Medicine
NW3 1SN London
England

Guodong Chen
Dept. Of Chemical Engineering
The University of Toledo
Toledo, OH 43606
U.S.A.

Li Ang Chen
Department of Chemical Engineering
North Carolina State University
Raleigh, North Carolina 27695
U.S.A.

Paul A. D'Orazio
Chiba Corning Diagnostics
63 North Street
Medfield, MA 02052
U.S.A.

Sylvia Daunert
Department of Chemistry and
Center of Membrane Sciences
219 Chemistry/Physics Building
University of Kentucky
Lexington, Kentucky 40506-0055
U.S.A.

Sergi Diez
Department de Quimica
Unitat de Quimica Analitica
Universitat Autonoma de Barcelona
08193-Bellaterra (Barcelona)
Spain

Edward J. Doherty
CytoTherapeutics, Inc.
2 Richmond Square
Providence, Rhode Island 02906
U.S.A.

Peter Edelman
Ciba Corning Diagnostics
63 North St.
Medfield, MA 02052
U.S.A.

Dwaine F. Emerich
CytoTherapeutics, Inc.
2 Richmond Square
Providence, Rhode Island 02906
U.S.A.

Janos H. Fendler
Department of Chemistry
Syracuse University
Syracuse, New York 13244-4100
U.S.A.

Ronald L. Fournier
Dept. Of Chemical Engineering
The University of Toledo
Toledo, Ohio 43606
U.S.A.

Sowmya Ganapathi
Dept. of Chemical and Materials
Engineering
University of Kentucky
Lexington, Kentucky 40506-0046
U.S.A.

Frank T. Gentile
CytoTherapeutics, Inc.
2 Richmond Square
Providence, Rhode Island 02906
U.S.A.

Kathi Glauner
Department Of Chemistry
University of Kentucky
Lexington, Kentucky 40506-0046
U.S.A.

Sudarram K. Guatam
Membrane Technology Lab
Chemical Engineering Department and
School of Biomedical Engineering
Indian Institute of Technology
Bombay - 400076
India

Parvez I. Haris
Interdisciplinary Research Centre in
Biomedical materials
Academic Department of Surgery,
Royal Free Hospital School of Medicine
NW3 1SN London
England

Richard J. Heaton
Interdisciplinary Research Centre in
Biomedical Materials
Academic Department of Surgery,
Royal Free Hospital School of Medicine
NW3 1SN London
England

Nathaniel G. Hentz
Department of Chemistry
University of Kentucky
Lexington, Kentucky 40506-0055
U.S.A.

Richard S. Hutchins
Department of Chemistry
University of Kentucky
Lexington, Kentucky 40506-0055
U.S.A.

Michael Jay
Division of Medicinal Chemistry and Pharmaceutics
College of Pharmacy and
Center of Membranc Scicnces
536 Pharmacy Bldg.
University of Kentucky
Lexington, Kentucky 40536-0082
U.S.A.

Douglass S. Kalika
Department of Chemical and Materials Engineering and
Center of Membrane Sciences
177 Anderson Hall
University of Kentucky
Lexington, Kentucky 40506-0046
U.S.A.

Nikolai M. Kocherginsky
University of Illinois
1102 S. Goodwin Ave.
Urbana, Illinois 61801
U.S.A.

Jinbo Lee
Department of Chemistry
University of Kentucky
Lexington, Kentucky 40506-0057
U.S.A.

Rebecca H. Li
CytoTherapeutics, Inc.
2 Richmond Square
Providence, Rhode Island 02906
U.S.A.

Ke Jian Liu
308 Strasenburgh Hall
Dartmouth Medical Center
Hanover, New Hampshire 03755
U.S.A.

Robert R. McCaffrey
Chiba Corning Diagnostics
63 North Street
Medfield, MA 02052
U.S.A.

Thomas C. Maley
Chiba Corning Diagnostics
63 North Street
Medfield, MA 02052
U.S.A.

Richard W. Mason
Chiba Corning Diagnostics
63 North Street
Medfield, MA 02052
U.S.A.

Colin D. Mattingly
Division of Medicinal Chemistry and Pharmaceutics, College of Pharmacy
University of Kentucky
Lexington, Kentucky 40536-0082
U.S.A.

Angela M. Moraes
Dept. Proc. Quimicos
FEQ
State University of Campinas
Campinas, SP 13083-0000
Brazil

David H. Rein
CytoTherapeutics, Inc.
2 Richmond Square
Providence, Rhode Island 02906
U.S.A.

Maria Helena A. Santana
Dept. Proc. Quimicos
FEQ
State University of Campinas
Campinas, SP 13083-000
Brazil

George A. Serad
Hoechst Celanese Corporation
P.O. Box 32414
Charlotte, North Carolina 28232-6085
U.S.A.

Rejesh R. Shinde
Department of Chemical and Materials Engineering
University of Kentucky
Lexington, Kentucky 40506-0046
U.S.A.

Eberhard Staude
Institut fur Technische Chemie
Universitat Essen
D-45117 Essen
Germany

Ram Subramaniam
Department of Chemistry
University of Kentucky
Lexington, KY 40506-0057
U.S.A.

Melissa Summers
Department of Chemical and Materials Engineering
University of Kentucky
Lexington, Kentucky 40506-0046
U.S.A.

Harold M. Swartz
308 Strasenburgh Hall
Dartmouth Medical Center
Hanover, New Hampshire 03755
U.S.A.

Manuel Valiente
Department de Quimica
Unitat de Quimica Analitica
Universitat Autonoma de Barcelona
08193-Bellaterra (Barcelona)
Spain

Sasidhar Varanasi
Dept. Of Chemical Engineering
The University of Toledo
Toledo, Ohio 43606
U.S.A.

Shekhar Vishwanath
Department of Chemical and Materials Engineering
University of Kentucky
Lexington, Kentucky 40506-0046
U.S.A.

Vesna Vukasinovic
Department of Chemistry
University of Kentucky
Lexington, Kentucky 40506-0055
U.S.A.

Andrew L. Zydney
Department of Chemical Engineering
University of Delaware
Newark, Delaware 19716
U.S.A.

# INDEX